INTEGRATED COASTAL AND OCEAN MANAGEMENT

INTEGRATED COASTAL AND OCEAN MANAGEMENT

Concepts and Practices

Biliana Cicin-Sain and Robert W. Knecht
with the assistance of Dosoo Jang and Gregory W. Fisk

Center for the Study of Marine Policy
Graduate College of Marine Studies
University of Delaware

UNESCO

With support from the
Intergovernmental Oceanographic
Commission, UNESCO; and the
Graduate College of Marine Studies,
University of Delaware

ISLAND PRESS
Washington, D.C. ◆ Covelo, California

Grateful acknowledgment is expressed for permission to include the following previously published material: Table 1.1, "Examples of Ocean and Coastal Use Models," and Table 1.5, "Global Distribution of ICM Efforts by Region," appear courtesy of Elsevier Science Publishers Co., Inc.; Table 1.4, "Stage-Based Model of Coastal Area Management," appears courtesy of Professor Adalberto Vallega (Genoa, Italy); Table 7.2, "Behaviors and Points of View Typically Associated with the Cultures of Science and Policy," courtesy of Taylor and Francis Publishing Co.; Table 9.3, "Tasks of the Mediator," courtesy of Harper Collins Publishers (originally appeared as Table 5.1 in *Breaking the Impasse*, by Lawrence Susskind and Jeffrey Cruikshank); Figure 1.1, "Relationship between Coastal Zone and Coastal Resource System," appears courtesy of the International Center for Living Aquatic Resources Management; Figure 1.2, "Cultural Ecology of Coastal Public Policymaking," appears courtesy of the National Academy Press, Washington, D.C. (originally appeared in *Science, Policy and the Coast: Improving Decisionmaking*); Figure 1.3, "Interaction of Uses of the Mediterranean Sea," courtesy of Professor Adalberto Vallega (Genoa, Italy); Figure 2.2, "The Six Stages of an ICM Process," courtesy of Elsevier Science Publishing Co., Inc.; Figure 3.1, "Maritime Zones," courtesy of Manchester University Press; Figure 9.1, "Tractability of Conflicts," courtesy of Chapman & Hall; Box 9.1, "Zoning of Coastal Land in Turkey," courtesy of Elsevier Science Publishing Co., Inc.; Box 9.3, "Marine Zoning in the Great Barrier Reef Marine Park, Australia," courtesy of the Great Barrier Reef Marine Park Authority; Box 9.4 "Set-back Rule for the North Carolina Coast, USA," courtesy of Professor David W. Owens (North Carolina, United States).

Grateful acknowledgment is also expressed for permission to reproduce the photographs found on the front cover: Red Sea artisanal fishermen in front of phosphate plant, courtesy of Ben Mieremet, National Oceanic and Atmospheric Administration, United States; Port of Singapore, courtesy of Port of Singapore Authority; marine recreation in Cancun, Mexico, courtesy of Dosoo Jang.

Library of Congress Cataloging-in-Publication Data

Cicin-Sain, Biliana.
    Integrated coastal and ocean management : concepts and practices /
    Biliana Cicin-Sain and Robert W. Knecht.
        p.   cm.
    Includes bibliographic references (p.   ) and index.
    ISBN 1-55963-603-3 (Cloth : acid-free paper). — ISBN
    1-55963-604-1 (Paper : acid-free paper)
    1. Coastal zone management. 2. Coastal ecology. 3. Marine
resources conservation. 4. Marine ecology. 5. Shore protection.
    I. Knecht, Robert W. II. Title.
HT391.C483   1998
333.91'7—dc21   98-12593
                                                                    CIP

Contents

··

Chapter 8. Formulation and Approval of an ICM Program 197

Chapter 9. Implementation, Operation, and Evaluation of ICM Programs 215

PART IV. COUNTRY CASE COMPARISONS AND LESSONS LEARNED

Foreword

··

I am pleased that the Intergovernmental Oceanographic Commission (IOC) was able to support the preparation of this book on integrated coastal management (ICM). Long involved in the coordination of global ocean research programs, the IOC has recently expanded its activities into the policy and management side. This development has been largely in response to the mandates of the United Nations Conference on Environment and Development (UNCED), held in Rio de Janeiro in June 1992. Agenda 21 of UNCED and several other international agreements clearly call for more integrated management of coastal and ocean resources, and integrated coastal management—the subject of this book—is the appropriate framework within which to achieve this goal.

With its thirty-seven years of service to large and small developed and developing nations and with its substantial scientific base, the IOC felt well positioned to assist nations in technically sound and better integrated management of their coastal resources. Hence, the IOC has undertaken more than a dozen workshops and capacity-building efforts devoted to a variety of coastal management issues in various parts of the world. Recent ICM workshops, for example, have been held in China, East Africa, western Africa, Korea, Madagascar, Pakistan, Indonesia, Brazil, and India. In the IOC's training and capacity-building activities, a strong effort is made to bridge the science-policy interface to better focus research activities on important management problems and needs.

The IOC sees this book as a significant contribution to its efforts in strengthening the theory and practice of ICM. Professors Cicin-Sain and Knecht cover the full range of topics relevant to the development and implementation of ICM programs. A strength of the book, from the IOC perspective, is the inclusion of case studies on ICM practices in nearly two dozen coastal countries. This comparative information should be helpful to countries in the process of developing ICM programs. In this connection, in cooperation with several other organizations, the IOC is planning the establishment of a World Wide Web site on the Internet devoted to the exchange of information on ICM.

I believe that other IOC programs, such as the Global Ocean Observing System (GOOS) and the Programme on Coastal Ocean Advanced Science and Technology, will also play a role in providing coastal zone managers with the coastal and ocean data they need to manage in a more informed manner. In fact, the coastal module of GOOS is being developed especially to meet these needs.

Coastal managers should ensure that their requirements are factored into the planning process for this program.

I am especially pleased to see this book published early in 1998—the International Year of the Ocean. The United Nations General Assembly gave the IOC a leading role in planning and overseeing activities related to the International Year of the Ocean. As is well known, a number of activities are planned for 1998—from the Expo '98, in Lisbon, Portugal, to the many other undertakings in different venues around the world. The value and usefulness of this book, however, will extend well beyond the International Year of the Ocean and, indeed, into the next century, providing advice and assistance to all those who work to improve our stewardship of the earth's valuable coastal zones.

Gunnar Kullenberg
Secretary, Intergovernmental Oceanographic Commission,
UNESCO, Paris, France

Acknowledgments

··

Many people assisted us in a variety of ways in the production of this book, and to them we owe our sincere gratitude. Many thanks go to Dr. Gunnar Kullenberg, Secretary of the Intergovernmental Oceanographic Commission (IOC), UNESCO, who responded with enthusiasm when we raised the possibility of spending our half-year sabbatical in 1996 at the IOC's offices in Paris working on this book. The financial support the IOC provided for preparation of the book and Dr. Kullenberg's gracious hospitality are acknowledged with deep appreciation. We met many good colleagues while at the IOC and have continued to enjoy their friendship since that time. Among our IOC colleagues, special thanks go to Natalie Philippon-Tulloch, Haiqing Li, and Salvatore Arico for their collegial support of this work, and to Silvia Vernizzi, who facilitated all aspects of our stay at the IOC.

We sincerely thank Dr. Carolyn A. Thoroughgood, Dean of the Graduate College of Marine Studies at the University of Delaware, for her unfailing support and encouragement of our work on integrated coastal management, and for the college's financial contributions in the preparation of this work. Many thanks go to the Delaware Sea Grant College Program for its funding, over a number of years, of our work on multiple-use ocean and coastal management, and to the Beijer Institute of Ecological Economics, Royal Swedish Academy of Sciences, for its support of our field work in Fiji and Western Samoa. The support and assistance of the secondary authors of the book, Dosoo Jang and Gregory W. Fisk, in all phases of the work—especially in conducting, analyzing, and interpreting the cross-national survey, in preparing the detailed case studies, and in refining the final product—were essential, and if the book is successful, much of the credit should go to their thorough and careful work. For keeping our Center for the Study of Marine Policy at the University of Delaware operating smoothly and effectively while we focused much of our attention on the book, credit goes to our Office Manager, Catherine C. Johnston. Our sincere thanks also go to senior editor Todd Baldwin of Island Press for his efficiency in handling all matters related to publication of the book, and to Pat Harris and Christine McGowan for their superb editorial skills.

Our analysis in the book relies in part on the results of a cross-national survey of ICM (integrated coastal management) practices in twenty-nine countries. Our sincere thanks go to the ICM experts who made time available to respond to the survey. They include Datin Paduka Fatimah Abdullah (Malaysia), Etty Agoes (Republic of Indonesia), Shahid Amjad (Islamic Republic of Pakistan), Ir. Ooi Choon Ann (Malaysia), Karen Anutha (Australia), Jack H. Archer (United States and Republic of Bulgaria), Luis Arriaga (Ecuador), Asher Edward (Federated States of Micronesia), Mohd Nizam Basiron (Malaysia), Gonzalo A. Cid (Chile), Christopher Dahl (Federated States of Micronesia), Rokhmin Dahuri (Republic of Indonesia), Leo de Vrees (Netherlands), Ana Laura Lara Dominguez (Mexico), Konstantin Galabov (Republic of Bulgaria), Victor A. Gallardo (Chile), Edgardo D. Gomez (Republic of the Philippines), Susan Gubbay (United Kingdom), Syed Mazhar Haq (Islamic Republic of Pakistan), Ampai Harakunarak (Thailand), Wayne Hastie (New Zealand), Marcus Haward (Australia), Renato Herz (Brazil), Larry Hildebrand (Canada), Abdul Aziz Ibrahim (Malaysia), Robert Kay (Australia), R. Krishnamoorthy (Republic of India), Ji Hyun Lee (Republic of Korea), Kem Lowry (Democratic Socialist Republic of Sri Lanka), Clarissa C. Magdaraog (Republic of the Philippines), Alain Miossec (France), Francisco Montoya (Spain), Chris Morry (Canada), Raj Murthy (Canada and Republic of India), Hiroyuki Nakahara (Japan), Sefa Nawadra (Fiji), Magnus Ngoile (United Republic of Tanzania), Stephen B. Olsen (Ecuador), Erdal Ozhan (Turkey), Sirichai Roungrit (Thailand), R. A. D. B. Samaranayake (Democratic Socialist Republic of Sri Lanka), Armando G. Sanchez Rodriguez (Chile), Hance Smith (United Kingdom), Triono Soendoro (Republic of Indonesia), Juan Luis Suarez de Vivero (Spain), Suraphol Sudara (Thailand), Carlos Valdes-Casillas (Mexico), Joeli Veitayaki (Fiji), Ying Wang (People's Republic of China), Alejandro Yanez-Arancibia (Mexico), and Huming Yu (People's Republic of China).

A number of colleagues in different countries reviewed either all or part of the book. Their thoughtful comments and suggestions were much appreciated and were taken into account wherever possible. They include Jack H. Archer (United States); Salvatore Arico, IOC, France; Leo de Vrees (Netherlands); Michael Fischer (United States); Susan Gubbay (United Kingdom); Larry Hildebrand (Canada); Paul Holthus, World Conservation Union (IUCN), Switzerland; Ji Hyun Lee (Republic of Korea); Haiqing Li, IOC, France (China); Kem Lowry (United States); Alain Miossec (France); Raj Murthy (Canada and the Republic of India); Magnus Ngoile, IUCN, Switzerland (United Republic of Tanzania); Stephen B. Olsen (Ecuador); Erdal Ozhan (Turkey); Adalberto Vallega (Italy); Carmen Rossi Wongtschowski (Brazil); and Huming Yu, International Maritime Organization, Philippines (People's Republic of China).

We should note, however, that any errors in the book are our sole responsibility. Although this book was supported, in part, by the IOC, UNESCO, the observations and recommendations contained herein represent our own analyses and perspectives and do not necessarily represent the official position of IOC, UNESCO.

A number of research assistants at the Center for the Study of Marine Policy worked long and hard on compiling the references and correcting and formatting various versions of the manuscript. Special thanks go to Nigel Bradly, Naomi Brown, Katherine Bunting-Howarth, Forbes Darby, Suzanna D. Donahue, Shannon Farrell, Deborah Goldstein, Jorge Gutierrez, Diane Jackson, Jeffrey Levinson, and Tracey Wiley for their help in these essential tasks.

Finally, our thanks go to our families for their love and support, especially our twelve-year-old daughter, Vanessa, and Biliana's mother, Lubica Cicin-Sain. Dosoo Jang is especially grateful to his wife, Dr. Young Joo Lee for her support, and Gregory Fisk thanks his wife, Diane Jackson Fisk, for her support.

Biliana Cicin-Sain and Robert W. Knecht
Newark, Delaware

List of Tables

List of Figures and Boxes

....

Introduction

....

Our intent in writing this book is to present an account of the concept of integrated coastal and ocean management (ICM) and to illustrate how it can be accomplished by describing ways in which particular nations or their subnational governments (provinces, localities) have implemented various aspects of it. The major goals of the book are to provide the following:

- A synthesis and analysis of international prescriptions for ICM;

- A presentation of the major concepts and methodologies of ICM;

- A practical guide to the establishment, implementation, and operation of ICM programs;

- An analysis of different patterns of ICM followed in different countries;

- Our own prescriptions for approaches that seem to be most successful, based on our experience, a cross-national survey we conducted, and the scholarly literature in the field.

In our view, the term *integrated coastal and ocean management* implies a conscious management process that acknowledges the interrelationships among most coastal and ocean uses and the environments they potentially affect. Hence, in a geographical sense, ICM typically embraces upland watersheds, the shoreline and its unique landforms (beaches, dunes, wetlands), nearshore coastal and estuarine waters, and the ocean beyond to the extent it is affected by or affects the coastal area. Given that many nations have claimed jurisdiction over 200 nautical-mile ocean zones off the shores of their coasts, the coastal area in some cases incorporates the entire offshore ocean zone.

ICM is a process by which rational decisions are made concerning the conservation and sustainable use of coastal and ocean resources and space. The process is designed to overcome the fragmentation inherent in single-sector management approaches (fishing operations, oil and gas development, etc.), in the splits in jurisdiction among different levels of government, and in the land-water interface. ICM is grounded in the concept that the management of coastal and ocean

1

resources and space should be as fully integrated as are the interconnected eco-systems making up the coastal and ocean realms. We wish to stress that ICM does not replace traditional single-sector resources management. For example, ICM is not intended to replace coastal water quality management and fisheries manage-ment programs but to ensure that all their activities function harmoniously to achieve agreed water quality and fisheries goals. Obviously, if a degraded coastal habitat affects the attainment of fisheries management goals, management of that habitat should be within the ambit of an integrated coastal management process. The fundamentals of ICM are explained in detail in chapter 2.

We believe that a book on ICM is especially needed at this time because of the recent recognition of the importance—indeed, the necessity—of employing more integrated approaches to management of the earth's resources. Better integrated resource management is a fundamental prerequisite of sustainable development. Mandates for the use of more integrated management approaches were prominent in the recommendations in Agenda 21, an action program emanating from the United Nations Conference on Environment and Development (UNCED), held in Rio de Janeiro, Brazil, in June 1992 (also known as the Earth Summit or the Rio Conference). Such mandates are also found in the 1982 United Nations Conven-tion on the Law of the Sea, which entered into force in November 1994. In addi-tion, measures for protecting marine biodiversity are called for in the 1992 Con-vention on Biological Diversity. To succeed, these measures require integrated approaches such as those embodied in the ICM concept. Moreover, ICM has been singled out by the Intergovernmental Panel on Climate Change (IPCC) as a key tool for dealing with the threat of accelerating sea-level rise in low-lying coastal areas. The Global Programme of Action on Protection of the Marine Environment from Land-Based Activities, emanating from the 1995 Washington Conference on Land-Based Activities Affecting the Marine Environment, also points out the importance of better integrated coastal management measures to control land-based sources of marine pollution.

Unrelenting pressures on the world's coastal areas due to ever increasing pop-ulations threaten the viability of coastal ecosystems and expose increasing num-bers of people to the very real hazards of living at the water's edge. Often unknowingly, coastal populations are at risk from potentially catastrophic effects of hurricanes and typhoons and the almost certain consequences of a slow but serious rise in sea level. Thus, measures to mitigate the natural hazards of the coastal zone must be comprehensive in scope and they must deal with all facets of the problem—meteorological aspects, effects of storms on the shore land and its development, evacuation and postdisaster planning, insurance programs, beach protection measures, and the like.

Given the extent of interest in integrated resource management, especially over the past half-dozen years or so, and the many reports and articles that have appeared in the literature, we saw the need for a substantial text that addresses each aspect of ICM in some detail and analyzes the practice of ICM in a range of circumstances and national settings. Even so, given the complexity of the subject, we know that gaps in our coverage of the topic are inevitable; we earnestly solic- it readers' suggestions of ways to improve this book in future editions.

Goals of the Book

Simply stated, the primary goal of this book is to provide coastal and ocean man- agers with essential information about integrated coastal management so that they can put functional and effective programs in place. We hope that the book will be relevant to coastal nations at various stages of economic development, from less developed to more developed. Similarly, we trust that what is offered here will be equally useful to nations with strong central governments and those with relative- ly undeveloped local governments as well as those with strong government capac- ity at three levels—national, provincial or state, and local or community. We anticipate that the book will be useful in all types of coastal and marine settings, from those in the tropics to those in more temperate climates.

A secondary goal of the book is to provide a clear description of the benefits of ICM to help policy makers in coastal nations decide whether and how to develop ICM programs. Policy makers need to know what is likely to be involved in implementing an ICM program: Will government reorganization be needed? Will new legislation or decrees be necessary? What levels of government will need to be involved? How much is the effort likely to cost? Thus, the book contains infor- mation policy makers will need as they consider the initiation, implementation, and operation of an integrated coastal management program.

We focus on those parts of the ICM process we believe to be most critical to a successful effort—that is, to the creation of a management process with the best chance of achieving the intended goals. Our experience has shown that the insti- tutional dimension often does not receive the attention it merits. Mechanisms that coordinate and harmonize the various sectoral programs are absolutely funda- mental to rational coastal management. Similarly, no ICM program can accom- plish its goals without satisfactory arrangements relating provincial and/or local governments and the central government and relating private and public actions. As a rule, all levels of government, the private sector, local communities, coastal and marine stakeholders, and nongovernmental organizations all play significant roles in determining the use of coastal lands and waters. To the extent that this is the case, each of these perspectives must be constructively involved in ICM.

It is our hope that the approaches outlined here will be relevant and useful regardless of the stage of coastal management present. Most nations have coastal management efforts of some sort already in place. Often, these efforts focus on a major problem in an important coastal area, such as coastal erosion or the need to protect a particularly threatened stretch of coral reef. Such efforts, though probably not comprehensive in scope, can be a reasonable starting point for a broader ICM program. We hope this book will provide the information necessary to construct a broader national effort on an existing base.

Finally, it is our intent that the book be practical. It was written for individuals and organizations responsible for the initiation, implementation, and operation of integrated coastal and ocean management programs in the coastal nations of the world. At the same time, we hope that academics, scholars, and others interested in the field of marine policy will find the book of interest.

Scope and Content of the Book

This book is not the first effort to set forth a framework or a set of guidelines for undertaking ICM. In various forms, information related to coastal management concepts and practices in a number of different national settings has been available for a number of years; see, for example, Sorensen and McCreary 1990; Kenchington 1990; Chua and Scura 1992; and Clark 1992 and 1996. Prior to UNCED in 1992, much of the literature on ICM was largely descriptive, reviewing approaches undertaken in various countries to manage shorelines and coastal areas for various purposes—to minimize pollution, control coastal erosion, promote coastal tourism, and the like. The earliest generic guidelines for coastal area management were disseminated by the Organization for Economic Cooperation and Development (OECD) in 1987. These guidelines, like those that were to follow in the 1992 Earth Summit, contained a greater prescriptive component. They suggested various policies that coastal management programs should implement and processes that the programs should include.

Coming out of UNCED were strong recommendations for coastal nations to develop and implement more integrated programs for managing their coastal and ocean resources, as discussed in detail in chapter 3. Not surprisingly, this mandate led to the preparation and dissemination of several additional sets of guidelines for integrated coastal management. The first such post-UNCED guidelines were prepared by the World Bank in 1993 in collaboration with the Food and Agriculture Organization of the United Nations (FAO) and the United Nations Environment Programme (UNEP). Another set of guidelines, first prepared by the Priority Action Programme of UNEP's Mediterranean Action Plan in the early 1990s, was subsequently published by UNEP in 1995. From the World Coast Conference in November 1993, sponsored by the Netherlands and attended by ninety nations, came a set of ICM principles and recommendations that were also prescriptive in nature. A number of other efforts are currently under way to fashion ICM training courses that in effect represent extended discussions of ICM guidelines, sometimes with simulations or training exercises built in. These include the Train-Sea-Coast courses being developed by the United Nations Division for Ocean Affairs

•

and Law of the Sea, courses being offered by the International Ocean Institute, courses by the Intergovernmental Oceanographic Commission, and others.

This book builds on this earlier and ongoing work and endeavors to extend it where possible and appropriate. In particular, the book presents the following:

- *Synthesis and analysis of the international prescriptions for ICM* evolving from processes associated with UNCED and the Law of the Sea Convention. Our intent is to provide an up-to-date account of thinking at the international level regarding what constitutes ICM and what are its core elements and principles.

- *Major concepts, processes, and methodologies involved in ICM.* We address central questions such as, What are the goals and functions of ICM? What triggers the need for ICM? What is being managed in ICM? Who should carry out ICM? What management methods and tools are available to coastal decision makers?

- *A practical guide for the coastal decision maker* reviewing options available at each major stage of the ICM process, including problem identification and analysis and formulation, implementation, operation, and evaluation of ICM programs. We review the major options available at each of these stages, difficulties decision makers may encounter, and ways of overcoming such difficulties.

- *Analysis of various patterns of ICM followed in different countries.* We describe variations in ICM practices in different nations of the world—developed and developing—and in all geographical regions. Because systematic comparative information about the ICM programs of different countries is notably unavailable, we conducted a cross-national survey of ICM practices in a sample of countries and prepared narrative accounts of various nations' efforts to plan for and implement ICM. The survey and case studies are described more fully later in this introduction.

- *Presentation of our prescriptions on approaches that seem most successful.* Drawing on the ICM literature, the cross-national survey and case studies, and our own experiences in working on ICM in various nations, we offer our own prescriptions of what approaches seem to work best under what circumstances.

Given our orientation as social scientists (especially in policy analysis and public administration), we pay special attention to the institutional aspects of ICM. For example, regarding the intersectoral integration inevitably needed in ICM, how can the many government agencies and other groups usually operating in coastal areas be made to work together in a coherent and harmonized fashion? We address in some detail the question of intergovernmental integration, also often a thorny issue in the ICM process: what is to be the relationship between national and subnational (provincial and local) authorities, how are they to relate, and through what mechanisms? A related question is how to determine appropriate conflict resolution strategies to address the many conflicts that typically occur among diverse ocean and coastal uses. Because scientific understanding is crucial for good ICM decision making, we also focus on the relationship of science and management, providing examples that illustrate how scientific information can be incorporated into the management process and discussing the obstacles often encountered in this effort. Similarly, we stress the need to monitor parameters indicative of the performance of the ICM program or one of its parts. In our view, corrections, modifications, and improvements are difficult, if not impossible, without a carefully designed and well-functioning monitoring component—one that not only reveals how well a program is performing but also promotes the kinds of learning that can improve the performance.

Major Sources of Information

Our discussion relies on information from a combination of sources. First, we conducted an extensive review of the literature, drawn both from academic circles and from reports of national governments, international organizations, and non-governmental organizations (NGOs), on the concepts and practice of ICM; this literature forms much of the basis for our discussion of concepts and methods in the first part of the book. Second, to describe the ICM experiences of a range of nations, both developing and developed, we drew on academic and in-country government and NGO sources and on our cross-national survey of ICM experts. We used these comparative data in two major ways: to illustrate the experiences of a range of nations with various aspects of ICM and to develop detailed case studies of such experiences, which we present in appendix 1. Finally, of course, we drew heavily on our personal backgrounds and experience in the design and implementation of ICM programs in various countries; on our experience in lecturing and working in various countries, especially Latin America, Europe, Asia, and the South Pacific; and on our experience as NGO participants in the international negotiations preceding and following the United Nations Conference on Environment and Development and, for author Robert W. Knecht, in the international negotiations concerning the Law of the Sea Convention. Knecht brought another particularly relevant set of experiences to bear in the writing of this book—as initial director of the U.S. coastal zone management program charged with implementing the world's first large-scale coastal management effort (1972–1981), he faced many of the decisions and choices that coastal decision makers around the world must address.

1996 ICM Survey of Nations

Although case studies on individual nations' experiences with ICM may be found in the literature, it is often difficult to compare ICM processes in different countries because the studies tend to use different variables and ask different questions. Thus, we believed it would be useful to survey a selected number of nations to ascertain patterns in ICM practice. The survey, conducted from February to June 1996 and reproduced in appendix 2, was sent via fax to one to five ICM experts in each of twenty-nine countries. In each country, we contacted key expert informants, individuals in the academic world or in relevant government agencies with close knowledge of the evolution and functioning of the country's ICM program (see appendix 2 for a list of respondents). Responses to the survey thus reflect expert opinion rather than official government positions. We chose the following countries for our sample, as shown in table I.1: *developed countries*: Canada, the United States, the United Kingdom, France, the Netherlands, Spain, Japan, the Republic of Korea, Australia, and New Zealand; *middle developing countries*: Mexico, Barbados, Brazil, Chile, Turkey, Thailand, Malaysia, Fiji, and South Africa; *developing countries and countries in transition*: Ecuador, Bulgaria, the People's Republic of China, the Republic of Indonesia, the Islamic Republic of Pakistan, the Republic of India, the Republic of the Philippines, Democratic Socialist Republic of Sri Lanka, the Federated States of Micronesia, and the United Republic of Tanzania.[1]

We chose this set of countries for the following reasons. First, we wanted examples from a range of different geographical regions. Among *developed countries*, we focused on those with colonial experiences (such as the United Kingdom), since a common historical pattern before worldwide decolonization beginning in the 1960s was the exportation of "metropolitan" institutions to colonies and, later, to newly independent countries. It is important to understand these historical institutional models and practices because in some cases they still influence government organization in now-independent states. Among middle developing countries, we sought examples from the major growing economies of Asia, Latin America, and Europe that, despite having achieved remarkable economic advances in recent years, still exhibit characteristics of underdevelopment. Among developing countries, we sought examples from different areas (Africa, Latin America, the Caribbean region, southern and eastern Asia, and the Pacific Islands)

1. There are many ways to categorize nations according to level of development; unfortunately, no readily available indicator encompasses the full range of economic, social, and political development factors that should be taken into account in estimating a nation's level of development. Here, we have arrayed nations by gross national product per capita, a readily available measure. This categorization, however, may mask important economic forces taking place in a particular nation. For example, although China's gross national product per capita places it in the "developing countries" category, its spectacular economic growth in recent years, particularly in its coastal areas, could place it in at least the "middle developing countries" category.

and included several small-island states because islands represent a special case in integrated coastal management. Finally, wishing to include at least one example of a country in transition from a communist economy to a free-market economy, we chose the Republic of Bulgaria, a country in which ICM efforts are ongoing as part of a World Bank project.

Table I.1. Countries Included in 1996 Survey
(by Level of Development and Region)

	North and South America	Europe	Asia	Oceania	Africa
Developed countries	Canada ($19,510)	United Kingdom ($18,340)	Japan ($34,640)	Australia ($18,000)	
	United States ($25,880)	France ($23,420)	Republic of Korea ($8,260)	New Zealand ($13,350)	
		The Netherlands ($22,010)			
		Spain ($13,440)			
Middle developing countries	Mexico ($4,180)	Turkey ($2,500)	Thailand ($2,410)	Fiji ($2,250)	South Africa ($3,040)
	Barbados ($6,560)		Malaysia ($3,480)		
	Brazil ($2,970)				
	Chile ($3,520)				
Developing countries and countries in transition	Ecuador ($1,280)	Republic of Bulgaria ($1,250)	People's Republic of China ($530)	Federated States of Micronesia[a]	United Republic of Tanzania ($140)
			Republic of Indonesia ($880)		
			Islamic Republic of Pakistan ($430)		
			Democratic Socialist Republic of Sri Lanka ($640)		
			Republic of India ($320)		
			Republic of Philippines ($950)		

Note: Figures in parentheses indicate gross national product (GNP) per capita, derived from the World Bank's World Development Report (1996b). GNP measures the total domestic and foreign value added claimed by residents. It comprises gross domestic product (GDP) and net factor income from abroad. GNP per capita indicators in parentheses are 1994 estimates.
[a]Estimated to be lower middle income ($726 to $2,895).

Our selection of countries was guided by two additional considerations. In each country chosen: (1) there had been efforts to establish and, in some cases, operate a system of ICM; (2) we could identify key expert(s) with close knowledge of the ICM effort who would be likely to respond to our survey.

The response rates for the survey are shown in table I.2. The high overall response rate (78 percent) is gratifying and reflects, we think, a high degree of interest in comparing ICM practices around the world. We should note that responses were received from all countries in the survey but Barbados and South Africa. Despite repeated contacts, respondents in these countries did not return their surveys.

The Need to Tailor ICM to Fit a Nation's Unique Circumstances

This book shows that ICM is not a "one size fits all" concept. It is not a fixed approach that can be applied in a wholesale fashion to all situations, and it is not a methodology based on any one nation's approach to coastal zone management. Furthermore, the use of zoning schemes to separate uses geographically is not necessarily an integral part of ICM, although such zoning may be recommended in certain situations to accomplish a particular purpose, as in management of a specific marine protected area. Rather, as we stress throughout the book, ICM is an ongoing process designed to ensure that all decisions and activities related to or affecting a country's coastal area are consistent with, and supportive of, agreed goals and objectives for the region and the nation.

Again, the ICM process does not replace the sectoral management programs

Table I.2 Rates of Response to Cross-National ICM Survey

Country Grouping[a]	Surveys Sent[b]	Responses Received	Response Rate (%)
Developed countries (10)	16 (1)	14	93
Middle developing countries (9)	24 (3)	15	71
Developing countries (10)	28 (1)	20	74
Total (29)	68 (5)	49	78

[a] Figures in parentheses indicate numbers of countries in groupings.
[b] Figures in parentheses indicate numbers of cases in which the survey was passed on to a more appropriate person or completed jointly by two respondents. In determining the response rate, these cases were subtracted from the total number of surveys sent.

virtually all countries already operate with respect to specific coastal and ocean resources, such as fisheries. In many cases, these programs are adequately managing the resources entrusted to them. Furthermore, such programs generally are staffed by the most knowledgeable and experienced people in the field in question. However, sectoral programs involving a single resource or use often are not equipped to handle conflicts with other uses and activities or to act in a manner supportive of overarching national coastal and ocean management goals. As stated earlier, the function of ICM is to ensure that such sectoral programs come within the ambit of a process that harmonizes multiple and diverse coastal and marine activities and ensures that they all operate in a manner consistent with the nation's agreed coastal and marine management goals. Hence, a properly functioning ICM program will consist of strong and well-run sectoral management programs operating as part of a larger system that includes institutional arrangements, processes, and procedures aimed at bringing about the necessary coordination and harmonization.

Thus, ICM must be tailored to meet each nation's unique situation. Virtually every country differs with respect to a number of variables:

- Physical, chemical, and biological characteristics of coastline and ocean
- Distribution, richness and diversity of natural resources
- Nature of cultural and religious traditions
- Existence (or lack thereof) of both indigenous and colonial descendants
- Level of economic development
- Nature of the legal system
- Nature of prevailing political system
- Nature and strength of central government
- Nature and function of provincial level of government
- Nature and strength of local and community levels of government
- Relative strengths of executive and legislative branches
- Nature of government bureaucracy
- Relative strengths of political parties and other interests, private and public
- Relative roles of local communities, nongovernmental organizations, and major coastal and marine stakeholders, including indigenous peoples
- Level of tourism development

Of course, this list could be much longer. The point is that nations and their governments can differ in countless ways, many of them important to the sound and effective management of the coastal zone and its resources. Clearly, to be successful, any government program must be adapted to these realities. This is especially true with ICM because a program's ultimate success depends on building positive working partnerships among the various levels of government and the sectoral programs active in the coastal zone.

It goes without saying, therefore, that the task of designing an ICM program appropriate for a given nation must be in the hands of those with a good understanding of the nation and its realities. Since this breadth of understanding is unlikely to reside in any single individual or government agency, a team of experts with varied backgrounds and experience is generally needed to design an ICM program to fit the unique context, needs, and realities of a particular nation. The composition of this team and the talents it brings to the task are central to the success of the resulting ICM program.

A Note on Terminology

Over the two and one-half decades of its existence, the concept of coastal management has attracted several names and corresponding acronyms. The 1972 legislation in the United States, the earliest national attempt at managing coastal zones, used the term *coastal zone management* (CZM). Early efforts in developing countries were often given the name *integrated coastal area management* (ICAM) because they were usually limited to a specific coastal area rather than the entire coastal zone. As the concept of coastal management gained greater recognition internationally, the phrases *integrated coastal zone management* (ICZM) and *integrated coastal management* (ICM) came into use. (The term *integrated* was included when it became clear that an integrated approach, rather than a single-sector approach, was essential for effective coastal management.)

More recently, in connection with implementation of the Convention on Biological Diversity, the term *integrated marine and coastal area management* (IMCAM) has begun to be used as well. In our judgment, these terms all refer to the same concept—that of integrated coastal management. For simplicity and consistency, throughout this book we use the term *integrated coastal management* (ICM).

Structure of the Book

This book is divided into four major parts. Part I, "The Need for Integrated Coastal Management and Fundamental Concepts," discusses the reasons why ICM is needed, focusing on the many conflicts that occur among coastal and ocean uses (chapter 1), and presents the fundamental concepts involved in integrated coastal and ocean management (chapter 2). Part II, "Evolution of International Prescriptions for Integrated Coastal Management," discusses in some detail the evolution of global perspectives on ICM emanating from both the Earth Summit and the Law of the Sea Convention (chapters 3 and 4). Part III, "A Practical Guide to Integrated Coastal Management," lays out important considerations for the design of ICM programs, including setting of policy and management goals (chapter 5); intergovernmental, institutional, and legal considerations (chapter 6);

and establishment of the proper scientific and technological base (chapter 7). Chapter 8 outlines the steps typically needed to formulate an ICM program and get it adopted. Chapter 9 focuses on implementation of ICM programs and on issues faced in their operation and evaluation. Part IV, "Case Comparisons and Lessons Learned," compares case studies illustrating how twenty-two different coastal nations are addressing specific aspects of ICM (chapter 10)[2] and provides with a summary of the book and highlights of successful practices in ICM programs (chapter 11). Appendix 1 presents the country case studies, and appendix 2 reproduces our cross-national survey and provides a list of respondents.

Tips on Reading the Book

Although we hope that readers of this book will be interested in all its parts, we suggest that the busy coastal professional interested primarily in establishing and implementing an ICM program focus especially on the following chapters: chapter 2, on definitions and concepts; chapters 5–9, on establishing, implementing, operating, and evaluating ICM programs and on institutional, legal, and informational considerations; and chapter 11, the summary and conclusions.

2. Case studies were prepared for twenty-two of the twenty-nine nations included in the cross-national survey. The absence of available published information on ICM programs in the remaining seven countries precluded us from preparing case studies for them.

PART I

THE NEED FOR INTEGRATED COASTAL MANAGEMENT AND FUNDAMENTAL CONCEPTS

....

Chapter 1

The Need for Integrated Coastal and Ocean Management

Introduction: The Coasts—Unique, Valuable, and Threatened

The place where the waters of the seas meet the land—the coasts—are indeed unique places in our global geography. They are unique in a very real economic sense as sites for port and harbor facilities that capture the large monetary benefits associated with waterborne commerce and as locations for industrial processes requiring water cooling, such as power generation plants. The coasts are highly valued and greatly attractive as sites for resorts and as vacation destinations, and they are valuable in many other ways as well. The combination of freshwater and salt water in coastal estuaries creates some of the most productive and richest habitats on earth; the resulting bounty in fishes and other marine life can be of great value to coastal nations. In many locations, the coastal topography formed over the millennia provides significant protection from hurricanes, typhoons, and other ocean-related disturbances. Hence, for most coastal nations, the coasts are an asset of incalculable value, an important part of the national patrimony.

But these values can be diminished or even lost. Pollution of coastal waters can greatly reduce the production of fish, as can degradation of coastal nursery grounds and other valuable wetland habitat. The storm protection afforded by fringing coral reefs and mangrove forests can be lost if the corals die or the mangroves are removed. Inappropriate development and accompanying despoilment can reduce the attractiveness of the coastal environment, greatly affecting tourism potential. Even ports and harbors require active and informed management if they are to remain productive and successful enterprises over the long term.

Beyond these values, and perhaps more important, the coasts are home to more than half of the world's population. Two-thirds of the world's largest cities are located on coasts and populations of coastal areas are growing faster than inland populations. For example, World Bank experts estimated in 1994 that two-thirds of the population of developing nations would be living along coasts by the end of the twentieth century (WCC 1994).

The presence of large and growing populations in the world's coastal areas creates major problems. In developed countries, needs are generated for ever larger sewage treatment plants, expanded landfills for the disposal of solid waste, and increased recreational facilities, to mention only a few. In developing countries, with less infrastructure in place, more people in the coastal zones means more pollution of coastal waters, more pressure on nearby natural resources (for example, mangrove forests for firewood and beach sand for construction), and more pressure on fishery resources. Clearly, the tendency for ever greater numbers of people to migrate to the world's coasts is exerting serious pressure on these areas that could put the value and productivity of many of them at risk. Unless effective steps to manage these areas are taken soon, losses of considerable consequence will occur.

But rational management of the resources of coastal areas is made complex by a number of inherent difficulties. Before the twentieth century, the oceans were used principally for two purposes: navigation and fishing. Except occasionally in the most congested ocean waters, conflicts between these uses were few and far between. Hence, traditional coastal and marine resource management has been characterized by a sector-by-sector approach. For example, fisheries have been managed separately from offshore oil and gas development, which is handled separately from coastal navigation. Yet these activities are now capable of affecting one another and do so with regular frequency. A second difficulty is that jurisdiction over various parts of coastal and ocean areas generally falls to different levels of government. The local government may control use of the shore land down to the water's edge and the state or provincial government may have jurisdiction over the territorial sea (typically extending 12 nautical miles from shore), with the national government having control over the Exclusive Economic Zone (EEZ) out to a distance of 200 nautical miles. In some cases, the jurisdiction of the national government begins at the shoreline and extends to the outer limit of the EEZ. Many coastal and ocean uses can affect all these zones and thus require the involvement of as many as three levels of government. A third difficulty involves the complexity of the ocean itself—its fluid and dynamic nature and the intricate relationships of the marine ecosystems and the environments that support them.

As a consequence of these difficulties, the traditional single-sector management approach, though quite satisfactory in the days of few ocean uses, frequently does not produce satisfactory results today. For example, an offshore oil development program may lead eventually to oil production, but if the decision-making process does not adequately take into account the effects of this development on other ocean uses and resources, the costs of the offshore oil production to the coastal nation could be very large indeed. Similarly, fisheries management regimes that deal only with fish catches and the harvesting process, and fail to protect the habitats critical to the well-being of those fisheries, cannot succeed over the long term.

In this chapter, we discuss the major reasons why an integrated approach to the management of coastal and ocean areas is desirable, describe several types of ocean and coastal uses and their interactions, and provide examples of conflicts among various ocean and coastal uses and their environmental implications. These examples underscore the need for integrated approaches to coastal and ocean management, a subject we turn to in chapter 2.

The Need for ICM

As noted by L. F. Scura and colleagues, the coastal zone represents the interface between the land and the sea, "but concern and interest are concentrated on that area in which human activities are interlinked with both the land and the marine environments" (Scura et al. 1992, 17), as illustrated by figure 1.1. The coastal zone has the following characteristics (Scura et al. 1992):

- Contains habitats and ecosystems (such as estuaries, coral reefs, sea grass beds) that provide goods (e.g., fish, oil, minerals) and services (e.g., natural protection from storms and tidal waves, recreation) to coastal communities.

- Characterized by competition for land and sea resources and space by various stakeholders, often resulting in severe conflicts and destruction of the functional integrity of the resource system.

Figure 1.1. Relationship between Coastal Zone and Coastal Resource System

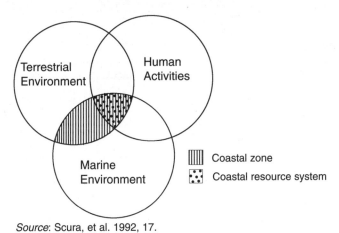

Source: Scura, et al. 1992, 17.

- Serves as the source or backbone of the national economy of coastal states where a substantial proportion of the gross national product depends on activities such as shipping, oil and gas development, coastal tourism, and the like.

- Usually is densely populated and is a preferred site for urbanization.

The coastal management system, in turn, can be thought of as a system of relationships among (1) people who live, use, or otherwise are concerned (in their beliefs or behaviors) with the coastal environment, (2) policy makers and managers whose decisions and actions affect the behavior of coastal peoples, and (3) members of the scientific community: natural scientists who study the coastal environment and social scientists who study human behavior in coastal zones (adapted from Orbach 1995). This system of relationships—the "cultural ecology of coastal public policy making," as M. Orbach calls it—is depicted in figure 1.2.

Ecological Effects and Multiple-Use Conflicts: Why ICM Is Needed

The major reasons why an integrated approach is needed for managing oceans and coasts are twofold: (1) the effects ocean and coastal uses, as well as activities farther upland, can have on ocean and coastal environments and (2) the effects ocean and coastal users can have on one another.

Figure 1.2. Cultural Ecology of Coastal Public Policy Making

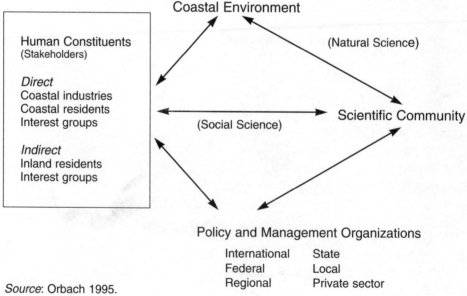

Source: Orbach 1995.

Coastal and ocean development activities (building of structures, mining, dredging, etc.) can significantly affect the ecology of the coastal zone and the functioning of coastal and ocean processes and resources. For example, development activities in beach and dune areas can change patterns of sediment transport or alter inshore current systems, and diking for agriculture can affect the functioning of wetlands through reduced freshwater inflows and through changes in water circulation. Similarly, industrial development in the coastal zone can decrease the productivity of wetlands by introducing pollutants, including heavy metals, and by changing water circulation and temperature patterns. Marine aquacultural activities in tropical areas often involve removal of mangrove forests to create aquaculture ponds, interfering significantly with the many functions mangrove systems perform, such as serving as buffers for coastal storms and nursery habitats for juvenile fishes. Activities such as port development and the dredging that inevitably accompanies it can significantly degrade coral reefs through the buildup of sediment. Activities farther inland, such as logging, agricultural practices (e.g., burning of cane sugar), and animal husbandry practices (e.g., pollution of streams by animal waste), represent important sources of damage to estuarine and ocean areas through increased flow of sediment, pesticides, and other pollutants into riverine and estuarine systems.

Different coastal and ocean uses such as fishing and offshore oil development, also often conflict with or adversely affect one another. Two major types of conflicts related to coastal and ocean resources can be noted: (1) conflicts among users over the use or nonuse of particular coastal and ocean areas and (2) conflicts among government agencies that administer programs related to the coast and ocean. By users we mean both direct, actual users of the coast and ocean (e.g., oil operators and fishermen), and indirect or potential users (e.g., environmental groups that promote the nonutilitarian values of the coast and ocean, members of the public who live in other areas, and future generations). Because most marine resources are public property and there is an important public, or societal, interest in the management of the land-side of the coastal zone, the rights and interests of such indirect users must also be taken into account (Cicin-Sain 1992).

Some typical manifestations of conflicts among users suggested by E. L. Miles involve: (1) competition for ocean or coastal space; (2) adverse effects of one use, such as oil development, on another use, such as fisheries; (3) adverse effects on ecosystems; and (4) effects on onshore systems, such as competition for harbor space (Miles 1991). Conflicts also occur among government agencies that administer programs related to the coast and ocean, including both interagency conflicts (among agencies at the same level of government, whether national, provincial, or local) and intergovernmental conflicts (or among different levels of government). Agency conflicts occur for a variety of reasons, including divergent legal mandates and different missions; differences in agency outlook and type and training of personnel; differences in external constituency groups; and lack of information or communication (Cicin-Sain 1992).

Models of Coastal and Ocean Uses and Their Interactions

Several efforts have been made to develop a typology of ocean and coastal uses and their interactions. A. Vallega (1996) presents an overview of the categories of ocean and coastal uses found in the literature (table 1.1). As can be seen in table 1.1, some authors, such as A. D. Couper in his global marine interaction model and Vallega in his coastal use framework, emphasize the "water side" of coastal and ocean uses, and others, such as J. C. Sorensen and S. T. McCreary (1990) and M. D. Pido and T. E. Chua (1992), emphasize the "land side" of the coastal zone (Vallega 1996). It should also be noted that none of the typologies presented in table 1.1 includes nonconsumptive uses of the marine environment and its

Table 1.1. Examples of Ocean and Coastal Use Models

Couper's Global Marine Interaction Model (Couper 1983)	Sorensen and McCreary (1990)	Pido and Chua (1992)	Vallega Coastal Use Framework (Vallega 1992)	Hawaii Ocean Resources Management (Example of CZM Approach) (1991)
1. Navigation and communication	1. Fisheries	1. Agriculture	1. Seaports	1. Research
2. Mineral and energy resources	2. Natural area and protection systems	2. Fisheries and aquaculture	2. Shipping	2. Recreation
3. Biological resources	3. Water supply	3. Infrastructure	3. Sea pipelines	3. Harbors
4. Waste disposal and pollution	4. Recreation development	4. Mining	4. Cables	4. Fisheries
5. Strategy and defense	5. Tourism	5. Ports and harbors	5. Air transportation	5. Marine ecosystem protection
6. Recreation	6. Port development	6. Industry	6. Biological resources	6. Beaches and coastal erosion
7. Research	7. Energy development	7. Tourism	7. Hydrocarbons	7. Waste management
8. Marine environmental quality	8. Oil and toxic spill contingency planning	8. Urban development	8. Metalliferous renewable resources	8. Aquaculture
	9. Industrial siting	9. Forestry	9. Renewable energy resources	9. Energy
	10. Agriculture	10. Shipping	10. Defense	10. Marine mammals
	11. Mariculture		11. Recreation	
			12. Waterfront structural development	
			13. Waste disposal	
			14. Research	
			15. Archaeology	
			16. Environmental protection and preservation	

Source: Vallega 1996.

resources. Examples of nonconsumptive uses are protection and promotion of nonutilitarian values of the ocean (the value of its mere existence and its value to future generations) and aesthetic uses (the human enjoyment and spiritual renewal that proximity to the ocean can provide). Also not included in these typologies is the crucial role of the ocean, from a global perspective, in regulating the earth's climate.

Drawing on these efforts, we present a revised list of major uses and activities of the coastal zone and ocean in table 1.2.

Table 1.2. Principal Coastal and Ocean Activities

Navigation and Communications
Shipping
Port and harbor development
Navigational aids
Communication cables

Living Marine Resources
Fishing (traditional, artisanal, industrial)
Aquaculture
Gathering of seaweed
Gathering of other marine creatures (e.g.,
  sea cucumbers, snails, shells, corals, pearls)
Tropical fish collection
Collection of marine mammals for con-
  sumption, display, or research
Watching marine mammals (e.g., whale
  watching)
Marine biotechnology applications; use of
  marine organisms or processes for product
  development

Mineral and Energy Resources
Hydrocarbon (oil and gas) exploration and
  production
Offshore drilling, pipeline laying, platforms,
  installations
Exploitation of sand and gravel aggregates
Exploitation of other minerals (gold, placer
  deposits, polymetallic sulfides, manganese
  nodules)
Other forms of ocean energy (e.g., wave
  energy, tidal power, ocean thermal energy)

Tourism and Recreation
Hotels, vacation homes
Tourism infrastructure (transportation,
  services)
Swimming and diving, underwater parks
Recreational fishing, boating
Nonconsumptive aesthetic uses

Coastal Infrastructure Development
Roads, bridges, other transportation
  infrastructure
Water supply and treatment
Reclamation or alteration of coastal waters
  (e.g., for building of human settlements,
  impoundment for aquaculture ponds, diking
  for recreational facilities)
Desalination facilities

**Waste Disposal and Pollution
Prevention**
Siting of industrial facilities
Sewage disposal
Dumping of dredged materials
Disposal of other wastes
Nonpoint sources of marine pollution
  (agriculture, runoff, river sedimentation)
Oil and toxic spill contingency planning

**Ocean and Coastal Environmental
Quality Protection**
Protection of the ocean's global role in
  regulating climate
Protection of the oceans from pollution
Protection of the oceans from transport and
  disposal of hazardous materials (radioactive, chemical, etc.)
Establishment of marine and coastal protected areas, parks to protect special areas or
  features (e.g., coral reefs, wildlife sanctuaries)
Marine mammal protection
Protection of cultural resources (e.g., religious sites, archaeological sites, ship-
  wrecks)
Protection of the oceans from transfer of
  alien species (e.g., through ballast waters)
Prevention and mitigation of harmful algal
  bloom phenomena

(continues)

Table 1.2 (*continued*)

Beach and Shoreline Management
Erosion control programs
Protection structures (against storms,
  waves)
Replenishment of beaches
Prevention and mitigation of coastal
  hazards (storms, inundation sea-level
  rise)

Research
Oceanography
Marine geology and coastal processes
Fisheries and marine mammal research
Marine biology, biodiversity, biotechnology
Archaeology
Studies of human uses of the ocean

Military Activities
Transit and maneuvers by navies
Military special areas (test ranges,
  exercise areas)
Enforcement of national ocean zones

It is important to keep in mind that upland uses can affect the coastal and ocean activities shown in table 1.2. Major upland activities are summarized in table 1.3.

Efforts have also been made to classify the relationships among ocean and coastal users—for example, according to whether they are conflictual or mutually beneficial. Various methods have been employed. Couper's global marine interaction model (1983) categorizes interactions among users as (1) harmful or conflicting, (2) potentially harmful, (3) mutually beneficial, or (4) harmful to one use but beneficial to another. Vallega (1990) applies this matrix to the Mediterranean context in figure 1.3.

Matrices such as the one in figure 1.3 are useful as starting points for coastal managers to think about the complementarity or lack thereof of alternative uses of a particular ocean and coastal area. Several shortcomings of these matrices, though, should be noted. They allow for consideration of interactions between only two uses, whereas coastal managers often must deal with the interactions of multiple (more than two) uses. Moreover, good data on interactions among uses often are not available and must be collected on a case-by-case basis. This has led authors such as Miles (1992) to argue that analysts must identify the sharpest points of conflict among uses and suggest ways of resolving them.

Table 1.3. Major Upland Activities Affecting Coastal and Ocean Activities

Agriculture	Nonpoint sources of pollution
Forestry	Alteration of wetlands

River diversions, damming,
  other alterations
	Construction of human
	  settlements and roads

Industrial (point source) waste	Mining

Figure 1.3. Interaction of Uses of the Mediterranean Sea

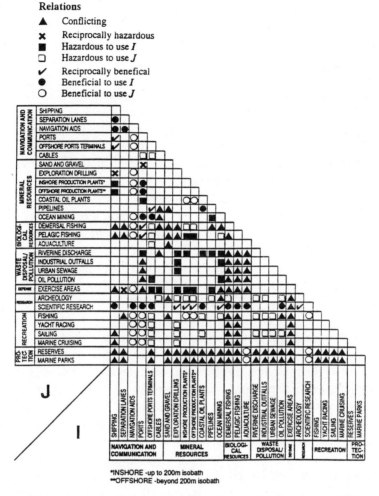

Source: Reproduced with permission from Vallega 1990; originally based on Couper 1983.

Examples of Interactions among Coastal and Ocean Uses and Their Environments

In this section, we discuss some examples of interactions among coastal and ocean uses drawn from both developed and developing countries. Most of the interactions fall in the "harmful" or "conflictual" category, although a couple of "mutually beneficial" interactions are noted at the end of the section.

Conflicts Related to Marine Transportation

Marine transportation is one of the oldest and most traditional uses of the sea. Most of the world's goods are transported by ship, making marine transportation a huge industry worldwide and of great importance to many nations. However, activities related to this use, such as navigational dredging and port expansion, can have significant negative effects on other coastal uses, such as fishing operations, and on environmental quality, as the following examples from China and the United States suggest.

Dalian Port, China: Port Expansion versus Fisheries and Mariculture

In recent years, fishery development efforts in China have concentrated on mariculture and distant-water fisheries. Mariculture industries in China produced 1.62 million metric tons (1.79 million short tons) in 1990 from an offshore area of 430,000 hectares (1.06 million acres). The mariculture farming area more than doubled from 1983 (180,013 hectares, or 444,810 acres) to 1988 (413,260 hectares, or 1.02 million acres).

This expansion took place primarily in estuaries and paralleled the rapid development of major seaports in China as trade and shipping activities increased. Container turnover by twenty-nine of China's major Pacific seaports almost doubled from 1986 (640,000 TEU [twenty-foot equivalent unit, the size of a standard container]) to 1989 (1.25 million TEU).

Dalian Port, one of the largest ports in China, needed more anchoring space in Dalian Bay, bordering the northern Yellow Sea. Dalian Bay was also an important area for fisheries, with annual exports valued at $195 million, about 20 percent of the value of Dalian's total exports.

In expanding the mooring area, the port authority in Dalian compensated fishermen for the use of some traditional fishery areas. However, the compensation did not totally stop the encroachment by mariculture rafts. The interaction between these two uses (fisheries and transportation) resulted in mutually harmful effects. On the one hand, the encroachment of mariculture rafts caused economic losses to the port authority by increasing the number of berthing days of visiting ships due to shrinkage of anchoring areas. In addition, during 1989–1991 twenty-nine vessels reported to Dalian Port that their rudders were entangled by mariculture rafts. On the other hand, fishermen complained of poor water quality caused by anchoring activities in the traditional fishing and mariculture areas.

Source: Yu 1994

Port of Miami, United States: Dredging and Filling versus Fishery Habitat

The Port of Miami is located on an artificial island in Biscayne Bay in southeastern Florida. The port serves the cruise and commerce markets of the Caribbean region and Latin America. In the late 1970s, the port planned a three-part expansion project to accommodate increasing passenger and cargo volumes, including enlargement of Dodge Island, filling and expansion of adjacent Lummus Island, and dredging to deepen navigational channels.

The dredging and filling destroyed approximately 2.4 hectares (6 acres) of mangrove forest, 33.2 hectares (82 acres) of sea grass, and 65.5 hectares (162 acres) of shallow-water fishery habitat. Mitigation was requested by the state and federal environmental protection agencies. According to the permit issued by the U.S. Army Corps of Engineers (COE) on October 6, 1980, the mitigation plan called for planting 101.6 hectares (251 acres) of unvegetated Biscayne Bay bottom with sea grass and at least 2.27 hectares (5.6 acres) of intertidal areas with mangroves within four years of the date of issuance of the permit. The COE stipulated that a 70 percent survival rate for the sea grass plantings should be attained.

During phase I of the sea grass mitigation effort, initiated in 1982, a total area of 15.4 hectares (38 acres) was planted at a cost of $781,245, with an average survival rate of 12 percent. During phase II in 1984 another 8 hectares (20 acres) of sea grass were planted, at a cost of more than $1.1 million. In fall 1985, the cost for planting the remaining acreage was estimated at $4 million. Despite the port's efforts to maximize the chances of sea grass survival, the mitigation project was delayed and sea grass was not planted at the required rate. The main reasons for these difficulties were the increasing planting costs for the port, uncertainty associated with the technical aspects of sea grass planting, and plantings made on benthic areas unsuited for sea grass survival. In July 1986, the port began preparing a proposal for alternative mitigation to fulfill its obligation.
Source: Wessel and Hershman 1988.

Conflicts Related to Coastal Land Reclamation

"Reclamation" of coastal areas by filling in wetlands, diking, and building dams and other barriers to exclude coastal waters has been practiced in many parts of the world for many years. The Netherlands, with its low-lying coastal zone, which is highly vulnerable to natural hazards, and its limited land area, represents the most advanced case of building coastal defense structures and of deploying many other means for gaining "land from the sea." Reclamation has also been a prominent feature of coastal development in industrialized and rapidly industrializing countries with limited developable coastal land and high population densities, such as Japan and the Republic of Korea (South Korea).

South Korea: Coastal Area Reclamation versus Fisheries
As of 1995, South Korea's total reclaimed area was about 109,000 hectares (270,000 acres) with most large-scale reclamation projects on the

western coast. The purpose of reclamation has been mostly agricultural, accounting for more than 92 percent, with industrial and residential purposes accounting for only 7 percent and 1 percent, respectively. Pursuant to the National Basic Plan for Reclamation of 1991, the South Korean government has launched new projects to convert 261 tidal estuarine areas of 123,510 hectares (305,200 acres) into agricultural, residential, industrial, power plant, and waste disposal sites by 2001.

In spite of the benefits of land expansion, reclamation of coastal areas that are important hatchery and nursery grounds for fishes results in conflicts with traditional uses such as fishing and mariculture. Coastal reclamation in the Siwha area, one of the major reclaimed areas on the western coast, provides a good example of such conflicts.

The reclamation of the Siwha area—11,303 hectares, or 27,930 acres (2,303 hectares, or 5,691 acres, of Siwha I and 9,000 hectares, or 22,000 acres, of Siwha II)—began in 1986 and was expected to be completed by 1997. The Siwha Bay area was important traditionally for mariculture of seaweeds, oysters, and shellfish. Not only did fishermen lose their traditional fish-farming areas, but also their fisheries declined due to pollution from agricultural pesticides and fertilizers used on the reclaimed area and industrial sites built there.

Source: Lee et al. 1993; Hong 1995.

Bangladesh: Estuarine Area Reclamation versus Navigation and Waterborne Commerce

In Bangladesh reclamation of land to increase agricultural area is characterized by two types of activities: (1) accretion promotion activities with closure of moribund or secondary channels, and (2) prevention of saltwater flooding by impoldering or impounding newly accreted areas with earthen dikes and drainage structures.

As part of land reclamation practices under the Coastal Embankment Project (CEP), large-scale closure employing cross-dam construction techniques has taken place in Bangladesh since 1956. The first such dam, Noakhali Cross-Dam No. 1, was built over a branch of the Meghna River, creating accreted siltation of about 21,000 hectares (52,000 acres). In 1964, a second dam was built in the dying Meghna River but farther downstream, causing an accretion of 31,000 hectares (77,000 acres). Two other small dams were built on Monpura Island during 1978–1989. These dams resulted in an accretion of 200 hectares (494 acres). The next step of the CEP was to impolder these existing and newly accreted areas by earthen dikes and drainage structures. The purpose of the polder construction was to prevent seawater from flowing over the newly reclaimed areas and to reserve freshwater for agriculture.

A crucial issue that was not scientifically addressed was the adverse effects created by these disturbances to the water system, including accelerated erosion at undesirable locations, unacceptable siltation patterns, siltation of drainage outlets, loss of fish spawning grounds, and hindrance of commerce dependent on navigation. Although these topographical changes have increased agricultural sites, they have had serious adverse effects on navigational safety and biological ecosystems.
Source: Barua 1993.

Conflicts Related to Offshore Oil Development

Although offshore oil development represents one of the most economically profitable uses of the sea, it can have significant adverse effects on other uses of the coastal zone and ocean, such as fishing operations, coastal tourism, recreation, aquaculture, and marine transportation. Some examples of conflicts surrounding offshore oil development may be drawn from Norway, China, and the United States.

Norway: Oil Development versus Marine Fisheries
Traditionally, Norwegian waters have provided abundant fishing grounds, enabling Norway to become the world's primary producer and exporter of Atlantic salmon. On the other hand, petroleum production activities within these waters have rapidly increased since the late 1960s have had negative effects on the fishing industry. Petroleum production activities can harm fishing activities in at least three different ways: through reduction of fishing space, through pollution, and by means of seafloor debris. Although the expanding offshore oil development and resulting pollution have decreased access to fishing grounds, the most direct negative effect has been the problem of seafloor debris. By the end of 1981, more than 3,000 incidents of loss or damage to fishing gear had been reported; estimated costs totaled nearly $5 million. The Norwegian authorities have pursued a strategy of integrated marine policy as a response to these interuser conflicts.
Source: Andresen and Floistad 1988.

China: Oil Development versus Estuarine Fisheries
China's offshore oil development has been concentrated in the shallow waters of the semi-enclosed Bohai Sea. The drilling muds (a sometimes toxic by-product of the drilling process) are discharged largely on-site and receive little treatment. The long-term effects of these discharges on the health of the coastal ecosystem of the Bohai Sea are causing con-

cern. In addition, in the Changjiang River estuary and Hangzhou Bay, which are part of the East China Sea, an oil concentration of 0.46 milligrams per liter was recorded, significantly exceeding the level (0.05 milligrams per liter) set by class 1 of the national seawater quality standard.

These estuarine areas represent China's largest fish spawning and feeding grounds for several species of high economic value. Fishermen have complained about the loss of their intertidal aquaculture areas to oil-related projects and about fish and shellfish mortality attributable to oil spills.

Source: Yu 1994.

Threats to Coral Reefs from Various Marine Uses

Coral reefs are among the world's most valuable ecosystems in terms of their biological diversity and their productivity and as a source of livelihood for many coastal communities. As noted by A. T. White and colleagues (1994), coral reefs provide food for a countless number of economically marginalized communities, they are a source of jobs and revenues (from the fishing and tourism sectors), are a source of recreation and enjoyment, and protect tropical coasts from storm damage, among other values. However, coral reefs face many threats from the destructive activities of other coastal users and from activities in upland areas, as the following examples from the Philippines, Egypt, and Sri Lanka suggest.

Philippines: Destructive Fishing Practices versus Coral Reefs and Fish Habitat

The residents of San Salvador Island in the Philippines face problems related to severe damage or destruction of fish habitat from the use of explosives and sodium cyanide for catching aquarium fish, the use of fine-mesh nets, and overfishing. Nearly one-fifth of the island's coral cover has been killed by recent misuses of sodium cyanide in fishing coupled with the physical destruction of large portions of coral reefs by scuba divers collecting aquarium fish. After introducing sodium cyanide into holes and tanneries, the divers remove entire pieces of the reef in order to reach fish that are hiding. The local community was poorly organized and generally unwilling to work cooperatively to address these destructive fishing methods. In 1988, the Marine Conservation Project for San Salvador (MCPSS) was established in the community, using education, community organization, and community participation to establish a municipal marine park. This approach seems to have put an end to coral reef destruction on the island.

Source: Buhat 1994.

Philippines: Deforestation from Logging versus Coral Reefs and Fisheries

Coastal development is rapidly expanding in the Bacuit Bay area in the Philippines, leading to conflicts over resources among the tourism, fisheries, and logging industries. Logging in the area has accelerated erosion and sediment input into the main river and bay, resulting in high sediment deposition in the bay and on the coral reefs. This has killed the corals and led to a subsequent decline of associated fisheries.
Source: Hodgson and Dixon 1988.

Egypt: Unregulated Tourism Development versus Coral Reefs

By 1991, tourism was the second largest source of foreign exchange in Egypt. As a result, the Hurghada–Port Safâga region on the Red Sea coast became host to numerous holiday villages, which are growing considerably faster than the government's capacity to provide infrastructure. Consequently, facilities for tourists were poorly regulated and sited. These poor management practices caused extensive degradation of coastal and marine habitats, especially coral reefs.

The Tourism Development Authority was created in 1991 to support environmentally sound, master-planned tourism. Although there were conflicts among the oil, tourism, and fisheries industries, coastal zone management and environmental assessment tools provided a framework for a planned approach to tourism, thus alleviating the problems in the Hurghada–Port Safâga region and providing a sound basis for complementary reef conservation and tourism along the Red Sea coast.
Source: World Bank 1994.

Sri Lanka: Mining of Coral Reefs versus Shoreline Stability

One of the utilitarian functions of fringing and patch-type coral reefs is to buffer adjacent tropical shorelines. The coral formations create a natural breakwater system, protecting coastlines from high-energy, storm-driven waves and swells. On the coastlines of Sri Lanka, erosion has been a critical problem.

The human activity most harmful to shoreline and beach stability is the severe mining of coral and coral sands for lime production, cement manufacture, and other purposes. Coral material is harvested from both coastal quarries and offshore areas using dynamite, crowbars, and boats. Erosion is further accelerated by the mining of natural replenishment sands for use as construction aggregate and for lime production and placers. In locations along Sri Lanka's southwestern coast such as Hikkaduwa, the net effect of these practices has been severe coastal erosion.
Source: DuBois and Towle 1985.

Conflicts Related to Tourism

Tourism, the world's largest economic enterprise, is frequently a major money earner for coastal nations. Although coastal and marine tourism is a significant economic activity for coastal nations, it can also have detrimental effects on other uses of the coast and on the coastal environment. Conflicts between tourism and fisheries are illustrated by the case of St. Lucia in the Caribbean Sea.

St. Lucia: Tourism versus Fisheries

Tourism and other recreational activities have increased rapidly in the smaller island states of the Caribbean Sea, such as St. Lucia, which are likely to experience conflicts over resource use. In Soufrière, on the western coast of St. Lucia, there was a serious conflict between the water-dependent activities of tourists (e.g., scuba diving and snorkeling) and those of commercial fishermen, such as spiny lobster and crab trap fishermen.

The natural attributes of the Soufrière area make it an important destination for foreign tourists; the region's luxuriant coral reefs make it the island's best diving destination. Commercial crab trap fishermen who had traditionally set their pots in or near coral reef areas complained about the behavior of conservation-minded divers who sometimes opened pots to release trapped fishes and lobsters. To resolve this conflict, the island's Department of Fisheries, through its 1986 declaration of marine reserves and fishing priority areas (FPAs), sought to exclude the fishermen from what were considered ecologically sensitive areas in the tourist-frequented area while providing them with unlimited access to other prime fishing locations.

Source: George and Nichols 1994.

Mutually Beneficial Interactions

Although in this section we have emphasized problematic relationships among various uses of the coastal zone and ocean—interactions in need of management—we should note that some interactions of marine and coastal uses can be mutually beneficial and supportive. This is one of the main challenges for coastal managers: to identify and encourage a mutually beneficial mix of activities in particular coastal and ocean areas. We cite two examples of mutually beneficial interactions, one from the United States and the other from Canada.

United States: Recreational Fishing and Offshore Oil Development

More than 90 percent of offshore oil production in the United States takes place in the Gulf of Mexico offshore from the state of Louisiana. There is also a strong recreational fishing and diving industry centered on the area's oil and gas platforms, which attract congregations of fish. It has been estimated that oil and gas platforms are the destinations of approximately 37 percent of all saltwater recreational fishing trips in the region. The oil and gas platforms are also a valuable resource for recreational divers, since there are few suitable hard bottom areas or natural coral reefs off the Louisiana coast.

Because of the importance of the oil and gas platforms to offshore fishing and recreational diving, and in consideration of their inevitable removal as they become obsolete, the Louisiana Artificial Reef Initiative (LARI) was formed in 1984 to develop an artificial reef program using the platforms. The average one-way distance traveled by user groups to reach the structures was at least 62 kilometers (38.5 miles). A location closer to shore was to be considered where new artificial reefs would be constructed using obsolete oil platforms, thus solving the oil industry's disposal problem and creating an effective and convenient fishing destination for the recreational fishermen.
Source: Stanley and Wilson 1989.

Canada: Sand Mining and Commercial Fishing

The Bay of Fundy is located in eastern Canada, bordering the provinces of New Brunswick and Nova Scotia. Strong tidal currents, ranging up to 6 knots in a water column averaging 75 meters (246 feet) deep, move in and out of the bay. The main sand wave field occurs at the mouth of Scots Bay in the Bay of Fundy. An estimated volume of 35 million cubic meters (1.23 billion cubic feet) of coarse-grained, well-sorted sand, in water 17 meters (56 feet) deep could be mined as a source of aggregate.

This sand wave field has developed on a basal transgressive gravel surface that occurs as a distinct layer over a thick sequence of estuarine sediments. Fishing activity is intense on the gravel lag surface, as evidenced by a dense pattern of trawl marks on the seabed. In contrast, fishing is not done on the sand wave field because of the intensity of sediment transport. Therefore, with removal of the sand in the wave field, a substantial area of similar habitat (e.g., gravel lag) to that presently being fished will be made available for fishing activities. This example shows how two seemingly competitive resource industries, in this case mining and fishing, may coexist and benefit each other.
Source: Lay et al. 1993.

Over the past few decades, it has become apparent that a more formal approach to management of the coastal and ocean region is needed—one suited to the realities of conflicts among various uses and the environmental effects of multiple coastal and ocean activities, as illustrated by the foregoing examples.

Early Efforts at Coastal Management

The history of coastal area management over the past thirty years is one of evolution from largely land-oriented approaches encompassing few uses to multiple-use approaches encompassing broader coastal areas both on the land and sea sides, and emphasizing ecosystem and multiple-use interdependencies. Such a progression is illustrated by A. Vallega's "stage-based model of coastal area management," reproduced here in table 1.4.

Some of the earliest efforts in this regard were made by developed coastal nations as coastal zones became degraded due to inappropriate development and

Table 1.4. Stage-Based Model of Coastal Area Management

Stage	Objective	Coastal Uses under Management	Geographical Coverage
1960s: rise	Use management addressing a single environmental issue socially perceived as important	One or a few uses (e.g., seaports, recreational uses)	The shoreline
1970s: implementation	Use management and environmental protection	Few uses (e.g., seaports, manufacturing plants, recreation and fishing)	Various alternative extents: • The shoreline • A coastal zone delimited according to arbitrary criteria • A coastal zone delimited according to administrative criteria
1980s: maturity	Use management and environmental protection	Multiple-use management	Various alternative extents characterized by the proclivity to move seaward to extend management to national jurisdictional zones
1990s: international primacy	Integrated coastal area management (ICAM)	Comprehensive use management, management of the coastal ecosystem	A zone extending • Landward according to various criteria • Seaward to the outer limit of the widest national jurisdictional zone

Source: Vallega 1996

poor planning. The United States initiated one of the first formal efforts with its Coastal Zone Management Act of 1972. Although the U.S. concept of coastal zone management (CZM) purports to be integrated, in fact most of the state programs that have emerged focus primarily on the management of shore land use and, so far at least, less so on coastal water-related issues. Nonetheless, a great deal of experience has been amassed as twenty-nine [thirty-one in 1997] coastal states and territories have developed and implemented coastal management programs in a wide variety of political, cultural, and physical settings (Archer 1988).

Shortly after initiation of the CZM program in the United States, coastal management efforts began in a number of other countries, many of them developing countries. Often, these programs were encouraged and supported by donor organizations or donor nations in an effort to ensure that development projects reflected good coastal planning and practice. The early work of this type generally dealt with a particular coastal region, a specific development project, or an important coastal problem such as erosion. Understandably, these early initiatives, sometimes aggregated under the term *coastal area management programs* (CAMPs), tended not to be comprehensive or well integrated. However, collectively, they do represent the beginning of the worldwide coastal management movement and can provide a basis on which to build broader programs. Good information about these activities can be found in work by J. C. Sorensen and S. T. McCreary (1990) and early editions of the CAMP newsletters (now the newsletter *Intercoast* published by the University of Rhode Island).

In 1993, Sorensen estimated that "approximately 142 integrated coastal zone management efforts have been initiated by approximately 57 sovereign or semi-sovereign states,"[1] as shown in table 1.5 (Sorensen 1993, 45). In conducting our

Table 1.5. Global Distribution of ICM Efforts, by Region

Region	Coastal States	Number of States with ICM Programs	Total Coastal States in Region	States in Region with ICM Programs (%)
North America	Canada Mexico United States	3	3	100
Caribbean and western Atlantic islands	Barbados* British Virgin Islands* Dominica* Puerto Rico St. Kitts–Nevis* St. Lucia* Turks and Caicos Islands* U.S. Virgin Islands	8	26	31

(continues)

Table 1.5 (*continued*)

Region	Coastal States	Number of States with ICM Programs	Total Coastal States in Region	States in Region with ICM Programs (%)
Central America	Belize* Costa Rica* El Salvador* Honduras	4	7	57
South America	Argentina Brazil* Colombia* Ecuador* Venezuela*	5	11	45
Europe and North Atlantic Ocean	Albania* Bulgaria* Croatia* France* Greece Ireland* Italy Netherlands* Portugal Spain* United Kingdom*	11	35	31
Near East	Egypt* Israel* Oman* Saudi Arabia* Syria* Turkey	7	15	46
Africa and western Indian Ocean	Côte d'Ivoire* Guinea-Bissau Nigeria Republic of South Africa Tunisia	5	38	13
Asia	Bangladesh* Brunei Darussalam* China Indonesia* Japan* Malaysia* Maldives* Philippines• Singapore* Sri Lanka• Taiwan• Thailand•	13	21	57

Table 1.5 (*continued*)

Region	Coastal States	Number of States with ICM Programs	Total Coastal States in Region	States in Region with ICM Programs (%)
Oceania	American Samoa	7	21	33
	Australia*			
	Federated States of Micronesia*			
	Guam			
	Hawaii			
	New Zealand			
	Northern Mariana Islands			

Note: An asterisk (*) marks those programs or projects that were directed by the national government either to manage all or most of the nation's coastal zone or done as pilot programs in a relatively small percent of the nation's entire coastal zone.

Source: adapted from Sorensen 1993.

cross-national ICM survey in 1996, we identified eight additional nations with some type of ICM program. They are Fiji and Western Samoa in the Oceania region; the United Republic of Tanzania and Madagascar in the region of Africa and the western Indian Ocean; Chile in South America; and Pakistan, India, and the Republic of Korea in Asia.

Birth of the ICM Concept

Preparation for the United Nations Conference on Environment and Development (UNCED), which began in the late 1980s, stimulated further thinking on the coastal management concept. Several United Nations organizations, most especially the Food and Agriculture Organization of United Nations (FAO), submitted concept papers on coastal management to the UNCED secretariat as input to the preparatory process (FAO 1991). Increasingly, analysts referred to the kind of management seen as necessary to manage the world's coasts as "integrated coastal management." Consultants and others working in the field began using the same terminology when referring to the increasingly more comprehensive management programs being designed and implemented; see, for example, publications by T. E. Chua describing work on ICM with six nations in Southeast Asia (Chua and Scura 1992; Chua 1993).

In 1988, the World Meteorological Organization (WMO) and the United Nations Environment Programme (UNEP) jointly created the Intergovernmental Panel on Climate Change (IPCC) for the purpose of conducting scientific assessments of global climate change and possible responses to it. The IPCC formed a

1. This number excludes the United States and its territories shown in the table.

subgroup on coastal zone management (CZM subgroup) to examine in particular the issue of sea-level rise and the threat it might pose to low-lying coastal nations. Subsequently, the CZM subgroup issued a set of recommendations that endorsed integrated coastal management as the appropriate framework within which to develop and implement specific measures to reduce vulnerability to accelerated sea-level rise. Indeed, the recommendations made it clear that the effectiveness of such measures depended upon their being implemented within a broader ICM framework (IPCC 1992).

As discussed in some detail in chapter 3, by the time of the UNCED conference itself (June 1992), the ICM concept had become an integral part of the ocean and coasts chapter (Chapter 17) of Agenda 21, the principal action program to emerge from UNCED.

Summary

This chapter discussed the major reasons why an integrated approach to the management of coastal and ocean areas is needed: (1) the many values (economic, ecological, social) present in coastal areas; (2) the effects that ocean and coastal uses, as well as activities farther upland, can have on ocean and coastal environments; and (3) the effects ocean and coastal uses can have on one another. A range of cases of conflicts over use of coastal areas was discussed, exemplifying the need for ICM to address and mitigate the negative consequences of such conflicts and to safeguard coastal values. The evolution of the ICM concept was briefly depicted—from limited coastal management programs in the 1970s to more multifaceted and complex programs in the 1990s, as called for in UNCED.

In chapter 2, we turn to a more extensive discussion of the central concept of integrated coastal management.

Definitions of Integrated Coastal Management and Fundamental Concepts

Introduction

The scholarly literature on integrated coastal management has tended to focus primarily on the land-sea interface and on approaches, options, and methods for controlling the uses of coastal land (see, e.g., UNDIESA 1982; Clark 1991; Sorensen and McCreary 1990; Chua 1993). A complementary but separate body of literature has examined the use and management of ocean areas under national jurisdiction, exploring such aspects as the organization of national agencies to address ocean issues and promotion and regulation of ocean uses (see, e.g., Vallejo 1993; Juda and Burroughs 1990; Vallega 1992; Cicin-Sain and Knecht 1985; Knecht, Cicin-Sain, and Archer 1988; Van Horn, Peet, and Wieriks 1985). This separation in the literature reflects, in effect, historical practice by nations around the world. Traditionally, "ocean" management issues—such as the extent of a nation's maritime zone and its maritime boundaries with other nations and the nation's stance in international fora on such issues as freedom of navigation and conservation of highly migratory species (e.g., whales and long-distance fisheries)—have been within the purview of national governments, particularly of foreign ministries and specialized fisheries and maritime agencies (traditionally dealing with maritime transportation and ports), and often, as well, of military agencies such as naval departments. Coastal management, in contrast, traditionally began on the land side of the coastal zone, focusing on issues related to the special interface between the land and the sea, such as shoreline erosion measures, protection of wetlands, siting of coastal development, and public access to the coast. Because these issues initially centered on control and regulation of coastal land, they tended to be handled by planning-type agencies at provincial and local levels of government, sometimes with national-level intervention and at other times without it.

Increased use of the oceans and coastal zones in the twentieth century—including the establishment and operation of offshore oil production installations and aquacultural facilities, all forms of coastal recreation and tourism, use of the ocean for waste disposal, ocean mining, and so on—has posed considerable challenges to this often dual system of management—ocean and coastal, and has called attention instead to the "seamless web" that ties ocean and coastal activities together. Over the past several decades, for example, we have come to understand the important influences of land activities on the quality of the water on which ocean-based activities such as fishing, aquaculture, and tourism depend; it is now well recognized that land-based activities account for more than 70 percent of all marine pollution. Similarly, all ocean activities—from fishing to marine transportation to offshore oil production to recreation—emanate from and have their bases on coastal land in the form of harbor and port space, onshore processing facilities, and the like.

Throughout this book, we draw on both the ocean and coastal management bodies of literature and experience, reflecting our strong belief that a major aspect of integrated coastal management, as envisioned in Chapter 17 of Agenda 21, is considering and addressing the mutual influences of land and sea. Given the divergent historical evolution of ocean and coastal affairs in most nations, however, ocean and coastal activities may not be within the purview of the same government agencies, thus often involving problems of coordination and harmonization of governmental activities in the coast and ocean. This is a typical problem in integrated coastal management and one we address in some detail in chapter 6.

Considerable work has been done in recent years in defining the major characteristics of integrated coastal management; see, for example, Clark 1991, Sorensen and McCreary 1990, Chua 1993, Scura et al. 1992, Vallega 1993, FAO 1991, Bower 1992, Van der Weide 1993, and Cicin-Sain 1993. Although different authors emphasize somewhat different aspects of integrated coastal management (partly as a result of diverse disciplinary backgrounds and partly as a reflection of their varied experiences with integrated coastal management in different parts of the world), there appears to be growing agreement on the outlines of a general model of integrated coastal management. Further, as discussed in chapter 4, the models of integrated coastal management that several major international entities, such as the World Bank, UNEP, the OECD, the World Conservation Union, formerly the International Union for the Conservation of Nature and Natural Resources (IUCN), and others have put forth in recent years in the form of international ICM guidelines are by and large consistent with one another and reflect growing international consensus on what ICM is all about.

In this chapter we thus present a simplified model of integrated coastal management by addressing a set of questions which decision makers are likely to pose:

What is integrated coastal management?
What are its goals?
What triggers the need for ICM?
What does *integrated* mean in ICM?

What is being managed (land, sea areas, resources)?

What are the functions of ICM—what does ICM actually do?

Where is ICM applied—what are the boundaries?

How is management carried out and what guiding principles are used?

Who should carry out the management—national or subnational levels of government or both?

What are the typical stages in developing ICM?

What capacity is needed for ICM?

What kinds of institutions should carry out ICM?

Although our discussion in this chapter is largely conceptual and presents basic definitions of different aspects of ICM, we provide examples from countries around the world.[1] Here, we rely largely on the survey we conducted in 1996 with key experts on ICM in twenty-nine countries in both the developed and developing worlds, as described in the introduction to this book.

Fundamental Concepts

In this section, we pose and answer a series of questions as a way of explaining the concepts that we believe are fundamental to an understanding of integrated coastal management.

What Is Integrated Coastal Management?

Integrated coastal management can be defined as a continuous and dynamic process by which decisions are made for the sustainable use, development, and protection of coastal and marine areas and resources. First and foremost, the process is designed to overcome the fragmentation inherent in both the sectoral management approach and the splits in jurisdiction among levels of government at the land-water interface. This is done by ensuring that the decisions of all sectors (e.g., fisheries, oil and gas production, water quality) and all levels of government are harmonized and consistent with the coastal policies of the nation in question. A key part of ICM is the design of institutional processes to accomplish this harmonization in a politically acceptable manner.

Integrated coastal management is a process that recognizes the distinctive character of the coastal area—itself a valuable resource—and the importance of conserving it for current and future generations. The coastal area, be it continental or island based, is a special area where land and sea meet that includes various characteristics:

• The coastal area is characterized by dynamic and frequently changing phys-

1. The conceptual discussion in this chapter relies in part on Cicin-Sain 1993.

ical features (e.g., changes in beaches and barrier islands due to the force of winds and waves).

• Valuable ecosystems of great productivity and biodiversity are present, such as mangrove forests, sea grass beds, other wetlands, and coral reefs—all of which provide crucial nursery habitat for many marine species.

• Coastal features such as coral reefs, mangrove forests, and beach and dune systems that serve as critical natural defenses against storms, flooding, and erosion.

• The area is generally of great value to human populations as they seek to settle in, use, and enjoy coastal marine resources and space.

• The coastal area provides the base for all human activities in the ocean—from marine recreation and fishing to marine transportation and offshore mineral development. Many of these activities represent significant economic benefits—both actual, for those resources already under exploitation, and potential for resources yet to be exploited. All such activities depend, to various extents, on the coastal area for their operation.

• Because the coastal area is often highly desired by various users and populations, coastal space is a finite resource over which there are often conflicts.

• Management of the two sides of the coastal area—land and sea—poses difficult challenges and complexities based, in part, on the public character of the ocean area and the generally mixed public and private character of the land area. Typically, the presence of "general-purpose" government authorities for land and "single-purpose" authorities for the ocean further complicates the governance issue.

Islands, unique in being surrounded and enclosed by the sea, represent the maximum coastal condition and thus require a high degree of integrated coastal management. For small islands, the coastal zone and ocean may be the only potentially developable assets. Consequently, planning and management for these resources require great care if a long-term pattern of sustainable development is to be achieved.

What Are the Goals of ICM?

The goals of integrated coastal management are to achieve sustainable development of coastal and marine areas, to reduce vulnerability of coastal areas and their inhabitants to natural hazards, and to maintain essential ecological processes, life support systems, and biological diversity in coastal and marine areas. Integrated coastal management is multipurpose oriented: it analyzes implications of development, conflicting uses, and interrelationships among physical processes and human activities, and it promotes linkages and harmonization between sectoral coastal and ocean activities.

What Triggers the Need for ICM?

The need to establish a program of integrated coastal management in a particular nation may arise for a number of reasons. Depletion of coastal and ocean resources (e.g., through overfishing or exploitation of corals for building materials) typically is a powerful trigger. Another important catalyst may be an increase in pollution that endangers public health, or poses threats to water-based industries such as aquaculture, fishing, and tourism. A desire to increase the economic benefits obtained from use of the coast and ocean (as through fostering marine tourism) may also point out the need for ICM planning and management. A related catalyst may be the desire to develop uses of the coastal and marine area previously not exploited in a particular country, such as extraction of offshore oil or other minerals, marine aquaculture, or new forms of fishing for underexploited stocks or in different areas.

Our survey asked respondents to describe the origin of ICM in their respective countries and to note whether any major catalyst had facilitated its initiation. Their responses are summarized in tables 2.1 and 2.2.

Regarding the reasons for initiating ICM, the majority of respondents (56 percent) reported some kind of environmental problem, such as depletion of resources, pollution, or ecosystem damage. Respondents from developing countries were most likely to designate an environmental issue as the reason for ICM initiation (75 percent), followed by respondents from middle developing countries (53 percent).

In contrast, respondents from developed countries were much more likely to cite some type of economic development as the reason for initiating ICM (43 percent) than were those from middle developed countries (26 percent) and developing countries (20 percent). This is somewhat counterintuitive, since prevailing assumptions would suggest that the main interest of developing nations lies in seeking economic development opportunities. What the data suggest, we think, is a growing realization in developing countries, especially the middle developing countries, of the environmental costs of rapid or inappropriate economic development.

Regarding catalysts for initiating ICM (table 2.2), the most frequent response was "national-level government initiative" (73 percent of all responses).

Table 2.1. Reasons for Initiating Integrated Coastal Management in Selected Countries

	All (%) (N = 49)	Developed (%) (N = 14)	Middle Developing (%) (N = 15)	Developing (%) (N = 20)
Depletion of resources	18	0	27	25
Pollution	20	21	13	25
Ecosystem damage	18	21	13	25
Economic benefits from coast and ocean	22	36	13	20
New economic opportunities on coast or in ocean	6	7	13	0
Damage from coastal hazards	10	14	7	10
Other	4	0	13	0
Uncertain	0	0	0	0

Source: Authors' 1996 cross-national survey.
Note: Percentages do not add up to 100% because respondents were allowed to mark multiple reasons.

Table 2.2. Catalysts for Initiating Integrated Coastal Management in Selected Countries

	All (%) (N = 48)	Developed (%) (N = 13)	Middle Developing (%) (N = 15)	Developing (%) (N = 20)
Environmental crisis	23	8	20	35
Proposal for new coastal or marine development	42	15	67	40
National-level government initiative	73	92	60	70

Table 2.2 (*continued*)

	All (%) (N = 48)	Developed (%) (N = 13)	Middle Developing (%) (N = 15)	Developed (%) (N = 20)
State or local initiative	33	23	33	40
Initiative from NGO	35	46	40	25
Initiative from international organization	25	8	33	30
Initiative from regional entity	19	15	13	25
External funding	25	0	27	40
Earth Summit prescriptions	31	23	27	40
Other	4	8	0	5
Uncertain	2	0	0	5

Source: Authors' 1996 cross-national survey.
Note: Percentages do not add up to 100% because respondents were allowed to mark multiple catalysts.

What Does Integrated Mean in ICM?

As discussed in the next chapter, Agenda 21 challenges us to think about the entire spectrum of area encompassing both the land and water sides through its call for "integrated management and sustainable development of coastal and marine areas, including Exclusive Economic Zones." It emphasizes as well the need for proper management of marine fishery resources under national control and the importance of the connection between land and sea, particularly regarding land-based sources of marine pollution.

Five main zones can be identified in the coastal-marine spectrum: *inland areas*, which affect the oceans mainly via rivers and nonpoint sources of pollution; *coastal lands*—wetlands, marshes, and the like—where human activity is concentrated and directly affects adjacent waters; *coastal waters*—generally estuaries, lagoons, and shallow waters—where the effects of land-based activities are dominant; *offshore waters*, mainly out to the edge of national jurisdiction (200 nautical-miles offshore); and *high seas*, beyond the limit of national jurisdiction.

Although natural processes in these five zones tend to be highly intertwined, it is difficult to integrate management regimes across the zones because the nature of property, the nature of government interests, and the nature of government institutions tend to differ in these zones, as summarized in figure 2.1. Roughly speaking, with regard to the nature of property in coastal areas, there tends to be a continuum of ownership: in inland areas, private property tends to predominate; on coastal lands, there tends to be a mix of public and private property; and in coastal and offshore waters, public property concerns are dominant. This generalization, of course, varies somewhat from country to country according to cultural conceptions of private and public property. In many South Pacific islands, for example, coastal lands and waters are often communally controlled by village-level councils of elders (see, e.g., South et al. 1994).

With regard to the nature of government interests, local or provincial interests tend to predominate in inland areas, whereas there tends to be a mix of local, provincial, and national interests on coastal lands and in coastal waters. Moving farther out, ultimately to offshore waters and the high seas, national and international interests become most important. The nature of government institutions also differs in the various zones. On land, there are often well-established "multiple-purpose" government institutions at the local and provincial levels to address such questions as control of land use and conflicts among uses. On the water side, there tends to be only "single-purpose" provincial or national agencies operating, each concerned primarily with a single use of the ocean, such as fisheries operations or oil and gas extraction. Given these differences, management of the five zones may require common and complementary, yet somewhat differentiated, approaches and institutions.

Several dimensions of integration need to be addressed as a part of an ICM process:

Figure 2.1. Nature of Property and Government Interests and Institutions in Coastal and Ocean Areas

	Inland Areas	Coastal Lands	Coastal Waters	Offshore Waters	High Seas
Nature of property	Private	Private or public	Predominantly public		
Nature of government interests	Local or provincial	Mix of local, provincial and national		Mainly national	Mainly international
Nature of government institutions	Multiple-purpose agencies		Single-purpose agencies		

Source: Cicin-Sain 1993.

1. *Intersectoral integration.* Integration among different sectors involves both "horizontal" integration among different coastal and marine sectors (e.g., oil and gas development, fisheries, coastal tourism, marine mammal protection, port development) and integration between coastal and marine sectors and land-based sectors that affect the coastal and ocean environment, such as agriculture, forestry, and mining. Intersectoral integration also addresses conflicts among government agencies in different sectors.

2. *Intergovernmental integration*, or integration among different levels of government (national, provincial, local). National, provincial, and local governments tend to play different roles, address different public needs, and have different perspectives. These differences often pose problems in achieving harmonized policy development and implementation between national and subnational levels.

3. *Spatial integration*, or integration between the land and ocean sides of the coastal zone. As discussed in chapter 1, there is a strong connection between land-based activities and what happens in the ocean involving water quality, fish productivity, and the like; similarly, all ocean activities are based or dependent on coastal land. And yet, as figure shown in 2.1, different systems of property ownership and government administration predominate on the land and ocean sides of the coastal zone, often complicating the pursuit of consistent goals and policies.

4. *Science-management integration*, or integration among the different disciplines important in coastal and ocean management (the natural sciences, the social sciences, and engineering) and the management entities. Although, as discussed in some detail in chapter 7, the sciences are essential in providing information for coastal and ocean managers, there often tends to be little ongoing communication between scientists and managers. (Here, the sciences are broadly construed to mean the natural sciences concerned with the oceans and coasts, such as oceanography, coastal processes, and fishery sciences; the social sciences, concerned with coastal human settlements and user groups as well as management processes that govern ocean and coastal activities; and coastal and ocean engineering, which focuses on all forms of coastal and ocean structures.)

5. *International integration.* Integration among nations is needed when nations border enclosed or semi-enclosed seas or there are international disputes over fishing activities, transboundary pollution, establishment of maritime boundaries, passage of ships, and other issues. Although in many instances, coastal and ocean management questions are within the purview of

national and subnational governments within national jurisdiction zones (200-nautical-mile EEZs, extended fishery zones), in many other cases, nations face ocean and coastal management problems vis-à-vis their neighbors and thus must seek internationally negotiated solutions. Typically, the national government plays the leading role in such negotiations.

What Is Being Managed in ICM?

Authors differ in terms of what areas, resources, and activities they include under the aegis of integrated coastal management. At its heart, ICM, in our view, is concerned with area management and with interactions among various resources and activities in specific coastal and ocean areas. Although, as discussed earlier, ICM must include both coastal lands and coastal waters because of the important reciprocal effects of processes and activities in these two areas, how far offshore and onshore an ICM regime should extend depends on the situation at hand; the topic is discussed further in a subsequent section of this chapter.

What resources and activities should come under the aegis of ICM? The many resources and activities that take place in coastal lands and waters—fisheries, nonrenewable resource extraction, tourism, agriculture and aquaculture,residential and commercial real estate development, marine transportation, recreation, and so forth—all represent specialized activities that are generally already within the purview of specialized agencies. In most cases, ICM would not supplant such specialized sectoral management but would instead supplement, harmonize, and oversee it. Thus, for example, fishery managers would continue to concern themselves with fishery allocations and the like, but an integrated coastal management entity would take primary responsibility for the effects of land-based sources of pollution on fishery nursing areas as well as with the links (both positive and negative) between fisheries and other uses.

What Are the Functions of Integrated Coastal Management?

In contrast to sectoral entities and processes, which tend to be concerned with only one use or resource of the coastal and marine environment, the ICM process is expected to address several important functions related to overall patterns of use, the well-being of marine and coastal areas, and the protection of key fisheries habitat, as set out in table 2.3.

Typical activities related to these functions are detailed in Table 2.4.

In our survey, we asked respondents which types of activities had been part of their country's ICM efforts; the responses are summarized in table 2.5.

Table 2.3. Major Functions of Integrated Coastal Management

Area Planning
Plan for present and future uses of coastal and marine areas; provide a long-term vision.

Promotion of Economic Development
Promote appropriate uses of coastal and marine areas (e.g., marine aquaculture, ecotourism).

Stewardship of Resources
Protect the ecological base of coastal and marine areas; preserve biological diversity; ensure sustainability of uses.

Conflict Resolution
Harmonize and balance existing and potential uses; address conflicts among coastal and marine uses.

Protection of Public Safety
Protect public safety in coastal and marine areas typically prone to significant natural, as well as human-made, hazards.

Proprietorship of Public Submerged Lands and Waters
As governments are often outright owners of specific coastal and marine areas, manage government-held areas and resources wisely and with good economic returns to the public.

Table 2.4. Typical ICM Activities

Area Planning
Studies of coastal environments and their
  uses
Zoning of uses
Anticipation of and planning for new uses
Regulation of coastal development projects
  and their proximity to the shoreline
Public education on the value of coastal and
  marine areas
Regulation of public access to coastal and
  marine areas

Promotion of Economic Development
Industrial fisheries
Artisanal fisheries
Mass tourism
Ecotourism
Marine aquaculture
Marine transportation
Port development
Marine recreation
Offshore minerals
Ocean research
Access to genetic resources

Stewardship of Resources
Conduct of environmental assessments
Conduct of relative risk assessments
Establishment and enforcement of environ-
  mental standards
Protection and improvement of coastal water
  quality (point sources, nonpoint sources)
Establishment and management of coastal
  and marine protected areas
Protection of marine biodiversity
Conservation and restoration of coastal and
  marine environments (mangrove forests,
  coral reefs, wetlands, etc.)

Conflict Resolution
Studies of multiple uses and their interac-
  tions
Applications of conflict resolution methods
Mitigation of unavoidable adverse effects on
  some uses

(continues)

Table 2.4 (*continued*)

Protection of Public Safety
Reduction of vulnerability to natural disasters and global changes (e.g., sea-level rise)
Regulation of development in high-risk areas through such methods as establishment of "set-back lines"
Construction of coastal defense measures (e.g., seawalls)
Creation of evacuation plans or other measures in case of coastal emergency

Proprietorship of Public Submerged Lands and Waters
Establishment of leases and fees for use of publicly held coastal and marine resources and spaces
Establishment of joint ventures to exploit non-renewable resources (e.g., offshore oil)

Table 2.5. Types of ICM Activities Conducted in Selected Countries

	All (%) (N = 49)	Developed (%) (N = 14)	Middle Developing (%) (N = 15)	Developing (%) (N = 20)
Area Planning				
Studies of coastal environments and their uses	92	79	93	100
Zoning of uses	76	79	73	75
Anticipation of and planning for new uses	57	71	53	50
Regulation of coastal development projects and their proximity to the shoreline	73	79	73	70
Public education on the value of coastal and marine areas	69	64	73	70
Regulation of public access to coastal and marine areas	65	71	73	55
Promotion of Economic Development				
Industrial fisheries	55	57	67	45
Artisanal fisheries	63	43	80	65
Mass tourism	53	50	67	45
Ecotourism	69	57	67	80
Marine aquaculture	88	86	93	85
Marine transportation	51	50	53	50
Port development	73	79	67	75
Marine recreation	80	93	87	65
Offshore minerals	45	43	40	50
Ocean research	61	50	67	65
Access to genetic resources	22	21	20	25
Stewardship of Resources				
Conduct of environmental assessments	90	86	80	100
Conduct of relative risk assessments	45	71	40	30

(*continues*)

Table 2.5 (*continued*)

	All (%) (N = 49)	Developed (%) (N = 14)	Middle Developing (%) (N = 15)	Developing (%) (N = 20)
Stewardship of Resources				
Establishment and enforcement of environmental standards	73	79	67	75
Protection and improvement of coastal water quality (point sources, nonpoint sources)	76	93	73	65
Establishment and management of coastal and marine protected areas	80	86	87	70
Conservation and restoration of coastal and marine environments (mangroves, coral reefs, wetlands)	78	71	73	85
Conflict Resolution				
Studies of multiple uses and their interactions	71	71	67	75
Applications of conflict resolution methods	47	50	53	40
Mitigation of unavoidable adverse effects on some uses	47	64	40	40
Protection of Public Safety				
Reduction of vulnerability to natural disasters and global changes(e.g., sea-level rise)	41	64	34	30
Regulation of development in high-risk areas through such methods as the establishment of "set-back lines"	49	50	53	45
Construction of coastal defense measures (e.g., seawalls)	61	64	60	60
Creation of evacuation plans or other measures in case of coastal emergency	31	50	27	20
Proprietorship of Public Submerged Lands and Waters				
Establishment of leases and fees for use of publicly held coastal and marine resources and spaces	49	64	47	40
Establishment of joint ventures to exploit nonrenewable resources (e.g., offshore oil)	35	29	33	40

Source: Authors' 1996 cross-national survey.
Note: Percentages do not add up to 100% because respondents were allowed to mark multiple activities.

Among the major findings were the following:

- The most significant ICM activity reported was *area planning*, with 92 percent of responses reporting conduct of studies of the coastal zone, 76 percent reporting zoning of uses, and 73 percent reporting regulation of coastal development projects and their proximity to the shoreline.

- Respondents also frequently reported activities to *promote economic development*, especially for marine aquaculture (88 percent of responses), marine recreation (80 percent of responses, with the greatest proportion from developed countries), port development (73 percent of responses), and ecotourism (69 percent of responses, with the greatest proportion from developing countries).

- With respect to *stewardship of resources*, 90 percent of respondents reported the conduct of environmental assessments and 80 percent reported the establishment and management of coastal and marine protected areas.

- Regarding *conflict resolution*, 71 percent of respondents reported the conduct of studies of multiple uses and their interactions, while only less than half (47 percent) reported the use of conflict resolution and mitigation approaches.

- Concerning *protection of public safety*, the most common activity was construction of coastal defense measures such as seawalls, mentioned by 61 percent of respondents.

- Finally, with regard to *proprietorship of public submerged lands and waters*, less than half of the respondents (49 percent) reported the use of leases and fees for ocean and coastal uses, and 35 percent reported the use of joint ventures for exploitation of nonrenewable resources.

Where Is ICM Applied?

One of the thorniest questions in integrated coastal management is how far inland and how far offshore an ICM regime should extend. In terms of the inland boundary, a watershed (containment area) approach permits better control of pollutants coming into a particular marine environment. However, watersheds often span large distances and encompass multiple jurisdictions. Hence, if too wide an area is included under the aegis of integrated coastal management, attention and resources may well be diverted away from the area constituting the heart of the land-sea interface. Thus, a watershed-based inland boundary may be appropriate for the specific purposes of controlling land-based sources of marine pollution and fresh water inflows but not for other ICM purposes.

Where to establish an offshore boundary for integrated coastal management is also a difficult question. Living marine resources do not respect human-made boundaries in coastal waters, nor do ocean processes. Nevertheless, as discussed earlier, different governmental units will be involved and somewhat different interests will be at stake as one goes farther offshore. How far from the coast an ICM regime ought to extend will depend in part on the characteristics of the physical system offshore, especially the continental shelf system, as well as on the relations between national and provincial governments in a particular nation. For example, the state of Oregon, which is in the forefront of ocean management efforts in the United States, has declared a state stewardship area encompassing the continental shelf offshore from the state (extending about 56–130 kilometers, or 35–80 miles, offshore) on the grounds that the resources and activities found in this area vitally affect the state's inhabitants (Bailey 1997).

As documented by J. C. Sorensen and S. T. McCreary (1990), there is great diversity among nations in the kinds of boundaries they have established for their coastal and ocean management efforts. Our 1996 survey confirmed this finding of diversity. As shown in table 2.6, more than half of the respondents (57 percent) reported that the landward boundary either varied according to use or was not yet determined. The same answer was given regarding seaward boundaries by 38 percent of respondents. Twenty-one percent of respondents reported that the boundary had been established at the 12-nautical-mile territorial sea limit, with the greatest proportion of respondents from developed countries (36 percent) reporting this boundary. Eight percent of respondents reported boundaries at the edge of the 200 nautical-mile EEZ, with the greatest number of respondents from middle developing countries (21 percent) reporting this boundary.

Table 2.6. Nature of ICM Landward and Seaward Boundaries in Selected Countries

	All (%) (N = 48)	Developed (%) (N = 14)	Middle Developing (%) (N = 14)	Developing (%) (N = 20)
Landward Boundary				
Up to 100 m (0.062 mi.)	4	0	14	0
100–500 m (0.062–0.311 mi.)	8	7	0	10
500 m to 1 km (0.311–0.62 mi.)	4	0	0	10
1–10 km (0.62–6.21 mi.)	10	0	7	15
Extent of local government jurisdiction (e.g., coastal city or county)	4	7	0	5
Watershed	6	0	14	10
Varies according to use	38	50	36	30
Not yet determined	19	21	29	15
Uncertain	6	14	0	5

(*continues*)

Table 2.6 (*continued*)

	All (%) (N = 48)	Developed (%) (N = 14)	Middle Developing (%) (N = 14)	Developing (%) (N = 20)
Seaward Boundary				
Mean low tide or mean high tide	2	7	0	0
Arbitrary offshore distance from tidal mark	17	0	14	30
3-NM[a] territorial sea boundary	6	7	7	5
12-NM territorial sea boundary	21	36	14	15
Edge of continental shelf	2	0	0	5
Limit of national jurisdiction/ 200-NM EEZ or fisheries zone	8	7	21	0
Varies according to use	23	21	21	25
Not yet determined	15	14	14	15
Uncertain	6	7	7	5

Source: Authors' 1996 cross-national survey.
[a] NM = nautical miles.

How Is Management Carried Out, and What Guiding Principles Are Used?

Integrated coastal management involves (1) a set of both substantive and procedural principles; (2) a management strategy that emphasizes adaptation and feedback; and (3) the use of particular approaches, methods, and techniques. In this section, we consider the issue of principles for ICM; the other two questions are addressed, respectively, in chapter 7 (methods) and chapter 9 (monitoring and evaluation).

Two broad categories of principles for guiding ICM can be identified: principles based on agreed international norms for environment and development that have emanated from the Earth Summit and key international agreements and principles specifically related to the special character of coasts and oceans.

Principles Related to Environment and Development

As discussed in some detail in chapter 3, the Rio Declaration on Environment and Development (Rio Declaration) is a set of twenty-seven principles to guide national and international actions on environment, development, and social issues approved by all nations participating in the Earth Summit conference. Some of these principles are new; others represent the reiteration of principles already established in international law; still others represent changes in established principles of international law (Van Dyke 1996). Overall, they provide a broad set of norms to guide nations in the pursuit of sustainable development. A full listing of the principles may be found in chapter 3; here, we briefly summarize the major ones.

1. *Principle of interrelationship and integration.* Although not explicitly stated as a principle in the Rio Declaration, the principle of interrelationship and integration forms the backbone of sustainable development and is the underlying theme of the Rio Declaration and Agenda 21 (UNDPCSD 1996). It means that we must address the interrelationships, or interdependence, among issues and sectors and between environment and development. In contrast to past thinking and past practices, environmental protection and development cannot be considered as separate activities; each one must incorporate the other.

2. *Inter- and intragenerational equity principles.* The principles of inter- and intragenerational equity relate to justice and fairness vis-à-vis questions of environment and development. The principle of intergenerational equity reflects the view that as members of the present generation, we hold the earth in trust for future generations (UNDPCSD 1996) and therefore we should not preclude the options of future generations (WCED 1987). The principle of intragenerational equity refers to the obligation to take into account the needs of other users (other members of society), especially regarding distribution of the benefits of development.

3. *Principle of the right to develop.* This principle relates to the basic right to life of every human being as well as the right to develop his or her potential so as to live in dignity. It is the first principle enunciated in the Rio Declaration.

4. *Environmental safeguards principle.* This principle relates to prevention of environmental harm through anticipatory measures to prevent harm rather than through post hoc efforts to repair it or provide compensation for it. Environmental safeguards go hand in hand with the precautionary principle (see below), and with two other Rio principles—the need for states to enact and implement effective environmental legislation and the principle of prevention of transboundary (across frontiers) environmental harm.

5. *Precautionary principle.* According to the precautionary principle, lack of scientific certainty is no reason to postpone action to avoid potentially serious or irreversible harm to the environment. Principle 15 of the Rio Declaration reads, in part, "Where there are threats of serious or irreversible damage, lack of full scientific certainty shall not be used as a reason for postponing cost-effective measures to prevent environmental degradation" (U.N. Document A/CONF. 151/26 (Vol. 1), 12 Aug 1992).

6. *"Polluter pays" principle.* This principle holds that it is important that the environmental costs of economic activities, including costs of prevention of potential harm, be internalized rather than imposed on society as a whole. The principle was originally developed by the Organization for Economic Cooperation and Development (OECD) to ensure that firms paid the full costs of controlling pollution and were not subsidized by the state. The principle is intended to apply within states rather than between states. Principle 16 of the Rio Declaration brings the "polluter pays" approach beyond a strictly developed country context; it calls on national authorities to "endeavor to promote the internalization of environmental costs and the use of economic instruments, taking into account the approach that the polluter should, in principle, bear the cost of pollution" (U.N. Document A/CONF. 151/26 (Vol. 1), 12 Aug 1992).

7. *Transparency principle and other process-oriented principles.* The transparency principle demands that decisions be made in an open, transparent manner, with full public involvement. This principle goes hand in hand with a number of related principles: encouragement of participation by all major groups, including women, children, youth, indigenous peoples and their communities, NGOs, local authorities, and others; the public's right to access to environmental information; and the importance of conducting environmental impact assessments to help ensure informed decision making and to provide for public participation and access to information.

Principles Related to the Special Character of Oceans and Coasts

There is no ready-made analog to the internationally agreed on Rio Declaration that applies to oceans and coasts; nevertheless, a number of important principles addressing the special character of oceans and coasts are contained in various publications. Drawing on the work of J. M. Van Dyke (1992), J. H. Archer and M. C. Jarman (1992), J. R. Clark (1992), B. Cicin-Sain and R. W. Knecht (1985; 1992), and Cicin-Sain, ed. (1992), we have put together a list of eleven major principles that we believe capture the essence of the uniqueness of oceans and coasts and can provide guidance for ocean and coastal management. These eleven principles are grouped into three main categories: (1) principles related to the public nature of the oceans, (2) principles related to the biophysical nature of the coastal zone, and (3) principles related to the use of coastal and ocean resources and space.

1. *Principles based on the public nature of the oceans.* As discussed in chapter 3, ocean resources have traditionally been thought of as part of the public domain, not to be exclusively owned or benefited from by any one group or person. This principle affirms the traditional public character of the oceans

and refers to the public trust doctrine (rooted in Roman law and part of the tradition of a number of countries). It holds:

> The Public Trust doctrine should, in nations where it applies, govern decisions in order to protect the interests of the whole community and the interests of intergenerational equity. This doctrine requires that conflicts be resolved in favor of keeping the oceans whole and protecting the interests of the public today and in the future. Managing resources as a commons should be preferred over privatizing such resources. If private developments are allowed, the public should receive financial benefits from such developments. (Van Dyke 1992).

2. *Principles related to the biophysical nature of the coastal zone.* These principles are derived from the special circumstances found at the land-sea interface. These special circumstances include the following:

- The coastal area is a distinctive resource system that requires special management and planning approaches. The major resource systems of the coast and coastal habitats (such as coral reefs and mangrove forests) are not only distinctive but also extremely productive of renewable resources; marine resources also are highly distinctive. The high mobility and interdependence of marine resources and processes also makes traditional land-based management not altogether suitable for coastal area management.

- Water is the major integrating force of coastal resource systems. Because it operates at the land-water interface, ICM relates to water in one way or another, whether making provisions for marine commerce, the ravages of sea storms, resource conservation, or pollution abatement. The water influence not only establishes special conditions but also dictates unusual and complex institutional arrangements (Clark 1992).

- Significant interactions take place across the land-water boundary and require that the whole system—upland, shore land, intertidal area, and nearshore waters—be recognized and managed as an integral unit.

Hence, the following principles are suggested:

   a. Since landforms fronting on the water's edge (sand dunes, mangroves, fringing coral reefs) play a key role in combating erosion and sea-level rise and contribute to long-term sustainability, they should be maintained.

  b. Care should be taken to maintain salt marshes, coastal wetlands, and other coastal habitats in their natural condition.

  c. Emphasis should be placed on "designing with nature"—for example, using special vegetation rather than physical structures for erosion control.

  d. In considering coastal development projects, interruption of the natural longshore drift system should be kept to an absolute minimum.

  e. Special protection must be provided for rare and fragile ecosystems and endangered and threatened species in order to ensure that the biodiversity of the ecosystem is not reduced or lost.

3. *Principles related to the use of coastal and ocean resources and space.* These principles relate to management of conflicts in coastal areas, development of guidelines for use, and public participation:

  a. Generally, protection of living resources and their habitats should be given priority over exploitation of nonliving resources; nonexclusive uses should be preferred over exclusive uses; and reversible exclusive uses should be preferred over irreversible exclusive uses. Potential conflicts should be identified early and in an orderly fashion, and equitable solutions should be developed by processes that protect and enhance public order (Van Dyke 1992; Archer and Jarman 1992).

  b. New developments in the coastal zone that are water dependent should have priority over those that are not. For example, construction of a new port facility should have priority over construction of a new office building.

  c. The historically based claims of indigenous peoples to ocean space and ocean resources should be recognized and their traditional practices of dealing with ocean resources from a perspective of kinship and harmony should be followed whenever possible (Van Dyke 1992).

  d. Based on recent assessment studies of climate change, adverse effects in the coastal zone, such as increased erosion, flooding, and saltwater intrusion, should be addressed within the framework of ICM.

e. When considering retreat as an adaptation option in dealing with accelerating sea level, efforts should be made to create or make provisions for new habitats for coastal resources (e.g., wetlands) and species that otherwise would be lost.

Who Should Carry Out the Management?

Most analysts would agree that a combination of national and provincial or local authorities is needed to carry out integrated coastal management, although analysts will vary on the extent to which they emphasize a "top-down" or a "bottom-up" approach. Local community concerns, even in centralized political systems, are always important in integrated coastal management processes, particularly with regard to inland areas and coastal lands. On the other hand, as discussed earlier, the national government's role becomes more and more dominant as one goes farther offshore.

Our 1996 cross-national survey asked respondents to report which was the primary level of government responsible for ICM in their country. As can be seen in table 2.7, some 61 percent of respondents reported that the national government is the primary level of government responsible for ICM, 16 percent reported the state or provincial government as the main party responsible, and 20 percent reported mixed responsibility between the different levels of government. Notwithstanding this finding of the national government as the main level involved, responses to other questions regarding actions taken at national, provincial, and local levels of government showed that in many nations, even though one level of government may have primary responsibility for ICM, other levels are involved as well. This makes the question of intergovernmental integration of policy a crucial one, as discussed in some detail in chapter 6.

Table 2.7. Primary Level of Government Conducting ICM Efforts in Selected Countries

	All (%) (N = 49)	Developed (%) (N = 14)	Middle Developing (%) (N = 15)	Developing (%) (N = 20)
Federal (national)	61	64	73	50
State (provincial)	16	21	7	20
Local	0	0	0	0
Other (mix of federal, state, and local)	20	14	20	25
Uncertain	2	0	0	5

Source: Authors' 1996 cross-national survey.

What Are the Typical Stages in Developing an ICM Program?

As with any other public policy, the policy for integrated coastal management generally goes through a number of predictable stages of development—issue identification and assessment, program preparation or formulation, formal adoption and funding, implementation, operation, and evaluation (see figure 2.2).

Although circumstances in different nations will, of course, differ, we detail here typical stages in the development of an ICM program; these are adapted, in part, from international guidelines prepared by the World Bank (1993). We emphasize that these steps are merely illustrative of one possible path of ICM development. Many different ways of developing ICM programs are in evidence around the world, reflecting each nation's particular physical, socioeconomic, cultural, and political conditions.

Stage 1. Identification and Assessment of Issues

• The need for management action is recognized as a result of such factors as an environmental crisis, deteriorating resource conditions, or perceived economic opportunities in the coast or in the ocean.

Figure 2.2. The Six Stages of an ICM Process

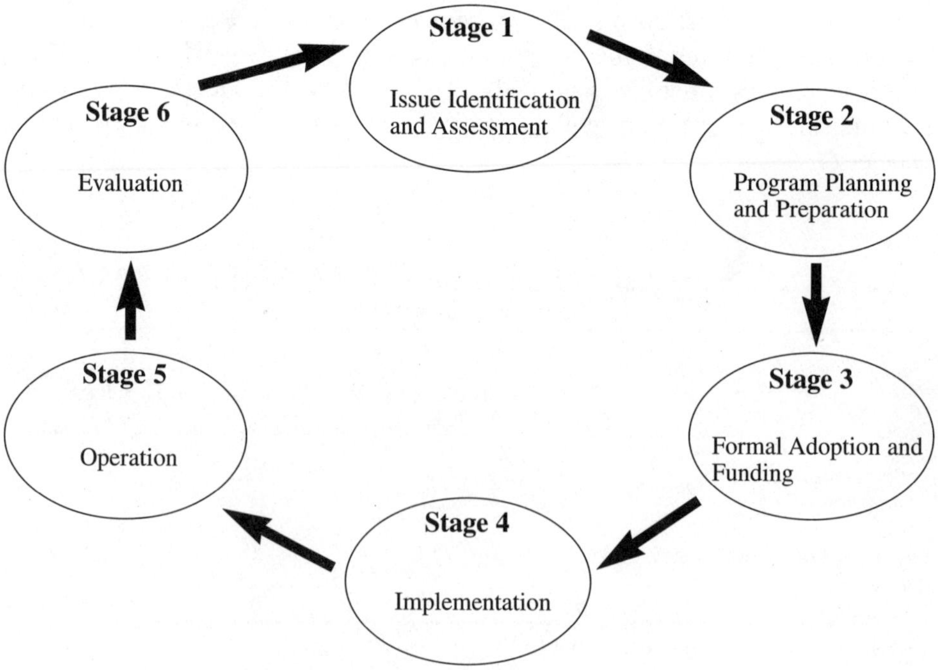

Source: Adapted, with modifications, from Olsen 1993.

- Consultative meetings with key agencies and stakeholders confirm the presence of problems and/or opportunities and the need for action.

- A concept paper outlining the need for ICM may be prepared.

- A team is created to formulate an ICM plan.

Stage 2. Planning and Preparation

- Necessary information and data on the physical, economic, and social characteristics of the coastal zone, as well as on existing political jurisdictions and on governance issues, are assembled.

- A plan for public participation in the ICM process is developed.

- Management problems (causes, effects, solutions) and development opportunities are analyzed.

- Priorities are set for addressing problems and opportunities, taking into consideration technical and financial feasibility and availability of personnel.

- Feasibility of new economic development opportunities is assessed.

- Appropriate coastal area management boundaries are considered. New management measures, such as zonation schemes, strengthened regulatory programs, and market-based incentives are considered.

- Institutional capacities are assessed. Options for development of suitable governance arrangements, including intersectoral and intergovernmental coordination mechanisms, are developed.

- Recommendations are made for policies, goals, and projects to include in the ICM management program.

- Appropriate monitoring and evaluation systems are designed.

- A timetable, a strategy, and a division of labor are established.

Stage 3. Formal Adoption and Funding

- Policies, goals, new management measures, and initial projects are adopted.

- Governance arrangements are established or improved, including establishment or strengthening of intersectoral and intergovernmental coordination mechanisms.

• Coastal management policies, principles, boundaries, zoning schemes, and so forth are adopted, often by legislative action.

• Staffing and required organizational changes are put into effect.

• Funding arrangements are put into effect.

Stage 4–6. Implementation, Operation, and Evaluation

• Governance body begins oversight of the ICM process and programs.

• New or revised regulatory programs come into effect.

• Individual sectoral line agencies continue to perform their regulatory and management responsibilities but now as part of the overall ICM program.

• Specific projects are designed and undertaken in connection with new economic opportunities in the coastal zone.

• A performance monitoring and evaluation program is initiated.

In our view, hence, it is important that ICM efforts not concentrate for too long on planning and delay implementation. This has been the tendency in some situations—to spend many years on studies, inventories, plans, and the like without moving to the stage of adopting, implementing, and enforcing an ICM program.

Our survey addressed this question, and, as can be seen in table 2.8, it is heartening to note that a large proportion (51 percent) of respondents reported having entered the implementation phase of a program in integrated coastal management.

What Capacity Is Needed for ICM?

Various kinds of "capacity" are needed to successfully carry out an integrated coastal management program:

• *Legal and administrative capacity*—for example, to designate a coastal zone, to develop and carry out coastal plans, to regulate development in vulnerable zones, and to designate areas of particular concern.

Table 2.8. ICM Steps Completed in Selected Countries

	All (%) (N = 49)	Developed (%) (N = 14)	Middle Developing (%) (N = 15)	Developing (%) (N = 20)
Conduct of studies				
Physical science	86	100	80	80
Resource inventories	90	93	87	90
Critical area studies	80	64	80	90
Socioeconomic studies	73	71	80	70
Studies on indigenous issues	47	50	47	45
Economic assessments	71	79	80	60
Policy studies	73	86	67	70
Conduct of coastal inventories	73	86	67	70
Preparation of an ICM plan	80	79	73	85
Formal government adoption of ICM policies	69	79	67	65
Enactment of new coastal law	53	64	33	60
Implementation of an ICM plan	51	50	47	55
Approval of funds for an ICM program	53	50	47	60
Establishment of monitoring program	55	57	53	55
Other	16	14	13	20

Source: Authors' 1996 cross-national survey.
Note: Percentages do not add up to 100% because respondents were allowed to mark all steps applicable to their countries.

- *Financial capacity*—adequate financial resources to carry out the planning and implementation of coastal management efforts.

- *Technical capacity*—information gathering and monitoring of coastal and marine ecosystems and processes, patterns of human use, and the effectiveness of government coastal management programs. Establishment and maintenance of coastal database and information system.

- *Human resources capacity*—personnel with interdisciplinary training in social sciences (including law and planning), natural and physical sciences, and engineering. Also, public awareness and understanding of the coastal ocean environment and the problems and opportunities it offers.

What Kinds of Institutions Should Carry Out ICM?

A fundamental part of most ICM programs is the institutional mechanism created to harmonize the various activities and programs that affect the coastal area and its resources. The proper functioning of such a mechanism is, of course, at the heart of a successful ICM process. To be effective, such a mechanism should have the following attributes:

- It must be authoritative; that is, it must have appropriate legal/legislative authority.

- It must be able to affect the activities of all agencies and levels of government that have decision-making authority relative to the coastal zone.

- It must be seen as a legitimate and appropriate part of the process.

- It must be capable of making "informed" decisions; that is, it must have access to appropriate scientific and technical expertise and data.

The institutional form of the harmonization mechanism will, of course, depend upon the particular nation involved and its government traditions and philosophy. Agenda 21 calls for states to consider establishing (or strengthening) appropriate coordinating mechanisms (such as a high-level policy-planning body), at both the local and national levels, to promote integrated management and sustainable development of marine and coastal areas and to consider strengthening (or establishing) national oceanographic commissions to catalyze and coordinate the needed research. Some options for developing coordinating mechanisms include creating an interagency committee, naming a lead agency, creating a new agency, and training agency personnel to instill an integrated, in contrast to a sectoral, perspective.

As we discuss in more detail in chapter 6, there are three features that tend to enhance the effectiveness of the integrated coastal management process: the coastal management entity and process should be at a higher bureaucratic level than the sectoral agencies to give it the necessary authority to harmonize sectoral actions; the effort should be adequately financed and staffed; and the planning aspect of integrated coastal management should be integrated into national development planning.

Keeping It Simple

Any conceptual discussion of integrated coastal management can be overwhelming—there are so many factors to consider and complexities to address. As with our earlier discussion of the meaning of integration, one must conclude that above all, the coastal manager must be realistic and avoid overselling "integrated coastal management" as some kind of crusade. "Integrated coastal management" is an ideal model that has yet to be fully implemented in any national context. Nevertheless, nations can take tangible steps in moving toward the ideal model depending on their specific needs and circumstances. This is clearly stated in the FAO report that served as an input in the deliberations for the oceans chapter at the Earth Summit (FAO 1991):

> A viable ICM program must be comprehensive but its content and complexity will vary from area to area according to development trends, conservation needs, traditions, norms, governmental systems and current critical issues and conflicts. Compatible multiple-use objectives should always be the main focus. If human and financial resources are limited, ICZM programs can be simplified to include only the following components: (i) harmonization of sectoral policies and goals; (ii) cross-sectoral enforcement mechanism; (iii) a coordination office and, (iv) permit approval and Environmental Impact Assessment procedures (EIA).

Summary

In this and the previous chapter, we established the need for integrated coastal management and defined the major elements of the ICM approach. In this chapter, we defined the meaning and goals of ICM and discussed typical reasons for developing an ICM effort; the functions of an ICM program and typical activities undertaken; the range of inland and seaward boundaries that may be chosen; principles to guide ICM decisions; the relative roles of national and local levels of government in carrying out ICM; typical stages in the development of an ICM program; and the importance of establishing some type of intersectoral coordination mechanism to harmonize policy actions vis-à-vis the coast and ocean. In chapters 3 and 4 we discuss in detail the evolution of international prescriptions for integrated coastal management, focusing in particular on the Law of the Sea Convention (LOS) and on the United Nations Conference on Environment and Development (UNCED).

As these chapters show, ICM has become the framework of choice in the major international pronouncements and agreements emanating from the Earth Summit and underlies the LOS as well. Because these international agreements offer guidance and, in some cases, funding for ICM, it is important for the coastal manager to understand the overall context of international agreements related to ICM. In chapter 3, we begin with a discussion of the evolution of ocean and coastal management and then focus specifically on two major sets of international agreements relating to ICM: the LOS and the set of agreements and prescriptions associated with UNCED.

In chapter 4, we focus particularly on efforts made since UNCED to implement and further operationalize and define the ICM concept. In chapters 5–9, we develop in more detail each of the major concepts discussed in this chapter and provide a "how-to" guide for formulating, implementing, operating, and evaluating ICM programs.

PART II

EVOLUTION OF INTERNATIONAL PRESCRIPTIONS ON ICM

Chapter 3

The Evolution of Global Prescriptions for Integrated Management of Oceans and Coasts

Introduction

As the need for improved management of coastal and ocean areas became clearer, various coastal nations embarked on management programs in their coastal zones. However, as discussed in chapter 1, these efforts tended to be limited in scope and application, often dealing only with a particular region of the coast or ocean or a specific problem such as coastal erosion. Academics and others, increasingly aware of the need for a broader kind of coastal management, began calling for more comprehensive, better integrated approaches (see, e.g., Sorensen and McCreary, 1990, Clark, 1991, Cicin-Sain and Knecht, 1985). This need began to be echoed at international conferences and in declarations and agreements coming out of such meetings. In this chapter, we focus on the way in which the ICM concept and its earlier manifestations have been reflected in formal international agreements, with special attention devoted to the pronouncements related to the United Nations Conference on Environment and Development (UNCED) held in Rio de Janeiro, Brazil, in June 1992; the earlier United Nations Law of the Sea (LOS) Conference (1973–1982), and the still earlier United Nations Conference on the Human Environment held in Stockholm in June 1972. In the twenty-year time span represented by these conferences, one can see the maturation of concepts of integrated management as the general idea is articulated in successive legal instruments and reports. We begin with a description of the way in which ocean issues and then environmental issues were addressed in the international arena.

The Evolution of International Regimes Involving the Ocean and the Environment

In this section, key international developments involving the ocean and the environment are described with special attention given to the manner in which they deal with the evolving integrated coastal management concept.

Third United Nations Conference on the Law of the Sea (1973–1982 and Subsequent Actions)

To describe the evolution of the Law of the Sea—the international "constitution" for ocean governance, we must take a brief excursion into history and provide a brief overview of the major paradigms that have governed the use of the world's oceans.

For about four-hundred years, until the middle of the twentieth century, the predominant paradigm for governance of the world's oceans was the notion of "Freedom of the Seas," pioneered by Hugo Grotius, a Dutch jurist, in 1609. Grotius's famed treatise, *Mare Liberum*, persuasively argues that because the world's continents are separated by the sea into a number of distinct land areas each of which could not develop without intercourse with the others, there is a "natural law" to the effect that the oceans should remain perpetually open for free trade and communication among nations (Friedheim, 1979). Moreover, Grotius argued, because it was, in practical terms, impossible to occupy, divide, or apportion the fluid and mobile sea, the ocean could not be considered "property" and owned as such.

The doctrine of freedom of the seas was developed in the 1600s as a reaction to efforts made earlier, in the 1400s, to divide up the oceans of the world. In 1493, Pope Alexander VI issued a papal bull decreeing that all lands west of the Azores would be Spain's and all lands east of that line (extending all the way around the globe) would appertain to Portugal. Soon, Spain was claiming exclusive navigation rights to the western Atlantic Ocean (challenging the exploratory voyages of the English) and Portugal was claiming the same rights for the South Atlantic and Indian Oceans. As a reaction to this effort to divide up the world, the freedom of the seas doctrine was developed by the Dutch, who had strong interests, through the Dutch East India Company, in trading with the East Indies, and who sponsored Grotius's treatise.

As coastal nations came to support the concept of freedom of the seas, they also saw the need to control the band of sea immediately adjacent to their shorelines. Without such a protective zone, armed ships could sail menacingly close to a country's shore and interfere with its commerce and security. Thus arose the legal concept of the territorial sea (set, by custom, in the late 1700s, at about three nautical miles offshore), holding that nations could establish control over the areas immediately adjacent to their coastlines. Within this zone, they could exert police powers, set customs, and control fishing; still allowing, however, for the "innocent passage" of foreign vessels through the zone.

Thus, from the 1600s to the end of World War II, the freedom of the seas doctrine prevailed. Coastal nations possessed relatively narrow territorial seas, and national activity beyond territorial waters was limited primarily to navigation, coastal fishing, and in a few cases, distant-water fishing and cable laying.

In contrast to this period of stability, since 1945 there has been almost constant activity, both internationally and nationally, with respect to ocean law and policy. The major new development during this time has been the "enclosure" by coastal nations of ocean space adjacent to their coastlines (Friedheim 1979), culminating in the 1980s with a new paradigm for international ocean governance: worldwide acceptance of national jurisdiction over 200-nautical-mile ocean zones.

It all started with a unilateral move by the United States in 1945 to assert jurisdiction over the resources of the continental shelf in response to important discoveries of oil and gas deposits offshore in the Gulf of Mexico. This move precipitated claims of extended jurisdiction by several Latin American nations, and by the late 1950s, many nations were claiming various forms of extended jurisdiction over ocean zones of various widths and powers (Hollick, 1981). Pressure built for the holding of an international conference to restore order and coherence to an increasingly chaotic situation.

Beginning in 1958, three Law of the Sea (LOS) conferences were held under the auspices of the United Nations to sort out the rights and duties of nations regarding the ocean. The first LOS conference, in 1958, drafted four conventions dealing with the high seas, the territorial sea and contiguous zone, fisheries, and the continental shelf, reaching no agreement, however, on the width of the territorial sea. The second LOS conference, held in 1960, also failed to reach agreement on the territorial sea, and expansive claims by coastal states continued (Hollick 1981). Finally, after the United States, the Soviet Union, and other maritime nations acceded to the demands of developing nations to include consideration of an international regime for the resources of the deep seabed, agreement was reached to hold a third United Nations conference on the Law of the Sea (UNCLOS III) in 1973.

In 1982, after nine years of negotiation, that conference did reach agreement on a new "constitution" for the world's oceans, giving legitimacy to the concept of 200-nautical-mile Exclusive Economic Zones under national jurisdiction while protecting most navigational freedoms and establishing 12 nautical miles as the maximum width of territorial seas. The 320 articles contained in the convention address virtually all ocean issues and establish international norms for ocean governance for years to come (United Nations, 1983). Analysis of the detailed provisions of the Law of the Sea as well as of the difficult negotiations leading up to these provisions, though of great interest, clearly lies outside the scope of this book; hence, we refer the reader to major works that undertake these analyses (Churchill and Lowe 1988; Friedheim 1993).

The convention came into force (after ratification by sixty nations) twelve years after conclusion of the conference in 1982, on November 16, 1994. Although significant differences between developing and developed nations over the regime established by UNCLOS III regarding governance of deep-seabed mineral resources delayed widespread acceptance of the treaty, a new agreement concluded in July 1994 (Agreement Relating to the Implementation of Part XI of the United Nations Convention on the Law of the Sea of 10 December 1982) has largely resolved these differences and paved the way for universal national ratification of the LOS Convention.

The major maritime zones sanctioned by the 1982 Law of the Sea Convention are depicted in figure 3.1: the territorial sea, the contiguous zone, the Exclusive Economic Zone, the continental shelf, and the high seas. Table 3.1 shows the numbers of nations claiming various extents of jurisdiction in these zones.

As E. M. Borgese (1995) notes, although the convention must be read as an integral whole, it consists of distinct building blocks, some of which (parts I–X) update and codify existing law and others of which (parts XI–XV) are constitutive, embodying new concepts, creating new law, and establishing new institutions. According to Borgese, the most innovative components of the convention are the following:

- The Exclusive Economic Zone (EEZ)
- The concept of "sovereignty" in the context of the EEZ
- The archipelagic state and the concept of archipelagic waters
- Formal recognition of the common heritage of mankind
- Establishment of the International Seabed Authority
- Comprehensive international environmental law
- A new regime for marine scientific research
- The most advanced framework for technological cooperation and development

Figure 3.1. Maritime Zones

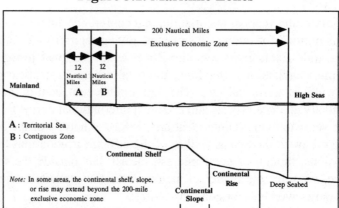

Source: Adapted from Churchill and Lowe 1983, 24.

Table 3.1. Maritime Zone Claims

Maritime Zone and Extent of Jurisdiction	Number of Nations Claiming Jurisdiction
Territorial Sea	
12 NM[a]	117
Less than 12 NM	11
More than 12 NM	16
20–50 NM: 5 nations	
200 NM: 11 nations	
Contiguous Zone	
24 NM	46
Less than 24 NM	8
More than 24 NM	1
Exclusive Economic Zone	
200 NM	85

Either to a line of delimitation determined by coordinates
  or without limits (4 nations claim a fishery zone of less
  than 200 NM and 15 nations claim one of 200 NM)
	8

Continental Shelf	
200-m isobath plus exploitability criterion	41
Outer edge of continental margin or 200 NM	23
200 NM	7
Other	13

Source: United Nations 1994b.
Note: An additional 16 nations claim status as archipelagic states, although not all of them have specified archipelagic baselines.
[a]NM = nautical miles.

- The most comprehensive and binding system of peaceful settlement of disputes
- Reservation of the deep seafloor for peaceful purposes

The 1982 Law of the Sea Convention, in summary, represents a constitution for the world's oceans. Not only are the rights of nations relative to the ocean and its resources carefully spelled out, but also the duties and obligations of nations are made clear. In a statement titled "A Constitution for the Oceans," Tommy T. B. Koh, who served as president of the Law of the Sea Conference during its final years, credited ambassador from Malta Arvid Pardo with contributing two seminal ideas to the work of the conference: "first, that the resources of the deep seabed constitute the common heritage of mankind, and second, that all aspects of ocean space are interrelated and should be treated as an integral whole" (Koh

1983). Indeed, both of these principles—common heritage and integrated management—are enshrined in the preamble to the convention.

However, although the Law of the Sea Convention delimits ocean zones under national jurisdiction and specifies the rights and responsibilities of nations in these zones, and although it emphasizes the concept of ocean "wholeness," it generally provides little guidance to nations on how to govern ocean resources in an integrated manner, how to deal with the effects of one use on other uses, and how to bring ocean and coastal management together. The Convention does establish the outer boundaries of national jurisdiction, set forth general principles for governing specific ocean uses (e.g., fishing operations and oil and gas development), and set standards for marine environmental protection, but it does not address conflicts among uses or deal with alternative institutional mechanisms for ocean and coastal management. Conflict resolution provisions, though one of the convention's major innovations, relate to international disputes rather than disputes among users within a given country.

Thus, although the Law of the Sea Convention provided the basic framework determining national jurisdiction and control over 200-nautical-mile ocean zones, it was not until the United Nations Conference on Environment and Development (UNCED), held in Rio de Janeiro in 1992, that more detailed guidance was provided to nations regarding options and management approaches for national ocean zones. To understand the background, content, and significance of UNCED, we must first go back to the early 1970s, when environmental concerns first surfaced on the international agenda.

United Nations Conference on the Human Environment (Stockholm, 1972)

Prior to the 1970s, environmental efforts at the international level were generally fragmented and reactive and tended to deal with relatively narrow problems or issues. Early international agreements, aside from those involving international lakes and rivers, tended to be of two kinds: those associated with migrating species of wildlife and those adopted in response to the increasing number and severity of spills from oil tanker accidents of one type or another (Caldwell 1990).

The first effort to be comprehensive and, indeed, even proactive, took place at a United Nations–sponsored conference on the environment held in Stockholm, Sweden, in 1972. The conference, formally called the United Nations Conference on the Human Environment, was held at the peak of a wave of environmental concern triggered by incidents such as the January 1969 oil spill offshore from Santa Barbara, California, in the United States and by a growing realization that environmental degradation was becoming a serious problem in many parts of the world.

The Stockholm conference was important in several respects. First, in preparation for the conference, many of the more than 100 nations attending prepared "national reports" on environmental conditions in their countries. This exercise led to increased environmental awareness in these countries and might have contributed to the fact that a number of them went on to establish permanent environmental departments or ministries. One of the most significant actions of the conference was the creation of a new institution to coordinate environmental activities within the United Nations system—the United Nations Environment Programme (UNEP) and its governing council. Formally established by the United Nations General Assembly during its 1972 fall session, UNEP has been an important actor in the international arena since that time, organizing such activities as the Regional Seas Programme and playing important roles in ozone depletion and biodiversity issues, among others.

The Stockholm conference also acted as a catalyst in accelerating the adoption of several pending international agreements dealing with ocean dumping, vessel-source pollution, and trade in endangered species. Moreover, the Stockholm declaration of principles and its accompanying 109-item action plan represented the first effort by the international community to deal with environmental problems on a comprehensive basis.

Although the Stockholm agenda and resulting action plan spanned the entire spectrum of environmental issues, in many respects ocean and coastal issues were emphasized throughout the conference. Nongovernmental organizations (NGOs) labored hard to focus world attention on the grave situation facing many of the world's stocks of great whales, which had experienced significant population decline. Earlier marine accidents such as the grounding of the tanker *Torrey Canyon* in 1965 off the coast of England and the Santa Barbara oil blowout of 1969 had highlighted the consequences of oil spills and the need for higher standards for coastal and marine environmental protection. As a result, some of the most important impacts of the Stockholm conference occurred in the ocean and coastal field. The London Convention (Convention on the Prevention of Marine Pollution by Dumping of Wastes and Other Matter), establishing the first global standards to govern the dumping of wastes into the oceans, was adopted in 1972 after having been given a considerable boost by the conference. Similarly, the first comprehensive convention on vessel-source pollution, the International Convention for the Prevention of Pollution from Ships (MARPOL), was adopted shortly after the Stockholm meeting. The governing council for the United Nations Environment Programme (UNEP) established the Regional Seas Programme in one of its earliest actions following its formal creation in December 1972. Through this program, UNEP has been instrumental in creating action programs in more than a dozen important enclosed or semienclosed seas around the world (Haas 1990).

United Nations Conference on Environment and Development (UNCED)

In the 1980s, several urgent new concerns were edging their way onto the international agenda, leading eventually to a decision to hold a second United Nations conference on the environment (and development) in 1992, on the twentieth anniversary of the Stockholm meeting. The first of these concerns involved a growing awareness that a significant number of the world's ills were inexorably linked—that the abject poverty and soaring birth rates of much of the developing world, together with the unsustainable production and consumption of many developed nations, were contributing directly to worsening environmental despoliation—and that reversing the situation required a shift to sustainable use of the world's resources—that is, sustainable development on the part of all nations. In 1983, the United Nations General Assembly appointed the World Commission on Environment and Development (WCED) and asked Norwegian Prime Minister Gro Harlem Bruntland to act as chairperson to conduct a serious examination of the interrelationships among these issues. The resulting report, issued in 1987— *Our Common Future*—was a significant guiding force in the preparations for the United Nations Conference on Environment and Development (UNCED), held in Rio de Janeiro, Brazil, in June 1992 (also known as the Earth Summit or the Rio Conference).

The second concern was related to the growing realization that industrialized societies appeared capable of doing significant damage to the earth's climate and life support systems. First, there was a concern that compounds such as chlorofluorocarbons (CFCs) could destroy the earth's ozone shield, thus exposing the biosphere and its living populations (including humans) to harmful levels of solar ultraviolet radiation. Similarly, important groups of scientists had become convinced that increasing emissions of "greenhouse gases," especially carbon dioxide, were threatening to trigger climate changes on a global scale, including increases in temperature that could alter weather patterns and cause potentially dangerous rises in sea level (IPCC, 1990). A second group of experts, under the leadership of the United Nations Environment Programme, was expressing grave concern over the increasing rate at which species were becoming extinct, signaling a serious decline in biological diversity worldwide (WRI, IUCN, and UNEP 1992).

These growing concerns, important as they were, might not have been sufficient in themselves to cause an event of the scope and significance of UNCED to take place. Indeed, a "window of opportunity" associated with a dramatic change in global politics allowed these issues to reach the top of the international agenda (on the concepts of agenda setting and windows of opportunity, see Kingdon, 1984). In the late 1980s, a political change of major consequence occurred with the collapse of the Soviet Union and the end of the cold war. This change was felt especially strongly in Europe, where it created an unprecedented opportunity for concerned governments (such as those of Germany, the United Kingdom, and the Nordic countries) to shift their attention to global environmental problems in a new and more focused way. The availability of a coherent set of policy recommendations in the form of the 1987 report of WCED also helped set the stage for action in this policy area. European leaders at the highest levels became versed in the issue of climate change, bringing the matter to the agendas of the annual meetings of the world's industrialized democracies (the G7 group) beginning with the Paris meeting in 1990.

Thus, the stage was set for these issues to be advanced to the top of the global agenda. The United Nations General Assembly adopted Resolution 44/228 at its eighty-fifth plenary meeting on December 22, 1989, formally deciding to convene the United Nations Conference on Environment and Development in Brazil in June 1992 (G.A. Res. 44/228, U.N. Doc. A/RES/44/228 [1989]). The resolution set forth the reasons prompting the decision to convene UNCED; noted that Brazil would host the conference; set the conference dates; affirmed (and reaffirmed) certain principles that should guide preparations for the conference; set out nine issues "of major concern in maintaining the quality of the earth's environment and especially in achieving environmentally sound and sustainable development in all countries"; listed twenty-three objectives for the conference; and established a Preparatory Committee, authorizing it to meet five times and charging it with preparing a provisional agenda and draft decisions for the conference. The "protection of the oceans, all kinds of seas, including enclosed and semi-enclosed seas, and coastal areas and the protection, rational use and development of their living resources" was the third of the nine issues highlighted in the resolution.

Preparatory Activities

The preparations for UNCED were, in fact, carried forward on four separate but parallel tracks, only one of which involved the work of the UNCED Preparatory Committee (PrepCom). Figure 3.2 illustrates the general nature of the multi-

Figure 3.2. Preparatory Activities Leading to the Earth Summit (UNCED)

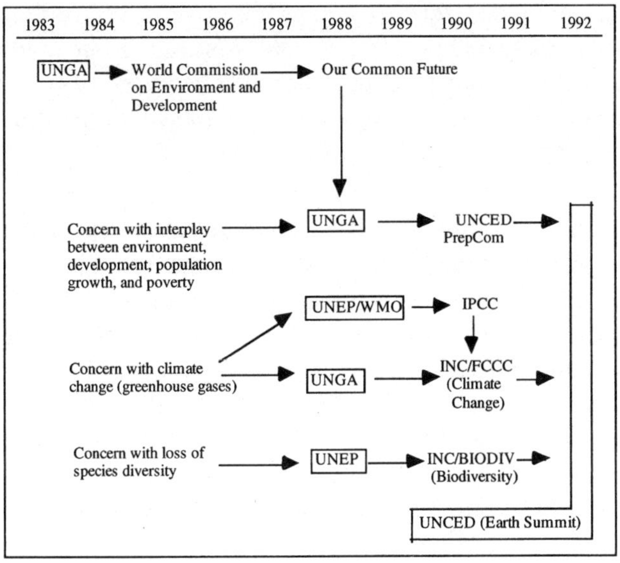

Source: Cicin-Sain and Knecht 1993

tracked preparatory process. Each of the major tracks is discussed in the following paragraphs.

UNCED Preparatory Committee (PrepCom). More than 150 nations participated in the five sessions of the PrepCom (any member nation of the United Nations General Assembly was eligible). Ambassador Tommy Koh of Singapore was elected chairman of the PrepCom. Ambassador Koh had served as president during the final years of the third United Nations Conference on the Law of the Sea and hence was very well acquainted with ocean and coastal issues. He was also a skilled negotiator and diplomat; much of the credit for the PrepCom's success must be given to his effective leadership. The PrepCom organized itself into three working groups: Working Group 1 dealt with atmospheric issues including climate change; Working Group 2 dealt with all other environmental issues, including coasts and oceans; and Working Group 3 dealt with institutional and legal matters, financial questions, and other cross-cutting issues. Following the initial organizing session in New York in March 1990, the PrepCom held four substantive sessions in the period 1990–1992 in Nairobi, Kenya; Geneva, Switzerland; and in New York.

The Intergovernmental Negotiating Committee for a Framework Convention on Climate Change (INC/FCCC). Given both the technical nature and the complexity of the issue, the United Nations General Assembly decided that negotiations on a new convention for climate change should be conducted in a separate forum. Hence, with Resolution 45/1212 in December 1990, the General Assembly established the Intergovernmental Negotiating Committee for a Framework Convention on Climate Change. As with the PrepCom, more than 100 nations chose to participate in the work of the INC/FCCC, which was headed by Jean Ripert of France. The committee formed two working groups: Working Group 1, on sources and sinks of greenhouse gases, and Working Group 2, on institutional responses. The INC/FCCC met six times between February 1991 and May 1992; by the conclusion of its final meeting, it had completed a draft convention on climate change.

Intergovernmental Negotiating Committee on Biological Diversity (INC/ BIODIV). This group, formed by the United Nations Environment Programme, was drawn from a group of experts who had been meeting for some time to discuss biological diversity issues. Again, it was decided that a special forum should be created to negotiate a draft convention on biological diversity for consideration at UNCED. Formal negotiating sessions were held in 1991 and early 1992, principally at UNEP headquarters in Nairobi. A number of controversial issues emerged, including protection of intellectual property rights (versus the sharing of technology with the "host" nation), sharing of financial benefits with the host nation, and some issues related to biotechnology.

Intergovernmental Panel on Climate Change (IPCC). Created by the World Meteorological Organization (WMO) and the United Nations Environment Programme (UNEP) with the assistance of the International Council of Scientific Unions (ICSU) in 1988, the IPCC (largely a scientific body) was charged with assessing the magnitude of the climate change problem and recommending appropriate responses. The IPCC's first formal assessment, issued in June 1990, was an important input into the work of the group negotiating the Framework Convention on Climate Change (INC/FCCC). In its first assessment, the IPCC predicted that global temperature could be expected to rise by about 2–3°C over the next 100 years due to the effects of greenhouse gases (Houghton, Jenkins, and Ephraums 1990). As part of the work of this group, an IPCC Subgroup on Coastal Zone Management developed a program for assessing the vulnerability of low-lying coastal nations to sea-level rise (Vellinga and Klein 1993).

Outputs of UNCED

The United Nations Conference on Environment and Development convened in Rio de Janeiro on June 3–14, 1992. It was, in many ways, a remarkable event. Virtually every nation on earth was represented (178 nations) and 114 heads of state were present, as were more than 1,000 official delegates, 9,000 media people, and representatives of 1,400 nongovernmental organizations. During the eleven days of the conference, the delegates completed work on the agreements that had been drafted during the preparatory process and, during the final two days, the heads of state signed the documents.

Five major outputs emerged from UNCED. They are (1) the Rio Declaration on Environment and Development, (2) the Framework Convention on Climate Change, (3) the Convention on Biological Diversity, (4) Agenda 21, and (5) a set of forest principles. Each is briefly described in the sections that follow. [1]

Rio Declaration on Environment and Development

The Rio Declaration on Environment and Development (adopted June 14, 1992, 31 I.L.M. 874), a set of twenty-seven principles to guide national and international actions on environment, development, and social issues, was approved by all nations attending the conference. As a "declaration of principles," the Rio Declaration is clearly of a nonbinding character; within the framework of international law, it can be considered as "soft law." As such, however, it can be seen as indicating the direction international law could take in the years ahead. Indeed, some of the principles might well find their way into legal instruments of one kind or another or, through common acceptance or acquiescence, ultimately become cus-

---

1. Discussion in these sections relies, in part, on Cicin-Sain and Knecht 1993 and Cicin-Sain 1993.

tomary international law, as has Principle 21 of the Stockholm declaration of principles, which deals with the responsibility of nations to refrain from actions that
damage the environment of neighboring countries (Caldwell 1990).

We have paraphrased the principles in the Rio Declaration and condensed them
into twenty principles, grouped in three categories—those that apply to development, those that apply to the environment, and those relating to process—as shown
in table 3.2.

The Convention on Climate Change

The overall objective of the Framework Convention on Climate Change (May 9,
1992, 31 I.L.M. 849) is to achieve the "stabilization of greenhouse gas concentrations in the atmosphere at a level that would prevent dangerous anthropogenic

Table 3.2. Principles Set Forth in the Rio Declaration on Environment and Development

On Development
- Human beings are at the center of concerns for sustainable development.
- States have the right to develop.
- Priority should be given to special needs of developing countries.
- States have a sovereign right to exploit resources but must ensure that their activities do not cause environmental damage to other jurisdictions.
- Nations have common but differentiated responsibilities for solving environmental problems.
- Unsustainable patterns of production and consumption must be reduced and eliminated.
- Appropriate demographic policies must be promoted.
- A supportive and open international economic system must be promoted.

On Environment
- Environmental protection must be an integral part of the development process.
- States shall enact effective environmental legislation (taking into account varying natural conditions).
- States shall develop national law regarding liability and compensation for the victims of pollution and other environmental damage.
- States shall discourage or prevent the relocation and transfer to other states of environmentally harmful activities.
- States shall apply the precautionary approach.
- Polluters should bear the cost of pollution.
- States shall employ environmental impact assessment.
- States shall be responsible for international notification of natural disasters, emergencies, and other activities with negative transboundary consequences.
- States must consider the environmental rights of peoples under oppression.
- Peace, development, and environmental protection are interdependent and indivisible.

On Process
- States need to promote the participation of all concerned citizens.
- States should be concerned with intergenerational equity and the rights of future generations.

Source: Report of the United Nations Conference on Environment and Development, Annex I, Rio Declaration on Environment and Development, A/CONF.151/26 (Vol. I) (August 12, 1992), abbreviated and
reorganized by the authors.

interference with the climate system . . . within a time frame sufficient to allow ecosystems to adapt naturally to climate change, to ensure that food production is not threatened and to enable economic development to proceed in a sustainable manner" (Art. 2). The agreement includes two principles: that the parties "should protect the climate system for the benefit of present and future generations of humankind" (Art. 3[1]) and that "the specific needs and special circumstances of developing country Parties . . . should be given full consideration" (Art. 3[2]).

In ratifying the Convention on Climate Change, nations commit themselves to do the following:

- Develop, periodically update, and publish national inventories of emissions by sources and removals by sinks.

- Formulate, implement, publish, and regularly update national programs containing measures to mitigate climate change.

- Promote and cooperate in the development, application, and diffusion of practices and processes that control, reduce, or prevent anthropogenic emissions of greenhouse gases.

Developed nations commit themselves to do the following:

- Adopt national policies and take corresponding measures to mitigate climate change by limiting emissions.

- Communicate, within six months of the convention's entry into force, detailed information on their policies and measures.

- Provide new and additional financial resources to meet the agreed full costs incurred by developing country Parties (Art. 4 [3]).

The Convention on Biological Diversity

The objectives of the Convention on Biological Diversity (June 5, 1992, 31 I.L.M. 818) are (1) conservation of biological diversity, (2) sustainable use of its components, and (3) fair and equitable sharing of the benefits arising from the utilization of genetic resources.

On ratifying this convention, each party commits itself to take the following actions:

- Develop national strategies, plans, or programs for the conservation and sustainable use of biological diversity.

- Integrate, as far as possible and appropriate, the conservation and sustainable use of biological diversity into relevant sectoral or cross-sectoral plans, programs, and policies.

• Identify components of biological diversity important for conservation and sustainable use.

• Monitor, through sampling and other techniques, the components of biological diversity, paying particular attention to those requiring urgent conservation measures and those offering the greatest potential for sustainable use.

• Establish a system of protected areas or areas where special measures need to be taken to conserve biological diversity.

• Develop, where necessary, guidelines for selection, establishment, and management of protected areas.

• Promote the protection of ecosystems and natural habitats and the maintenance of viable populations of species in natural surroundings.

Ratifying nations also commit themselves to a number of actions involving "ex situ" conservation, including the adoption of measures for recovery and rehabilitation of threatened species and reintroduction into their natural habit under appropriate conditions. Developed-country parties commit themselves to provide new and additional financial resources to enable developing-country parties to meet the agreed full incremental costs of implementing measures.

Agenda 21

Agenda 21, a forty-chapter action plan, was intended to serve as a kind of road map pointing the direction toward sustainable development. It represents an ambitious effort to provide recommendations across the entire spectrum of environment, development, and social issues confronting humankind today. In terms of social and economic issues, it addresses poverty, overconsumption and production, population, and human development problems.

In the areas of natural resources and the environment, Agenda 21 deals with the atmosphere, land resources, deforestation, desertification and drought, mountain ecosystems, agriculture and rural development, biological diversity, biotechnology, oceans and coastal areas, freshwater resources, toxic chemicals, hazardous wastes, solid wastes, and radioactive wastes. It has chapters devoted to the roles of major groups, including women, children and youth, indigenous peoples, nongovernmental organizations, local authorities, workers and trade unions, business and industry, the scientific and technological community, and farmers. Finally, concerning means of implementation, it discusses financial resources, transfer of technology, the roles of science, education, public awareness and training, capacity building, institutional arrangements, legal institutions, and information for decision making.

Like the Rio Declaration, Agenda 21 is a nonbinding document. Yet in signing the document, governments indicated a willingness to be part of the international consensus seeking to move toward a more sustainable society along the lines set forth in Agenda 21.

Statement of Forest Principles

The Statement of Principles for a Global Consensus on the Management, Conservation and Sustainable Development of All Types of Forests (adopted June 13, 1993, 31 I.L.M. 881), a prelude to a possible future convention on forests, was agreed to at UNCED. Not surprisingly, the forest issue was one of the most intractable at the conference. Industrialized nations of the North see the expansive rain forests of the South as a necessary and valuable "sink" for greenhouse gas emissions. In contrast, nations of the South—Brazil and Malaysia are good examples—see the rain forests as entirely theirs to develop or conserve as they see fit. Broadening the discussion to include all forests rather than just tropical rain forests helped to move the dialogue in a positive direction, but it was impossible to reach agreement at UNCED on anything more than a set of principles. However, the agreed principles may be helpful in guiding post-UNCED discussions toward of a future convention on the protection of forests.

Central Concepts at UNCED: Interdependence, Integrated Management, and Sustainable Development

All the major actions that came out of the Earth Summit—the Rio Declaration, Agenda 21, the Convention on Climate Change, the Convention on Biological Diversity, and the statement of forest principles—reflect a fundamental shift in thinking, a shift in paradigm: the understanding that henceforth, nations, groups, and individuals must address questions of environment and development and relations between North and South in a fundamentally different way from the way they have in the past.

Although the actions taken at Rio are numerous and complex (involving 2,500 recommended actions in 115 program areas), in our view two central concepts underlie the major outputs of the Earth Summit: interdependence and integration.

The Reality of Interdependence

As was documented in the report of the World Commission on Environment and Development, *Our Common Future* (WCED 1987), and emphasized in the Earth Summit negotiations, there has been growing realization in recent years that the world is facing a series of environmental crises—some global in nature, threatening the very future of life on earth, and some more local in nature, threatening the attainment of development and quality of life. These environmental crises are linked to patterns of economic development prevalent in the North and in the South and in the economic relationships between North and South. These relationships are presented, in a simplified manner, in table 3.3 and figure 3.3.

Table 3.3 shows the major environmental stresses the world is experiencing and indicates where they are mainly manifested, who is thought to bear primary responsibility for them, and who is thought to have the major responsibility or capability for solving them.

Figure 3.3 summarizes the economic patterns thought to contribute to these environmental stresses. In the developed countries, two major factors—patterns of production (excessive use of natural resources and excessive generation of waste) and patterns of consumption (excessive consumption)—create a series of environmental stresses, some global (greenhouse gases, ozone depletion) and others (such as toxic pollution) more localized in nature. The global environmental problems, generated mainly in the North (e.g., greenhouse gases), are seen by many as threatening the ability of the South to develop or not allowing the South "sufficient and adequate 'environmental space' for its future development" (Griffith 1992). In the developing countries, the combination of poverty and overpopulation leads to a number of environmental stresses, mainly local in nature, such as air and water pollution.

The international trade system and the international finance system are thought to contribute to these effects. The international trade system has emphasized the export of natural resources from the South to the North and the import by the South of manufactured goods from the North, contributing to overexploitation of

Table 3.3. Major Environmental Stresses

Environmental Stresses[a]	Where Effects Are Manifested	Who Bears Primary Responsibility	Who Should/Can Solve the Problem
Greenhouse gases	North, South	North	North
Ozone depletion	North, South	North	North
Acid rain	North	North	North
Nuclear waste	North, sometimes South	North	North
Environmental toxins	North, South	North	North
Water and air pollution	North, South	North, South	North, South
Deforestation	Mainly South	North, South	North, South
Desertification	Mainly South	North, South	North, South
Species depletion	North, South	North, South	North, South

Source: Cicin-Sain 1993.
[a]Environmental stresses are linked to one another, as are environmental stresses and patterns of economic development.

Figure 3.3. Interdependence of Environment and Development

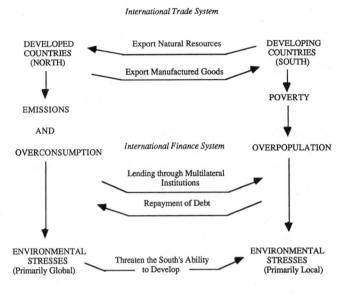

Source: Cicin-Sain 1993

the South's natural resource heritage. In recent years, too, many believe that given high interest rates, the imperative of debt repayment has contributed to the maintenance of poverty conditions in the South.

The point of this brief discussion is to underscore the inescapable fact of interdependence—between environment and development, among sectors, and among nations. As the Brundtland Commission put it, "Until recently, the planet was a large world in which human activities and their effects were neatly compartmentalized within nations, within sectors (energy, agriculture, trade), and within broad areas of concern (environment, economics, social). These compartments have begun to dissolve. This applies in particular to the various global "crises" that have seized public concern, particularly over the past decade. These are not separate crises: an environmental crisis, a development crisis, an energy crisis. They are all one" (WCED 1987, 4).

The Imperative of Integration

The reality of interdependence, hence, necessitates integration: integration between environment and development (sustainable development), integration among sectors, integration among nations (especially North and South). Although the concepts of integration and sustainable development make sense intuitively, they are not well defined in the Earth Summit outputs; yet operationalization of these key concepts is a prerequisite to proper implementation of the Earth Summit prescriptions. The following section briefly summarizes our interpretation of the concept of sustainable development, while in chapter 6, we discuss integrated policy as a necessary means for achieving it.

Definition of Sustainable Development

Although definitions of the term *sustainable development* abound in the literature, we believe that a good understanding of its basic meaning of can be gleaned by aggregating and extrapolating on various points made in *Our Common Future*, the report of the World Commission on Environment and Development (WCED 1987), which provided much of the intellectual underpinning for the Earth Summit negotiations.

Sustainable development may be thought of as a new paradigm, a new mode of thinking that serves as a guide to action. Achieving sustainable development entails a *continuous process of decision making* whereby certain questions are asked and the "right" choices and decisions are made. Thus, there is never an "end state" of sustainable development, since the equilibrium between development and environmental protection must constantly be readjusted.

The general goal of sustainable development is "to meet the needs of the present without compromising the ability of future generations to meet their own needs" (WCED 1987, 8); or in other words, sustainable development is a process of change in which the exploitation of resources, the direction of investments, the orientation of technological development, and institutional change are made consistent with future as well as present needs (46). Two ideas are underlined in this definition: (1) the concept of needs, especially those of the developing world, and (2) the idea of limitations on the environment's ability to meet present and future needs. Sustainable development requires that "societies meet human needs both by increasing productive potential and by insuring equitable opportunities for all" (44).

The means to sustainable development is *decision making* that (1) *is guided by a set of principles* (such as those contained in the Rio Declaration and other documents; (2) *is integrated in nature*, following an integrated management approach (discussed in part III of this book); and (3) *has the capacity to craft sustainable development* (e.g., can employ sufficient technological know-how, sufficient natural resources, and appropriate human resources). The major actors in sustainable development are national governments and the multilateral international organizations in which they participate, although participation by businesses, nongovernmental organizations, subnational governments, and all citizens, including women, youth, indigenous groups, and so forth, is strongly encouraged.

In summary, sustainable development involves three major emphases:

- *Economic development to improve the quality of human life*—"Human beings are at the centre of concerns for sustainable development. They are entitled to a healthy and productive life in harmony with nature" (Principle 1 of the Rio Declaration).

- *Environmentally appropriate development*—development that is environmentally sensitive and makes appropriate use (and sometimes nonuse) of natural resources: development that protects essential ecological processes, life support systems, and biological diversity

- *Equitable development*—equity in the distribution of benefits from development:

> *Intrasocietal equity* (e.g., among groups in society, respecting the special rights of indigenous peoples and other groups).

> *Intergenerational equity* (not foreclosing the options of future generations).

> *International equity* (fulfilling obligations to other nations and to the international community, given the reality of global interdependence).

These three main ideas can readily be translated into questions decision makers must ask regarding every environment and development decision: "How will it improve the quality of human life? How will it affect natural resources and the environment? Is there equity in the distribution of benefits from development?" Although it is sometimes used interchangeably with the concept of sustainable development, the concept of sustainable use, in our view, should be reserved for helping to answer the second of these questions regarding effects on natural resources and the environment.

In summary, sustainable development can be seen as a continuous process of decision making guided by a basic philosophy emphasizing development that improves the quality of human life (ensuring equity in the distribution of benefits flowing from development) and development that is environmentally appropriate, making proper use (and sometimes nonuse) of natural resources and protecting essential ecological processes, life support systems, and biological diversity. A necessary element of sustainable development is an integrated decision-making or policy process. Other elements that enhance the likelihood of sustainable development, as suggested by the World Commission on Environment and Development, include a political system that secures effective citizen participation in decision making; an economic system that can generate surpluses and technical knowledge on a self-reliant and sustained basis; a social system that provides for solutions to tensions arising from disharmonious development; a production system that respects the obligation to preserve the ecological base for development; a technological system that fosters sustainable patterns of trade and finance; and an administrative system that is flexible and has the capacity for self-correction (WCED 1987, 65).

An important theme, especially in the UNCED deliberations, concerns the importance of transparency in decision making and the effective participation of such major groups as women, indigenous peoples, local authorities, and farmers. In ocean and coastal areas where indigenous people are present, it is important that their knowledge and interests be fully considered in policy and decision making.

Ocean and Coastal Issues in Chapter 17 of Agenda 21

Given the general background outlined in the previous sections, we now review the major prescriptions for ocean and coastal management included in Chapter 17 of Agenda 21 (titled "Protection of the Oceans, All Kinds of Seas, Including Enclosed and Semi-Enclosed Seas, and Coastal Areas and the Protection, Rational Use and Development of Their Living Resources"). The reader should note that this is the longest chapter of Agenda 21 (forty-two pages) and one of the most complex. The introduction to the chapter (Para.17.1) stresses both the importance of oceans and coasts in the global life support system and the positive opportunity for sustainable development that ocean and coastal areas represent. The latter was a point highlighted at various times in the UNCED process: in contrast to other areas where the marrying of environment and development will require mostly regulatory measures to protect already degraded environments, ocean and coastal areas often present excellent opportunities in development, particularly for developing countries—opportunities that, if conducted in a sustainable development mode, can yield significant economic and social benefits for coastal human settlements while protecting environmental integrity. The introduction stresses a key point that underlies the entire chapter: that new approaches to marine and coastal area management will be needed, approaches that "are integrated in content and precautionary and anticipatory in ambit." The introduction also underscores that the United Nations Convention on the Law of the Sea "provides the international basis upon which to pursue the protection and sustainable development of the marine and coastal environment and its resources." The fact that Chapter 17 builds on and refines, rather than supplants, provisions of the Law of the Sea treaty was also highlighted frequently in the negotiations and was a point strongly argued by developed countries, most prominently the United States and members of the European Community. However, although references to the importance of the Law of the Sea treaty as a basis for action on oceans and coasts abound in Agenda 21, nowhere is there an explicit discussion of how the UNCED provisions on oceans and coasts relate to or refine the treaty.

Seven major program areas are included in Chapter 17: (1) integrated management and sustainable development of coastal areas, including Exclusive Economic Zones, (2) marine environmental protection; (3) sustainable use and conservation of living marine resources of the high seas; (4) sustainable use and conservation of living marine resources under national jurisdiction; (5) the addressing of critical uncertainties in management of the marine environment and climate change; (6) the strengthening of international, including regional, cooperation and coordination; and (7) sustainable development of small islands.

Integrated Management and Sustainable Development of Coastal and Marine Areas, Including Exclusive Economic Zones

In this section, coastal nations commit themselves to "integrated management and sustainable development of coastal areas and the marine environment under their national jurisdiction" (Par. 17.5). The text stresses the need to reach integration (e.g., identify existing and projected uses and their interactions and promote compatibility and balance of uses); the application of preventive and precautionary approaches (including prior assessment and impact studies); and full public participation (17.5).

The text calls for integrated policy and decision-making processes and institutions ("Each coastal State should consider establishing, or where necessary strengthening, appropriate coordinating mechanisms (such as a high-level policy planning body) for integrated management and sustainable development of coastal and marine areas, at both the local and national levels. . . . ") (Par. 17.6). It also provides a series of suggested actions such coordinating institutions should consider undertaking, such as preparation of coastal and marine use plans (including profiles of coastal ecosystems and of user groups), environmental impact assessment and monitoring, contingency planning for both human-induced and natural disasters, improvement of coastal human settlements (particularly in terms of drinking water and sewage disposal), conservation and restoration of critical habitats, and integration of sectoral programs (such as fishing and tourism) into an integrated framework (17.6). Also called for is cooperation among states in the preparation of national guidelines for integrated coastal management (17.11) and the undertaking of measures to maintain biodiversity and productivity of marine species and habitats under national jurisdiction (17.7).

This section also highlights the need for information on coastal and marine physical systems and uses, information on both natural science and social science variables (Par. 17.8), education and training in integrated coastal and marine management (17.15), and capacity building, including building of human resource capacity, support of pilot demonstration programs and projects in integrated coastal and marine management, and establishment of centers of excellence in the area (17.17). As in other parts of Agenda 21, there is a strong affirmation in this section of the need to include traditional ecological knowledge and sociocultural values as an input to management and of the importance of coastal areas for indigenous peoples (e.g., 17.15, 17.3, 17.6). Also emphasized is the need for international cooperation on both a bilateral and multilateral basis to support national efforts by coastal states in the objectives and activities noted earlier.

Marine Environmental Protection

Chapter 17 calls for coastal nations to increase their efforts to deal with both land-based and sea-based sources of marine pollution. Although land-based sources account for as much as 70 percent of the pollution, there is currently no global scheme to address such pollution (Par. 17.18). This section of chapter 17 stresses the need for a precautionary and anticipatory, rather than reactive, approach to prevent degradation of the marine environment (17.21). Nations commit themselves "to prevent, reduce and control degradation of the marine environment so as to maintain and improve its life-support and productive capacities" (17.22).

The section on controlling land-based sources of pollution points to the need to relate control of land-based sources to integrated coastal and ocean management (programme area A); to consider updating, strengthening, and extending the 1985 Montreal Guidelines for the Protection of the Marine Environment Against Pollution from Land-Based Sources; and to promote new regional arrangements where appropriate. Among other items, it also calls on UNEP to convene an intergovernmental meeting on protection of the marine environment from land-based sources of pollution (Pars. 17.24, 17.25). Priority is given to the control of sewage effluent (a paramount problem in many developing countries) and to other sources of pollution such as synthetic organic compounds, sediments, organohalogen compounds, and anthropogenic inputs of nitrogen and phosphorus (17.28). Watershed management practices are encouraged to prevent, control, and reduce degradation of the marine environment (17.29).

Similarly, building on the international framework already established through the International Maritime Organization and other relevant international organizations, additional measures to reduce degradation of the marine environment from sea-based activities are called for, including control of pollution from shipping, dumping, offshore oil platforms, and ports (Par. 17.30).

More systematic observations to measure marine environmental quality are called for, such as expansion of the mussel watch program, establishment of a clearinghouse for marine pollution control technologies, and establishment of a global profile and database (Par. 17.35). The section also stresses the need for technology transfer of clean or cost-effective pollution control technology (17.37) and education and training in this area (17.37, 17.38).

Sustainable Use and Conservation of Living Marine Resources of the High Seas

As the world has witnessed a significant expansion of fishing activity on the high seas in the past decade, a number of high seas resources have become overutilized and problems in the management of high seas fisheries have become evident. These include unregulated fishing, overcapitalization, excessive fleet size, reflagging of vessels to escape controls, insufficiently selective gear, unreliable databases, and lack of sufficient cooperation among states (Pars. 17.44, 17.45). In this section of chapter 17, therefore, "States commit themselves to the conservation and sustainable use of marine living resources on the high seas" (17.46) and reference is made to development of fisheries to meet nutritional needs as well as social, economic, and development goals. Also mentioned are protection of endangered marine species and maintenance or restoration of marine species at levels that can produce maximum sustainable yield (as qualified by "relevant environmental and economic facts taking into account relationships among species") (17.46). Regarding marine mammals, this section defers the question of management of cetaceans to the appropriate international organizations (17.47), recognizing the competence of the International Whaling Commission and its scientific work on large whales and other cetaceans, as well as the work of other international organizations, but goes no further (12.62).

The 1982 Law of the Sea Convention is referred to as the basis on which high seas fisheries should be managed, and nations are called on to give "full effect to these [LOS] provisions" with regard to fisheries populations whose ranges lie both within and beyond exclusive economic zones (straddling stocks) and highly migratory species (Par. 17.49). Nations are called on to convene an intergovernmental conference under the auspices of the United Nations to promote "effective implementation of the provisions of the Law of the Sea treaty on these species" (17.50). Other measures that nations are called on to implement include resisting the reflagging of vessels as a means of avoiding compliance with applicable conservation and management rules; prohibiting destructive fishing practices; fully implementing General Assembly Resolution 46/215, on large-scale pelagic driftnet fishing; and reducing waste, postharvest losses, and discards (17.54–17.56). The section also underscores the importance of international cooperation in fisheries management, particularly in enclosed and semienclosed seas and within global or regional fisheries bodies.

Sustainable Use and Conservation of Living Marine Resources under National Jurisdiction

Fishery yields have increased significantly in the past four decades in areas of national jurisdiction, where 95 percent of the world's total fish catch (80–90 million tons of fish and shellfish per year) is taken. This section of chapter 17 addresses management issues related to fisheries in Exclusive Economic Zones and other areas under national jurisdiction, reaffirming that the Law of the Sea Convention provides the basic framework for management of these areas (Par. 17.70). Fisheries management problems discussed in this section include local overfishing; unauthorized incursions by foreign fleets; ecosystem degradation; overcapitalization and excessive fleet sizes; undervaluation of catch; insufficient-

ly selective gear; unreliable databases; and increasing competition between artisanal and large-scale fishing and between fishing and other activities (17.72). In
addition to fisheries-related problems, problems related to coral reefs and other
marine and coastal habitats, such as estuaries and mangroves forests, are highlighted (17.73).

States therefore commit themselves to a number of objectives of both a development and a conservation nature, such as to develop fisheries to meet nutritional needs, taking into account traditional knowledge and small-scale fisheries; to
use selective fishing gear to minimize bycatch of nontarget species; and to protect
endangered marine species and ecologically sensitive areas (Par. 17.75). Among
the major management activities called for in this section are development of
fisheries and marine aquaculture, with special emphasis on small-scale artisanal
fisheries (17.79, 17.82); strengthening of legal and regulatory frameworks for
fisheries management; use of selective gear to reduce waste, postharvest losses,
and discards; use of environmentally sound technology, including assessment of
the environmental effects of major new fisheries practices (17.79); encouragement
of recreational and tourism activities based on marine living resources (17.81);
and protection of marine ecosystems exhibiting high levels of diversity and productivity, such as coral reefs, estuaries, wetlands, sea grass beds, and the like
(17.86).

The section also calls for transfer of environmentally sound technologies to
develop fisheries and aquaculture, particularly to developing countries (Par.
17.93), and for a number of measures designed to raise capacity for fisheries management, such as support to local fishing communities (17.95).

Addressing Critical Uncertainties for the Management of the Marine Environment and Climate Change

Rational use of coastal and marine areas will require better information on the
present state of these systems and for predicting future conditions. Better information is needed as well to assess the role of the oceans in driving global systems.
A long-term cooperative research commitment is called for to provide the data
required for global climate models and to reduce uncertainty. Meanwhile, precautionary measures should be undertaken to diminish the risks and effects of
global climate change, particularly on small islands and in low-lying and coastal
areas of the world (Pars. 17.97, 17.98).

Nations therefore commit themselves to improve understanding of the marine
environment and its role on global processes by promoting scientific research and
systematic observation, exchange of data and information (from both scientific
and traditional ecological knowledge sources), and development of standard inter-
calibrated procedures, measuring techniques, data storage, and management capabilities (Par. 17.100).

Among the activities listed in this section for states to consider are improved

forecasts of marine conditions; development of methodologies for coastal vulnerability assessment; research on the effects on marine systems of increased ultraviolet radiation caused by ozone depletion (Par. 17.101); utilization of studies by
the Intergovernmental Oceanographic Commission (IOC) and others on the role
of the oceans as a carbon sink (17.102); support of the role of the IOC and others
in collection, analysis, and distribution of ocean data (including data from the proposed Global Ocean Observing System) (Par. 17.103); improvement of regional
and global databases (17.106 and elsewhere); capacity building in the marine sciences for coastal and island states (17.104, 17.113, and elsewhere); and strengthening (or establishment, where necessary) of national oceanographic commissions (17.114). There are also calls in this section for research conducted in the
Antarctic to be made freely available to the international community and widely
disseminated (17.105).

Strengthening International, Including Regional, Cooperation and Coordination

This section of chapter 17 underlines the proposition that the role of international cooperation is to support and supplement national efforts and that given the
existence of numerous national and international organizations with competence
in marine issues, there is a need to improve coordination and strengthen links
among them and to ensure, as well, that a multisectoral approach to marine issues
is presented at all levels (Par. 17.116).

Nations therefore commit themselves to "promote institutional arrangements
necessary to support the implementation of the programme areas in this chapter"
for purposes of integrating sectoral activities; exchanging information and
improving linkages among bilateral and multilateral national, regional, and other
organizations; promoting regular intergovernmental review within the United
Nations system of ocean and coastal issues; and promoting effective cooperation
among UN agencies (Par. 17.117).

Management-related activities called for at the global level include requesting
the UN secretary-general and executive heads of UN entities to improve coordination among relevant UN agencies dealing with oceans and coasts, other specialized UN agencies, and other international, subregional, and regional coastal
and marine programs; and developing a centralized system to provide information
and advice on implementing of legal arrangements on oceans and coasts (Par.
17.118).

Management-related activities called for at the subregional and regional levels
include strengthening, and extending where necessary, intergovernmental regional cooperation, the Regional Seas Programme of UNEP, regional and subregional
fisheries organizations and regional commissions, and other measures to improve
coordination among UN entities and other multilateral organizations at regional

and subregional levels, including consideration of collocation of their staffs (Par. 17.120).

Paragraph 17.119 in this section also addresses the practice of establishing trade sanctions for environmental reasons. Tacitly aiming a criticism at the use of such sanctions by the United States to protect marine mammals beyond the limits of its jurisdiction, the paragraph emphasizes that unilateral trade policy measures for environmental protection outside the jurisdiction of the importing country should be avoided. Trade policy measures addressing international environmental problems should, as far as possible, be based on an international consensus and be guided by certain suggested principles, such as nondiscrimination, transparency, and consideration of the special circumstances of developing countries.

Sustainable Development of Small Islands

Because of their special characteristics (e.g., small size, limited resources, geographical dispersion and isolation from markets, ecological fragility and vulnerability, and high degree of biodiversity), small-island developing states represent a special challenge in integrating environment and development. For small-island developing states, oceans and coasts are of paramount importance as an avenue for development, both for strategic reasons and for the environmental challenges facing their coastal zones (e.g., storms, sea-level rise) (Pars. 17.124–17.126). To meet the challenges of planning for and implementing sustainable development, small-island developing states will need the cooperation and assistance of the international community (17.127).

In chapter 17 of Agenda 21, states therefore commit themselves to address the problems of sustainable development of small-island developing states and, inter alia, to adopt measures that will enable these states to address environmental change, mitigate adverse environmental effects, and reduce threats posed to marine and coastal resources (Par. 17.128).

Small-island developing states, with the assistance of the international community, are called on to conduct studies of island environments; determine carrying capacity; prepare plans for sustainable development; adapt coastal area management techniques to the special characteristics of small islands; review and, where appropriate, reform existing institutional arrangements to provide for effective implementation of sustainable development plans; design and implement responses to effects of climate change; and promote environmentally sound technology for sustainable development (Par. 17.129). International organizations are called on to recognize the special development requirements of small-island developing states and to give them adequate priority in providing assistance (17.132).

Interisland regional and interregional cooperation and information exchange should be strengthened. A call was made to convene a first global conference on sustainable development of small-island developing states in 1993 (Par. 17.131).

To build scientific and management capacity, the section also calls for the creation (or strengthening) of centers on coastal ocean science, technology, and management on a regional basis (17.134); training programs in integrated coastal area management and development (17.135); and public awareness programs (17.135).

Significance of the UNCED Agreements for Oceans and Coasts

Some commentators have criticized the UNCED recommendations in Chapter 17 of Agenda 21 as being too soft—as setting forth vague and general goals and advocating strategies that in some cases are not well defined or completely articulated. Indeed, many of the recommended approaches will have to be more fully fleshed out at future UN conferences of the type explicitly called for in Chapter 17.

But these shortcomings should not detract from UNCED's very significant achievements relative to oceans and coasts. Under the aegis of the UNCED process, broad international consensus was reached on major problems in the ocean and coastal arenas and on principles to guide concerted action to address these problems. The UNCED process also gave political legitimacy to the concept of integrated ocean and coastal management, underscoring the importance of *integration* in the process, something that heretofore had been argued mainly by academics (see, e.g., Sorensen and McCreary 1990; Clark 1991; Cicin-Sain and Knecht 1985).

It can be argued, too, that UNCED gave additional political impetus to the 1982 Convention of the Law of the Sea by highlighting the thorough and carefully balanced legal framework established by that convention, on which UNCED builds.

UNCED's Agenda 21 also highlighted the positive opportunity for sustainable development of oceans and coasts and the special importance of these resources for small-island developing states. UNCED's ocean and coastal recommendations stress the connections between the land and sea aspects of ocean and coastal management and emphasize the importance of dealing with land-based sources of marine pollution within the ambit of integrated coastal management.

As previously noted, UNCED gave rise to at least four other fora for continued discussion and refinement of ocean and coastal issues (coasts, islands, straddling and migratory fishery stocks, land-based sources of pollution). UNCED also underscored the desirability of strengthening regional arrangements for ocean and coastal management and highlighted the special challenges faced by small-island states.

With respect to some of UNCED's broader effects, we would call attention to three developments:

1. Recognition of the explicit link between environment and development has been solidly forged.

2. Global networks have been created among interest groups and NGOs that will change the dynamics of international negotiations for years to come.

3. There is now a clear awareness and understanding that issues such as poverty, underdevelopment, population growth, environmental quality, and sustainable resource use are interrelated and interdependent and must be dealt with as such.

In a sense, UNCED outlined a vision of what ought to be done to move toward a more sustainable and equitable global society and sketched out a road map for getting there. Hence, it created opportunities for interested individuals, organizations, and governments to begin pursuing these goals. UNCED must thus be seen as a beginning and not as an end, the marking of a new era of positive cooperation between the nations of the North and those of the South in the name of a better life for all the earth's inhabitants.

Summary

This chapter traced the evolution of international governance regimes for oceans and the environment and emphasized the importance of the United Nations Convention on the Law of the Sea and the United Nations Conference on Environment and Development (UNCED) as the major frameworks guiding nations in the governing of coasts and of ocean areas under national jurisdiction. The Law of the Sea Convention, which came into force in 1994, provides a basic constitution for the oceans, delimiting how nations may establish national control in ocean areas 200 nautical miles offshore from their coasts and establishing the rights and duties of nations in those zones and on the high seas. UNCED, held in Rio de Janeiro in 1992, twenty years after the initial UN Conference on the Human Environment in Stockholm, built on the LOS framework and provided further guidance to nations. In Chapter 17 of Agenda 21, UNCED emphasizes the central concepts of sustainable development and integrated management of coastal and marine areas and resources and underlines the concept that management of oceans and coasts must be "integrated in content and precautionary and anticipatory in ambit."

···

Chapter 4

Earth Summit Implementation: Growth in Capacity in Ocean and Coastal Management

Introduction

In April and May 1996, almost four years after the Earth Summit, the United Nations Commission on Sustainable Development (CSD), the major UN entity charged with monitoring implementation of the Earth Summit agreements, assessed the progress achieved in implementation of Chapter 17 of Agenda 21. Each year since its creation in 1993, the CSD has been evaluating progress in implementation of various chapters of Agenda 21 and of the two international conventions also adopted at the Earth Summit—the Framework Convention on Climate Change and the Convention on Biological Diversity. Along with "atmosphere," "oceans and coasts" were the last substantive areas whose implementation was reviewed prior to the UN General Assembly's overall review of Earth Summit implementation in June 1997 (Cicin-Sain 1996).

Implementation of Chapter 17 recommendations involves myriad actors. Those at the international level, in international organizations and lending institutions, often play a leading role in interpreting concepts and providing incentives to governments and others to implement commitments agreed to internationally. Those in nongovernmental organizations, through continual prodding, attempt to keep international, national, and local institutions true to their commitments. Those in national and local governments most often are the key to whether international agreements will truly be implemented or will largely remain mere "words on paper."

To be sure, there have been disappointments in the aftermath of UNCED. The "new and additional" funding expected by developing nations to help them implement the agreements has been slow in coming, in part because developed nations have been more inclined to meet the new commitments with existing funds than to allocate new money for them. Further, the United Nations Commission on Sus-

---

This chapter relies in part on Cicin-Sain, Knecht, and Fisk 1995.

tainable Development has not been allocated the funding, the staff, or the mandate to perform the oversight of UNCED implementation that many observers believe is necessary.

Yet there have been a number of important accomplishments. International financing institutions, the World Bank and the Global Environment Facility (the major source of new funding for post UNCED programs, as discussed later), have undergone reform, striving to achieve more transparency in their operations and to give a greater voice to developing nations in funding decisions (Cicin-Sain 1996). A new program of action to assist small-island developing states (SIDS) was adopted in Barbados in May 1994, and a significant new agreement on management of straddling and highly migratory fish stocks was adopted in August 1995. Also of considerable importance, a Global Programme of Action to deal with land-based activities that affect the marine environment was adopted at a UNEP-sponsored global conference in Washington, D.C., in November 1995 (Cicin-Sain and Knecht 1995).

Change has also been taking place within UN organizations concerned with oceans and coasts. For example, UNEP undertook a major reorganization of its programs to meet the mandates entrusted to it by Agenda 21. UNDP responded by creating a new capacity-building program called Capacity 21 (Cicin-Sain 1996). Regional institutions such as those that are part of UNEP's Regional Seas Programme are also responding to the UNCED decisions. Noteworthy is the response of the countries bordering the Mediterranean, which incorporated sustainable development and other mandates into the revised Barcelona Convention, adopted in 1995 (Vallega 1995). The South Pacific Regional Environment Programme (SPREP) has also responded significantly, establishing capacity-building programs in environmental management. A number of new training programs in ICM have also been established in the past few years; the Division for Ocean Affairs and the Law of the Sea (UNDOALOS), with UNDP assistance, created the Train-Sea-Coast program, and UNESCO's Intergovernmental Oceanographic Commission has developed an ICM training initiative as well.

At the national level, where much of the responsibility for implementing the UNCED agreements resides, at least seventy-one nations (of ninety responding to an NGO-conducted survey in 1993) have created new national institutions (e.g., national commissions on sustainable development) to oversee implementation of Agenda 21, and sixty have reported some identifiable policy or action being taken as a result of the Earth Summit (Cicin-Sain, 1996). There is a general view, however, that some nations have been slow in implementing the commitments they made in Rio.

With regard to efforts at the international level since the Earth Summit, we can identify three major developments that are contributing, or that we expect will contribute, to implementation of the concept of integrated coastal management:

1. The ICM concept has come to be embraced as a central organizing concept in a number of UNCED-related conferences and in international agreements

formerly lacking a coast and ocean emphasis, such as the Convention on Biological Diversity.

2. The ICM concept has been further interpreted and operationalized in several efforts by international entities to develop ICM guidelines to assist national governments and others in formulating and implementing ICM programs.

3. There have been a number of efforts by international organizations and others to build capacity in ICM by establishing new teaching and training programs.

These developments indicate growth in capacity at the international level for formulating and carrying out integrated coastal management programs.

Adoption of the ICM Concept Widens

Although the 1982 Law of the Sea Convention and Chapter 17 of UNCED's Agenda 21 contain the bulk of global prescriptions for ICM, other legal instruments and conference pronouncements have emerged that relate to and help define it. Typically, these agreements embrace ICM as an appropriate and even necessary means of addressing a rather wide range of issues, from the effects of climate change on low-lying coastal areas to control of land-based sources of marine pollution to protection of coral reefs. The most important of these agreements and pronouncements are highlighted in the sections that follow.

Framework Convention on Climate Change

The major objective of the Framework Convention on Climate Change (FCCC) (opened for signature at the Earth Summit and entered into force in March 1994) is to achieve the "stabilization of greenhouse gas concentrations in the atmosphere at a level that would prevent dangerous anthropogenic interference with the climate system" (Art. 2). The Convention was negotiated between January 1991 and June 1992 by an intergovernmental negotiating committee (INC) appointed by the United Nations General Assembly. The INC completed its work just prior to UNCED, although not without considerable controversy over the extent to which the new convention should set precise "targets and timetables" for stabilization and eventual reduction of greenhouse gas emissions. In the end, a group of developed nations led by the United States prevailed, and definitive targets and timetables were not included.

The work of the INC was closely supported by the Intergovernmental Panel on Climate Change (IPCC), established in 1988 by the World Meteorological Organization (WMO) and the United Nations Environment Programme (UNEP) to assess the magnitude of human-induced climate change and to recommend appropriate response options to various threats related to climate change. In 1990, the IPCC's working group on responses created a subgroup on coastal zone management (CZM) specifically to explore the threat of sea-level rise and recommend appropriate adaptive strategies (Carey and Mieremet 1992; Vellinga and Klein 1993). After developing a common methodology for vulnerability assessments and conducting of a number of case studies, the CZM subgroup concluded that successful adaptation to the threat of sea-level rise required that efforts at vulnerability reduction be undertaken within the context of integrated coastal management (IPCC 1992). The IPCC's finding was based on the recognition that a freestanding program responding solely to the problem of sea-level rise would be limited in effectiveness, given the interconnectedness of activities and environments in the coastal zone.

The IPCC's position on the key role of ICM was incorporated into the text of the Framework Convention on Climate Change by the negotiators. In Article 4 of the convention, nations commit themselves, inter alia, "to develop integrated plans for coastal zone management." Thus, the Framework Convention on Climate Change reinforces the more general prescriptions concerning ICM contained in Chapter 17 of Agenda 21 and shows how this management concept can relate to a particular issue—in this case, adaptation to the effects of climate change.

In 1995, the IPCC issued its second assessment of the issue of climate change (the first having come out in 1990) (IPCC 1990, 1995). In this assessment, the IPCC slightly reduced its earlier projection regarding the magnitude of sea-level rise expected by the year 2100 (from about sixty centimeters, or twenty-four inches, to about forty centimeters, or sixteen inches) but increased the certitude of its assessment that greenhouse warming (caused by human activities) was becoming a reality. In February 1997, an international workshop was convened in Taipei, Taiwan (China), to further develop ways to fit climate change adaptation planning into the framework of ICM. Guidelines for dealing with climate change within an ICM framework were formulated (Cicin-Sain et al. 1997).

Convention on Biological Diversity

The Convention on Biological Diversity (CBD) was opened for signature at UNCED in 1992 and entered into force in December 1994 after ratification by thirty nations. Like the Convention on Climate Change, this agreement was negotiated by a special intergovernmental negotiating committee, in this case established by UNEP. UNEP had been facilitating the work of a group of experts studying biodiversity since the mid-1980s. As with the Convention on Climate Change, some aspects of the negotiations proved controversial, especially those dealing with protection of intellectual property and sharing of benefits from the use of the products of biodiversity in biotechnology applications.

At the first meeting of the Conference of Parties held in the Bahamas in November–December 1994, it was decided that priority attention should be given to coastal and marine ecosystems, given their dominant contribution to global bio-diversity. The second Conference of Parties, held in Jakarta, Indonesia, on November 6–17, 1995, gave further attention to issues related to coastal and marine biodiversity (de Fontaubert, Downes, and Agardy 1996). Although the convention itself does not make specific reference to ICM, a principal outcome of the second Conference of Parties was Decision II/10, on "Conservation and Sus-tainable Use of Marine and Coastal Biological Diversity." Among other recom-mendations, this statement, now called the Jakarta Mandate,

> encourages the use of integrated marine and coastal area management as the most suitable framework for addressing human impacts on marine and coastal biological diversity and for promoting conservation and sustainable use of this biodiversity [and] encourages Parties to establish and/or strengthen, where appropriate, institutional, adminis-trative, and legislative arrangements for the development of integrated management of marine and coastal areas, and their integration within national development plans. (UNEP 1995b).

As the Conference of Parties and the Subsidiary Bodies of the CBD continue their implementation of the convention, ICM is being seen as an important tool in protecting coastal and marine biodiversity. For example, a group of experts met in Jakarta in March 1997 to discuss, among other things, the formulation of guide-lines for protecting coastal and ocean biodiversity within the framework of ICM. It has become increasingly clear that biodiversity per se cannot be protected by specialized measures taken in isolation; rather, biodiversity protection will need to be built into a broader, comprehensive management framework, such as that provided by ICM, for a successful outcome.

Global Programme of Action on Protection of the Marine Environment from Land-Based Activities

Chapter 17 of Agenda 21 invited UNEP's governing council to convene an inter-governmental meeting on protection of the marine environment from land-based activities. In calling for this meeting, Chapter 17 drafters stressed the urgent need to deal more effectively with marine pollution associated with land-based activi-ties, the cause of the bulk of pollution found in marine waters today. After sever-al preparatory meetings, the conference, sponsored by UNEP and hosted by the United States, took place in Washington, D.C., in October–November 1995. The conference adopted the Global Programme of Action on Protection of the Marine Environment from Land-Based Activities and the Washington Declaration, which highlights major aspects of the Global Programme of Action.

The text emerging from the conference makes clear that integrated coastal management is seen as an important tool in accomplishing the goals of the program at the national level. Indeed, the first action listed is as follows:

> 19. States should . . . focus on sustainable, pragmatic and integrated environmental management approaches and processes such as integrated coastal area management, harmonized, as appropriate, with river basin management and land use plans. (UNEP 1995c)

Although the Global Programme of Action falls within the category of "soft law" and is not legally binding on states, it does deal in a comprehensive and definitive way with a wide range of land-based activities and their effects on the coastal and marine environment. Furthermore, as L. A. Kimball has pointed out (1995), the program does contain four key elements that, depending on the effectiveness of their implementation, could significantly increase its prospects for success. These are (1) periodic scientific assessments of the health of the coastal and marine environments; (2) means to organize and expedite exchange of information, experience, and expertise; (3) means to coordinate the efforts of the many relevant international agencies; and (4) an intergovernmental review mechanism to consider progress in implementation of the program.

In our judgment, the extent to which the Global Programme of Action becomes a strong driving force for ICM will depend on three factors: (1) the extent to which national governments give serious and sustained attention to land-based activities as a source of coastal and marine pollution; (2) the extent to which nations in general, and developed nations in particular, support funding and technical assistance by the Global Environment Facility (GEF), multilateral banks, and other donors to address these problems; and (3) the extent to which UNEP, as the Secretariat, provides aggressive leadership in this area.

Programme of Action for the Sustainable Development of Small Island Developing States

The Global Conference on the Sustainable Development of Small Island Developing States, one of several conferences recommended in Chapter 17 of Agenda 21, was held in Barbados in April–May 1994. Its purpose was to explore the special problems of small-island developing states (SIDS), such as those related to their size, limited resources, special environmental problems, and vulnerability to newly recognized threats such as accelerating sea-level rise. From the conference came a comprehensive Programme of Action for the Sustainable Development of Small Island Developing States, currently in the implementation stage. Three of nine substantive issues addressed in the Programme of Action (climate change and sea-level rise, coastal and marine resources; and tourism resources) call for the formulation of new policies and programs in the context of integrated coastal area management. A strengthened capacity for integration of economic and environmental policy in national planning and across sectors was also called for in the section of the Programme of Action dealing with national institutions and administrative capacity. In addition, international organizations and donor nations were asked to support SIDS in responding to the call by the IPCC for vulnerable coastal nations to develop integrated coastal zone management plans including the development of adaptive response measures to the impacts of climate change and sea level rise (United Nations 1994a).

International Coral Reef Initiative

The International Coral Reef Initiative (ICRI), built on existing programs and expertise, combines national and international efforts to conserve and manage coral reefs and their related ecosystems, including mangrove forests and sea grass beds. The founding nations (the United States, Japan, Australia, Jamaica, France, the United Kingdom, the Philippines, and Sweden) announced the initiative at the first Conference of Parties of the Convention on Biological Diversity in December 1994 and at the high-level segment of the April 1995 session of the United Nations Commission on Sustainable Development. Accompanying the announcement was an invitation for other nations and organizations interested in coral reef protection and management to join the initiative. Current "partners," in addition to the nations just listed, include UNEP, UNDP, UNESCO, the World Bank, the Inter-American Development Bank, SPREP (South Pacific Regional Environment Programme), the IUCN, and AOSIS (Alliance of Small Island States). The ICRI does not seek to create new agencies or bureaucracies and is not a funding entity. Rather, it seeks to raise global and local awareness and to obtain national, regional, and global commitments to conserve and sustainably use coral reefs and their associated ecosystems.

In cooperation with the government of the Philippines, the ICRI sponsored a major global workshop titled "Partnership Building and Framework Development" in Dumaguete City from May 29 to June 2, 1995, at which forty-four nations were represented. Participants adopted a Call to Action and developed a Framework for Action. These documents provided background and starting points for six more focused regional workshops exploring regional needs and priorities, held from November 1995 to February 1996. It is hoped that these regional workshops will catalyze the development of national coral reef initiatives.

The Call to Action adopted at the Philippines workshop clearly endorses integrated coastal management as a "framework for achieving the sustainable use of, and maintaining the health of, coral reefs and associated environments." Indeed, one of the six principles contained in the Framework for Action states, "Integrated coastal management, with its special emphasis on community participation and benefit, provides a framework for effective coral reef and related ecosystem management" (ICRI 1995, 4).

R. B. Mieremet (1995) has described the ICRI as a "seed from the Earth Summit tree which now bears fruit." He traces its roots to the first meeting of the
UNCED PrepCom, where, among items suggested by national delegations for the
UNCED agenda, was one submitted by the U.S. delegation that embraced coastal
zone management and coral reef protection. Mieremet details the continuing
interest of the United States in these topics and outlines its present effort to support ICRI, quoting several parts of the report of the Philippines workshop.
Included is a statement by John McManus of the International Center for
Resources Management, referring to ICM:

> *The need to consider community-based management or co-manage-*
> *ment of coral reefs.* Problems in the coastal zones of developing coun-
> tries are generally complex and multifaceted. Solutions require inter-
> disciplinary analysis and planning in the framework of integrated
> coastal zone management. In most cases, a strong emphasis on vil-
> lage-led community development is needed so as to improve the equi-
> ty of resource distribution, to facilitate the management of resources by
> making their misuse socially unacceptable, and to generally improve
> the enforcement of regulations. (Mieremet 1995, 9)

Thus, in the context of coral reef management, ICM is seen as important—in
part, at least, because it protects local community interests through its call for full
participation of all those affected by coastal resource management decisions and
because it embodies the concept of equitable sharing of the benefits of ocean and
coastal resource use.

The ICM Concept Is Further Defined: Development of International Guidelines

Although Chapter 17 of Agenda 21 lucidly described the need for integrated
coastal management, as is common in such negotiated texts, there was little discussion of the meaning of ICM, its main elements and methods, the range of
forms it can take to address various local physical and sociopolitical contexts, and
how it can be put into place—the formulation, adoption, implementation, and
operation of an ICM program. Thus, a number of efforts have been made by international entities to further define, interpret, and operationalize the ICM concept.
Parallel efforts have taken place in the academic community; see, for example, the
special issue on the subject published in *Ocean & Coastal Management* in 1993
(Cicin-Sain, ed., 1993).

We can identify five major efforts to develop international guidelines for integrated coastal management: World Bank 1993; WCC 1994; OECD 1991; UNEP
1995a; and IUCN 1993 (see Pernetta and Elder 1993). Some of these efforts, such
as the OECD's work and UNEP's work, are based on work or experience prior to
UNCED.

International guidelines for practices such as ICM are often important for they can be viewed as setting standards of an international model or norm for countries to follow. In some cases, a country's adherence to such international standards, or lack thereof, can be used by international funding agencies as a basis for approving or disapproving program funds. To our knowledge, none of the guidelines referenced in the previous paragraph has attained this status. Nonetheless, it is our impression that these guidelines are influential and are utilized by both donor agencies and countries commencing ICM programs as guidance in the design and implementation of ICM efforts.

The World Bank's guidelines grew out of a joint effort initiated by the World Bank, UNEP, and FAO in fall 1992 in response to mandates from UNCED. A technical workshop on guidelines for coastal zone management was held at the World Bank's headquarters in November of that year. Subsequent to that meeting, several consultants working with World Bank staff completed the guidelines, which were then tabled by the Bank at the World Coast Conference (WCC), held in the Netherlands in November 1993. The conference statement, representing the views of the ninety or more participant nations, contains a strong statement about the urgency and the benefits of integrated coastal zone management (ICZM). The statement also lists the key elements of ICZM and outlines the obstacles nations often encounter as they endeavor to implement the concept.

The guidelines of the OECD (Organization for Economic Cooperation and Development) and UNEP's guidelines were developed prior to UNCED, generally as parts of other programs. The OECD's guidelines, developed from an analysis of seventeen country reports and sixteen case studies, were aimed specifically at identifying measures to achieve better integration in coastal zone management and, especially, integration of economic and environmental considerations. UNEP's guidelines, on the other hand, emerged from experience gained in administering the Mediterranean Action Plan (MAP) for the Mediterranean Regional Seas Programme. The guidelines were initially drafted by the Priority Action Programme of MAP, since integrated management of coastal areas was recognized by countries bordering the Mediterranean as having the highest priority.

The guidelines of the IUCN (International Union for the Conservation of Nature and Natural Resources, now the World Conservation Union), produced by the IUCN's Marine and Coastal Areas Programme as a contribution to the deliberations on coastal zone management, emphasize ecological and planning considerations in sustainable coastal development.

After we analyzed these five sets of guidelines (as described in the paragraphs that follow), two additional sets of guidelines became available. A guide to good ICM practices is contained in the report of an international workshop on lessons learned in ICM held in Xiamen, People's Republic of China, in May 1996 (Chua 1996). As mentioned earlier in this chapter in the discussion of the FCCC, draft guidelines addressing climate change within the framework of ICM were developed at an international workshop on climate change and ICM held in Taipei, Taiwan, China, in February 1997 (Cicin-Sain et al. 1997).

In table 4.1, we compare the five major sets of guidelines according to ten vari-

ables that we view as important in the design and implementation of ICM programs: (1) scope and purpose (major aspects covered), (2) principles, (3) definition of the management area, (4) functions and goals of ICM, (5) legal basis for
ICM, (6) horizontal integration (mechanisms for intersectoral coordination), (7)
vertical integration (mechanisms for intergovernmental integration), (8) financial
arrangements, (9) prescriptions for the use of science, and (10) capacity building.

Examination of these five sets of guidelines reveals rather more agreement
among them than might be expected given the different motivations and backgrounds underlying the drafting of each set. This seems especially true regarding
variable (1), the scope and purpose of ICM; variable (2), the key principles underlying the concept; variable (4), the scope and functions of ICM; variables 6 and 7,
horizontal and vertical integration in ICM; and variable (9) the use of science in
informing the ICM process.

From our analysis of the five sets of ICM guidelines, it is possible to distill a
set of guidelines that would generally be consistent with all five. Hence, we offer
the guidelines in table 4.2 as "consensus guidelines," which we believe to be a reasonable characterization of the ICM concept as it is understood today. It is not our
intention to suggest that the consensus guidelines have a status different from that
of the five sets from which they are derived. They clearly do not. Rather, they
can be thought of as representing the core of the ICM concept as it is understood
today at the international level. As time goes on and more nations become
involved with the concept, we expect to see a further convergence in thinking
regarding ICM.

As discussed earlier, the ICM concept has already been endorsed as an important and relevant tool in connection with implementation of the Framework Convention on Climate Change, the Convention on Biological Diversity, the Global
Plan of Action for the Protection of the Marine Environment from Land-Based
Activities, the Programme of Action for the Sustainable Development of Small
Island Developing States, and the International Coral Reef Initiative. It seems
likely that forces will exist for these programs to adopt similar definitions and
working guidelines for ICM in order to simplify the implementation efforts by
national authorities attempting to respond to all of these agreements and programs. This suggests to us that the concept of ICM will become more fully elaborated over the next several years and that by the end of the decade, internationally agreed specifications for ICM will have been formally adopted.

Capacity-Building Efforts in ICM Increase

Acknowledgment of the need to build in-country capacity for integrated management runs through all the decisions reached at the Earth Summit and is strongly
emphasized in Chapter 17 of Agenda 21. We have seen considerable progress
since UNCED in the area of capacity building in ocean and coastal management.
This is manifested in major ways: (1) the major UN entities involved in oceans

Table 4.1. Comparison of International Guidelines for ICM

World Bank Guidelines	World Coast Conference Report	UNEP Guidelines	OECD Guidelines	IUCN Guidelines
1. Scope/Purpose				
Guidelines are designed to "ensure that development and management plans for coastal zones are integrated with environmental (including social) goals and are made with the participation of those affected." Included are sections on institutional roles and responsibilities and on triggering of the need for ICZM. A three-stage process of plan formulation, program implementation, monitoring, enforcement, and evaluation is recommended.	Guidelines stress the urgent need for coastal states to strengthen their capabilities for ICZM; and the need to develop strategies and programs by the year 2000. ICZM is described as the most appropriate process for anticipating and responding to long-term concerns and needs while addressing present-day challenges. It is to be achieved through a planning process involving data collection and analysis, monitoring and evaluation, and an implementation process. Report states that the "quest for a unique recipe for ICM was misguided" and that several strategies can be effective.	Guidelines state that "integrated management of coastal areas is required to lay the foundation for sustainable development." ICZM guidelines are given in three main stages: initiation, planning (preparation of ICZM master plan), and plan implementation. Many of the challenges to sustainable development (population pressure, industry growth, coastal tourism) are illustrated in the context of the Mediterranean region.	Guidelines are the result of investigations using country information papers prepared by respondents in several countries and a survey instrument. The main stress is on ecologically sustainable development of the coastal zone. Recommendations are given for structure and processes for ICM, including creation of the institutional body, assessment of information, preparation of alternative plans, selection of a final plan, and monitoring and evaluation.	Guidelines are provided for the development of a coastal area plan that can be applied at a national level, through a review of coastal problems and the need for integrated cross-sectoral management (ICSM). The process by which ICSM can be achieved (CICAP—cross-sectoral integrated coastal area planning) is discussed; seven steps include problem definition, assessment and analysis, issues and options, formulation, adoption, implementation, and monitoring and evaluation.
2. Principles				
• Precautionary principle • "Polluter pays" principle • Use of proper resource accounting • Transboundary responsibility • Intergenerational equity	Follows the principles set out in the Rio Declaration—in particular, sustainable development and its long-term focus and the precautionary principle.	• Multidisciplinary approach • Problem solving, not problem transfer • Priority on prevention rather than cure • Precautionary approach	Guidelines stress that integrated coastal zone management is well suited for ecologically sustainable development (ESD). Implies principles of long-term sustainable growth, utilization of modern economic tools, etc.	Guidelines state that the starting point of ICSM is sustainable development. Therefore, the guidelines endorse a long-term and intra- and intergenerational focus. To be effective, the effort should be holistic, that is, interdisciplinary, combining social and natural sciences.
3. Definition of Management Area				
Ideally, the management zone should include all of the coastal resources of interest and all activities that are capable of affecting the resources and waters of the coastal zone (implies from the watershed to the 200 mile limit).	Report does not recommend where nations should set boundaries. Case studies illustrate a wide diversity of strategies, from only the land side of the coastal zone to somewhat integrated coastal and ocean management.	Guidelines do not recommend specific coastal boundaries. However, the first step of the planning stage is precise definition of coastal area boundaries.	Guidelines suggest an ecosystem approach that should cover part of the hinterland and include the "associated aquatic ecosystems" and seaward, potentially to the 200-nautical-mile limit.	Recognize that the significant variation in local natural and socioeconomic systems make setting up a management program for a specific scientific scale difficult. When the appropriate scale is selected, the CICAP must be consistent with that scale.

(*continues*)

Table 4.1 *(continued)*

World Bank Guidelines	World Coast Conference Report	UNEP Guidelines	OECD Guidelines	IUCN Guidelines
		4. ICM Functions/Goals		
• Strengthen and harmonize sectoral management • Preserve and protect the productivity and biological diversity of coastal ecosystems • Promote rational development and sustainable utilization of coastal resources	• Anticipate and respond to long-term concerns while addressing present-day challenges and opportunities • Stimulate sustainable development of coastal areas • Promote increased economic development and benefits	• Provide guidelines for legal and institutional strategy in area management and planning • Form ad hoc committees for dispute resolution • Combine land use controls and economic tools for pollution control and conservation	• Achieve ESD and • Maintain critical stocks of resources • Maintain or enhance critical environmental quality • Preserve certain natural resources (usually public goods) • Maintain the amenity value of the coastal zone	• Set national development objectives for ICSM • Initiate the CICAP for long-term benefits and an evolutionary ICSM process
		5. Legal Basis		
Agreed-on goals and objectives by line agencies and subnational governments. Legislation for boundary setting and zoning of the coastal area but not necessarily a comprehensive coastal zone act.	There is no one measure for success. New institutional and legal framework must take into account social, economic, and political parameters of each specific situation.	Endorses use of the National Coastal Management Act or other legal means of ensuring enforcement of various sectoral laws. Guidelines utilize a "top-down" approach.	Legislation is needed to create institutional bodies or the management council (lead agency) and to allow for coordination. Of importance is "a consistent government policy at national or regional level that provides clear direction and support for integration."	Within the CICAP steps, legal and administrative options are shown to be "essential components" of a coastal or ocean planning program. However, the guidelines provide no specific recommendations or suggestions for setting up those legal and institutional arrangements; they only emphasize the need for integration among the legal and administrative, socioeconomic, and biogeophysical components and among the public, scientists, managers, and users, including some form of dispute resolution.
		6. Horizontal Integration		
Alternatives include • Use of a national planning agency • Formal establishment of an interagency or interministerial council • Creation of a special coordinating commission or committee • Formal designation of a lead agency	There is no one recipe for success. Case studies document successful ICZM efforts created from both top-down and bottom-up arrangements, allowing for the need for increased intersectoral and intergovernmental integration. Depends on specifics of the case.	Recommends designating a lead agency for coastal management at the national level and creating an interdepartmental committee for coastal management consisting of the major relevant ministries.	Stresses that effective ICM coordination is achieved through "improving the linkages" among sectors and development of regulatory and economic tools. A coordinating body should be created along with management councils and subgroups. No specific recipe is provided.	No specific recommendations are provided; guidelines state only that mechanisms for integration are needed.

There is little discussion of this dimension. Guidelines do suggest the following: • Define the primary level of implementation (not necessarily federal) • Have all government levels and stakeholders participate in the "concept paper" for program formulation • Include local representation in the interagency coordinating mechanism	There is no one recipe for success. Case studies document successful ICZM efforts created from both top-down and bottom-up arrangements, allowing for the need for increased intersectoral and intergovernmental integration. Depends on specifics of the case.	Suggests a degree of national planning to inform regional and local authorities of the intention of the national authorities on coastal development policies. Recommends implementing a process to establish consistency between local activities and regional and national activities.	Makes little or no mention of the intergovernmental aspects of ICM. There is some discussion of the benefits of regional-level ICM planning. Regarding the coordinating body, the guidelines state that in many cases, local coordinating bodies that fit into the national system may be needed.	No specific recommendations are provided; guidelines state only that mechanisms for integration are needed.

Funding to put ICZM program in place can come from budgets of existing government agencies. New funding for research needs and new institutional arrangements should come from the national government, with assistance from international agencies.	Funding is to come from international and national funding institutions (including development banks) that take into account economic development, environmental, and social considerations. Some of the burden is to be borne by coastal users' groups and the general public.	• For administrative structure: national and regional government budgets • For infrastructure and pollution control: use fees, taxes, economic incentives • For conservation: funds from interested environmental groups, park entry fees, etc.	Guidelines state that a major factor of successful ICM is how well the government implementing the program can raise its own funds, either independently or with assistance from the national government. This flow of financial resources must be dependable and ongoing.	Guidelines do not mention strategies or recommendations regarding financial arrangements or economic tools for funding ICSM.

Attention is given to the value of natural coastal resources and the fact that the coastal zone is a "dynamic area with frequently changing biological, chemical and geological attributes."	Given the strong tie between the IPCC and the World Coast Conference, natural science aspects are stressed, with emphasis on the role of vulnerability assessments in ICZM and the effects of sea-level rise.	There is a strong reliance on effective risk assessment studies. Emphasis is given to economic valuation techniques. A section of the planning stage is devoted to estimating: • Current rates of resource use • Waste being discharged into the system • Effects of socioeconomic activities on natural ecosystems. Additional emphasis is placed on water pollution issues.	It is emphasized that scientific analysis, natural resource monitoring, and impact assessment play a crucial role in successful ICM.	There is a strong prescription for using science in the assessment and analysis stage of the CICAP and for understanding the short- and long-term concerns fueling the need for ICSM.

One aspect overseen by the interagency coordinating mechanism is human resources development. Guidelines also recognize the role of stakeholders and the general public in formulation and implementation of the program. However, no mention is made of indigenous peoples and their role in ICM.	Use of funding for capacity building is stressed, as are education and training in the implementation process through: • Interdisciplinary curricula • Establishment of ICZM training centers • Coordination of existing training efforts • Networks to facilitate information exchange	No mention is made of capacity building or attention to indigenous management.	No specific mention of capacity building is made in the main text, though there are references to it in the case studies.	Guidelines state that feedback and consensus building are needed among all interested parties in the coastal area. They state that communication and common linkages should be facilitated, though no formal measures for capacity building are outlined.

Source: Cicin-Sain, Knecht, and Fisk 1995.

Table 4.2. A Consensus Set of ICM Guidelines

Purpose of ICM

The purpose of ICM is to guide coastal area development in an ecologically sustainable fashion.

Principles

ICM is guided by principles in the Rio Declaration on Environment and Development with special emphasis on the principle of intergenerational equity, the precautionary principle, and the "polluter pays" principle. ICM is holistic and interdisciplinary in nature, especially with regard to science and policy.

Functions

ICM strengthens and harmonizes sectoral management in the coastal zone. It preserves and protects the productivity and biological diversity of coastal ecosystems and maintains amenity values. ICM promotes the rational economic development and sustainable utilization of coastal and ocean resources and facilitates conflict resolution in the coastal zone.

Spatial Integration

An ICM program embraces all coastal and upland areas whose use can affect coastal waters and the resources therein and extends seaward to include that part of the ocean that can affect the land of the coastal zone. An ICM program may also include the entire ocean area under national jurisdiction (exclusive economic zone) over which national governments have stewardship responsibilities under both the Law of the Sea Convention and UNCED.

Horizontal and Vertical Integration

Overcoming the sectoral and intergovernmental fragmentation that exists in today's coastal management efforts is a prime goal of ICM. Institutional mechanisms for effective coordination among various sectors active in the coastal zone and among the various levels of government operating in the coastal zone are fundamental to the strengthening and rationalization of the coastal management process. From the variety of available options, the coordination and harmonization mechanism must be tailored to fit the unique aspects of each particular national government setting.

The Use of Science

Given the complexities and uncertainties that exist in the coastal zone, ICM must be built on the best science (both natural and social) available. Techniques such as risk assessment, economic valuation, vulnerability assessment, resource accounting, cost-benefit analysis, and outcome-based monitoring should be built into the ICM process, as appropriate.

and coasts have expanded their own capacities in integrated coastal and ocean management and have also sponsored efforts to build capacity at national, regional, and subnational levels; (2) a number of new training efforts in integrated coastal and ocean management have been created at the international level, and examples of similar efforts at regional and national levels are also in evidence.

UN Organizations and Their Work in ICM

A number of UN organizations have made substantial contributions to the field of ICM, especially during the period following the 1992 Earth Summit.

The Intergovernmental Oceanographic Commission (IOC). In its thirty-seven years of existence, the IOC, a UNESCO organization, has supported many scientific research programs of great relevance to international environmental policy, spanning the full range of the ocean sciences and involving development and application of a rich variety of ocean technologies. In addition to its traditional emphasis on global ocean processes in the open sea, the IOC's work has increasingly come to encompass questions related to coastal areas and management issues. In response to the UNCED follow-up conference on small-island developing states (Global Conference on the Sustainable Development of Small Island Developing States, Barbados, April–May 1994), the IOC has undertaken a number of projects related to their needs. It has also conducted several workshops and training and technical exchange programs on coastal processes. Its Programme on Coastal Ocean, Advanced Science and Technology (COASTS), for example, provides an international framework within which national and regional programs and projects may be coordinated and synthesized to contribute to a global understanding of the fundamental properties and variability of the transition zone from the land to the open sea.

In November 1994 the IOC convened the Second International Conference on Oceanography in Lisbon. Titled "Toward Sustainable Use of Oceans and Coastal Zones," the conference brought together several hundred ocean scientists and marine policy makers from all parts of the world and reaffirmed the long-term commitment of the international marine science community to achieving sustainability in ocean use. First among the objectives to be met was the integrated management of marine and coastal environments including the maintenance of biodiversity (IOC 1994b).

The IOC is currently developing an overall strategy for its TEMA (training, education and mutual assistance in marine sciences) program and for its activities in capacity building. As part of this effort, eight workshops on integrated coastal management have been held at both national and regional (supranational) levels, with more planned in 1997 and 1998. As part of its efforts in ICM training, the IOC supported, in part, the preparation of this book.

United Nations Division for Ocean Affairs and the Law of the Sea (UNDOALOS). As part of the UN Secretariat, this office oversaw the development of UNCLOS III and remains intimately involved in its implementation. The office also served as the Secretariat for the UN negotiations related to straddling and highly migratory fish stocks, which led to adoption of a landmark agreement on an international regime for management of these species (Nandan 1995). The office stimulated some of the early work on integrated coastal management and promoted the conduct of case studies on ICM around the world. UNDOALOS has recently taken the lead in stimulating discussion and action on training needs related to the implementation of UNCLOS III and UNCED. A significant training program called Train-Sea-Coast is now in the implementation phase under the leadership of this office, as discussed later in this chapter.

Food and Agriculture Organization of the United Nations (FAO). The conservation and management of international fisheries are the responsibility of the fisheries element of FAO. Since the well-being of marine fisheries depends on healthy coastal habitats and clean marine waters, FAO is interested in effective coastal management techniques and has produced training and educational materials related to this topic. FAO sponsored a report on integrated coastal management (FAO 1991) during the UNCED preparatory process that was influential in the adoption of the ICM concept in Chapter 17 of Agenda 21. Over the past several years, FAO has been centrally involved in the development of a code for responsible fishing practices and in the negotiations of the post-UNCED United Nations Conference on Straddling Fish Stocks and Highly Migratory Fish Stocks. Moreover, FAO's legal office has recently completed a study on the legal aspects of ICM, focusing in particular on institutional issues (Boelaert-Suominen and Cullinan 1994).

United Nations Environment Programme (UNEP). The environmental problems, pressures, and stresses facing the coastal areas of the world are among the most severe anywhere. Hence, UNEP has a significant ICM effort under way as a part of its Regional Seas Programme. Much of UNEP's work in ICM has been conducted in the context of the Mediterranean Action Plan, particularly through the UNEP office located in Split, Croatia. Guidelines on integrated coastal management, first developed in 1993 in the context of the Mediterranean program, were published by UNEP in 1995 as a guidance document (UNEP 1995a).

UNEP's leading role in implementing the Convention on Biological Diversity also gives it a strong interest in protecting marine biodiversity. Similarly, UNEP has taken the lead in development of the Global Programme for Action on Protection of the Marine Environment from Land-Based Activities and thus has been quite involved with ICM as a framework through which control of land-based activities may be achieved. Recently, UNEP has initiated a training effort called NETTLAP to support environmental training, including coastal management, at the tertiary (university) level in Asia-Pacific (Bandora 1995) (discussed later in this chapter).

United Nations Development Programme (UNDP). UNDP has undertaken a significant program in capacity building as its response to the mandates of UNCED. Its program, called Capacity 21, involves substantial support for a wide range of efforts in integrated management, including integrated coastal management. UNDP provides funding for the Train-Sea-Coast program and the ICM training program of the International Ocean Institute (IOI), both of which are discussed on the following pages.

International Maritime Organization (IMO). IMO has traditionally been concerned with matters related to shipping, vessel-source pollution, safety at sea, and ocean dumping of wastes. For training and education, IMO operates the World Maritime University (WMU) at Malmö, Sweden, with a planning course on sea use among its various offerings. In recent years, IMO has devoted increased attention to efforts to ensure navigational safety and clean seas from the perspective of multiple marine uses. It serves as executive agency for the Regional Programme for the Prevention and Management of Marine Pollution in the East Asian Seas. This program, a cooperative effort with the GEF and UNDP, operates demonstration projects for application of ICM systems in prevention and management of marine pollution, especially pollution from land-based sources.

GESAMP. Originally the Group of Experts on the Scientific Aspects of Marine Pollution, GESAMP, jointly funded by IMO, FAO, UNESCO, WHO, WMO, the International Atomic Energy Agency (IAEA) and UNEP, is a relatively new entrant in the field of coastal management, having been mainly concerned with scientific assessment of marine pollution. Following the Earth Summit, GESAMP has broadened its mission to incorporate general concerns regarding protection of the marine environment, including the provision of advice on policy and managerial options. Reflecting this change in emphasis, GESAMP changed its name to Group of Experts on the Scientific Aspects of Marine Environmental Protection. GESAMP conducted a study to provide guidance on applications of social and natural sciences in integrated coastal management. The results of the study, including a number of case studies, were submitted to the twenty-sixth session of GESAMP in Paris in March 1996 (Khalimonov 1995; GESAMP 1996).

World Bank. The World Bank became increasingly involved in ocean and coastal issues as it expanded its environmental programs in the wake of UNCED. Its active environmental portfolio now stands at $11.5 billion for 153 projects in sixty-two countries; of this amount, $7.2 billion has been committed since UNCED. As reported by Andrew Steer, director of the World Bank's Environment

Department, the organization's environmental agenda is twofold: to support countries as they seek to reform their environmental management and to consider environmental issues in all World Bank activities. The Bank has spearheaded ICZM projects in the Mediterranean Sea, the Baltic Sea, the Danube River basin, the Black Sea, the Aral Sea, the Red Sea and the Gulf of Aden, the Caspian Sea, and Lake Victoria. ICZM was introduced on a pilot basis into the Bank's portfolio of coastal investment projects in Mexico, Ghana, Thailand, Albania, and Indonesia. World Bank–funded projects in coastal and marine protection include the Seychelles Biodiversity Conservation and Marine Pollution Abatement Project, the Ship Waste Disposal Project in China, and the Water Supply and Coastal Pollution Management Project in Brazil. The World Bank's Aral Sea Environment Program and Haaspsalu and Matsalu Bays Environment Project in Estonia concern the use of constructed wetlands for wastewater treatment (World Bank 1995b).

In addition to its regional work, the World Bank has supported a number of global initiatives in the coastal and marine area. It has published or funded several documents on integrated coastal management including *Africa: A Framework for Integrated Coastal Zone Management* (World Bank 1995a), which sets the stage for the identification of future investment in the coastal areas of Africa. In collaboration with the Swedish International Development Authority's Department for Research Cooperation, the Bank is also assessing marine biotechnology in developing nations. The Bank's Land, Water, and Natural Habitats Division has formulated a set of training modules on ICZM that include seminars on the Red Sea and the Gulf of Aden, North Africa, and the Gaza Strip. In association with the Swedish International Development Agency and the Swedish Agency for Research Cooperation with Developing Countries (SIDA/SAREC), the World Bank conducted national seminars on ICM for Tanzania and Mozambique (World Bank 1996a).

Global Environment Facility (GEF). Along with UNDP and UNEP, the World Bank is an implementing agency for the GEF, which funds programs and projects of global environmental benefit. The GEF has provided significant funding for projects and studies in ocean and coastal management, including the Red Sea Coastal and Marine Resource Management project, the Ghana Coastal Wetlands Management Project, and the Lake Victoria Environmental Management Project. In 1996, as part of the Red Sea and Gulf of Aden Program, the GEF conducted an innovative study on management of navigation risks, the primary threat to the marine environment in that region.

New and Existing Teaching and Training Efforts in ICM

The years that have elapsed since UNCED have seen a flurry of activity in training and teaching of integrated coastal management—conferences and workshops have been held (e.g., the conference convened by UNDOALOS and UNDP in Sardinia in June 1993 and the workshop convened by Canada and UNEP at the Canada CZ '94 Conference in Halifax in September 1994) and training programs have been initiated at the international, national, and regional levels. So much activity is taking place that we cannot do it justice here—a paper devoted solely to these developments is needed. In fact, a number of surveys of ICM training institutions are currently under way. One such effort is being coordinated by the United Nations University at UNESCO, Paris; another is a joint effort of the University of Dalhousie, the Marine Affairs and Policy Association, and the University of Delaware.

Respondents to our cross-national survey also reported engagement in a wide variety of ICM training efforts, both in-country and through international institutions, as shown in table 4.3.

To illustrate some of the new training programs that have been created in the field, we focus on two major examples: the United Nations Train-Sea-Coast program and the activities of the International Ocean Institute.

Train-Sea-Coast. Train-Sea-Coast is a program organized by the United Nations Division for Ocean Affairs and the Law of the Sea (UNDOALOS) with the support of the United Nations Development Programme's Science, Technology and

Table 4.3. Capacity-Building Measures in Selected Countries

	All (%) (N = 49)	Developed (%) (N = 14)	Middle Developing (%) (N = 15)	Developing (%) (N = 20)
Specialized in-country training of existing staff	73	57	73	85
Specialized training at overseas institutions	67	21	80	90
Participation in UNDP, UNEP, FAO, IOC, or other training programs	65	29	80	80
Hiring of new staff with appropriate qualifications and training	51	36	53	60
Establishment of new graduate programs in ICM at the university level	41	64	53	15
Other	10	14	13	5

Source: Authors' 1996 cross-national survey.
Note: Percentages do not add up to 100% because respondents were allowed to mark all categories that applied.

Private Sector Division (UNDP/STAPS), in collaboration with a number of UN and non-UN entities. The program was launched in 1993 to develop a coordinated global training program in ocean and coastal management in response to the prescriptions of Agenda 21. The idea for the program emerged from discussions held at the Consultative Meeting on Training and Integrated Management of Coastal and Marine Areas for Sustainable Development. The meeting, held in Sassari, Sardinia, Italy, on June 21–23, 1993, was sponsored by UNDOALOS with the support of the UNDP's Division for Global and Interregional Programmes (UNDP/DGIP)). Train-Sea-Coast is a decentralized international program for coordinated development and sharing of high-quality standardized course materials using a common methodology (Vallejo 1995).

Ten academic institutions located in nine countries in all major geographical areas of the world (Brazil, Costa Rica, Fiji, India, the Philippines, Senegal, Thailand, the United Kingdom, and the United States) are founding members of Train-Sea-Coast, joined in a network coordinated by the United Nations Division for Ocean Affairs and the Law of the Sea. These institutions, chosen for participation after field visits by UNDOALOS, have committed their own resources and those of their governments to support the program. Each of these institutions is developing a set of courses in ocean and coastal management, using a very detailed common methodology. Coordination within the network is expected to ensure coverage of a wide range of important topics while minimizing duplication and overlap. When courses are created and detailed course materials are developed, they are made available to all members of the Train-Sea-Coast network. The first training workshop, teaching course developers the common methodology, was held in Stony Brook, New York, on January 23–February 3, 1995.

Thus far, funding for the program has come from UNDOALOS, UNDP/STAPS, the government of Japan (in the form of contributions to the first workshop), and member institutions and their governments. Future plans call for establishment of an Internet network, application of long-distance learning methods, and fostering of collaborative research projects among the training centers.

International Ocean Institute Network. The International Ocean Institute (IOI), an independent international nonprofit nongovernmental organization with headquarters in Malta, has grown from two to eight operational centers through funding support by the Global Environment Facility and others. The eight centers are located in Canada, China, Costa Rica, Fiji, India, Japan, Malta, and Senegal. In 1995, arrangements were under way to establish three additional centers, in Qatar, South Africa, and Romania. The IOI conducts training programs, publishes books and newsletters (e.g., Ocean Yearbook and Across the Oceans), holds conferences (e.g., Pacem in Maribus), and has a variety of memoranda of understanding for joint work with educational and government entities in various areas of the world.

The institute's activities are largely, but not exclusively, aimed at organizations,

institutions, and persons in developing countries. The stated objective of the training program is to focus attention on the oceanic environment with special reference to land-sea-air interactions, the problems of small-island states, and sustainability of the oceans, especially enclosed and semi-enclosed seas. The approach
is to bring together various specialists involved in ocean management (i.e., science, technology, economics, law, and management) and help them transcend
their specializations and "create a common language leading to a common culture
or purpose" (Wood 1995, p. 134). As described by the IOI's founder, Elizabeth
Mann Borgese (Borgese, 1994), courses should impart information, enhance skills,
and, above all, change attitudes. Courses vary in duration from one to ten weeks.

Currently, all training programs are conducted in English; however, the centers
in Senegal and Costa Rica intend to offer training in French and Spanish, respectively, and the center in Qatar will provide training in Arabic. To take advantage
of the most advanced teaching methodologies, the IOI is developing a series of
teaching modules consisting of texts, audiotapes, videotapes, and simulations that
can be exchanged within the IOI network; the Internet system is used as well. The
IOI has received significant funding from the Global Environment Facility (GEF)
to support its work—U.S.$2.6 million, administered through UNDP (UNDP and
GEF 1993).

Other Examples. UNCED and the Law of the Sea Convention have also spawned
a number of ICM training efforts at the regional level. Among these are UNEP's
NETTLAP Program, an effort to coordinate environmental training programs at
the tertiary (university) level in the Asia-Pacific region (Bandora 1995); the
CANADA/ASEAN marine science program, which aims to build the technical
capacity of marine scientists in the ASEAN region; and the MEDCOAST training
programs, designed to develop ICM expertise in countries bordering the Mediterranean and Black Seas. MEDCOAST initiatives generally include conference
series and workshops, training programs, and collaboration in international
research (Ozhan 1995; Ozhan and Culhaogie 1995).

Many other institutions are involved in ICM training, particularly in developed
countries such as the United States, the United Kingdom, and Italy. In the United
States, significant efforts to teach ICM as part of the field of marine affairs and
policy began in the early 1970s and have been growing ever since. Among the pio-
neers in the field are the University of Washington, the University of Delaware,
and the University of Rhode Island. In the early 1990s, the two dozen teaching
and research programs in marine affairs and policy in the United States joined
together to form a professional association, the Marine Affairs and Policy Association (MAPA), which now number more than 250 individual and institutional
members.

MAPA publishes an extensive directory of individuals and organizations
involved in ICM and has established an Internet service. The University of Rhode
Island conducts a summer training institute in integrated coastal management

designed especially for participants from developing countries (Crawford, Cobb, and Friedman 1993) and operates an Internet service with profiles of coastal management programs around the world. The United Kingdom offers significant programs in ICM, particularly at the University of Cardiff and the London School of Economics and Political Science. In Italy, the International Centre for Coastal and Ocean Policy Studies (ICCOPS), was created in 1992, analyzes the theory and practice of integrated ocean and coastal management. ICCOPS has held a number of workshops on the subject and sponsored a series of publications, at both the global and regional (Mediterranean) levels.

In the developing world, the number of new institutions related to integrated coastal management appears to be growing rapidly—for example, the government of Malaysia has created the well-supported Malaysian Institute for Maritime Affairs (MIMA), and in Thailand, a new educational center in the field, CORIN (the Coastal Resources Institute), has been established at Prince of Songkla University. This list is not exhaustive, of course, but indicates the rapid growth in ICM education and training.

Summary

Clearly, UNCED catalyzed considerable ICM-related activity at the international level. Beyond a doubt, ICM has emerged as the framework of choice for realizing many of the goals of Chapter 17 of Agenda 21 and for implementing important aspects of both the Convention on Biological Diversity and the Convention on Climate Change. We believe that significant elaboration, refinement, and implementation of the ICM framework can be anticipated in the coming years.

This chapter reviewed three major manifestations of the growing international capacity in ICM: adoption of the ICM framework in widely different fora related to oceans and coasts, a growing convergence in the development of international guidelines and standards in ICM, and a healthy proliferation of training and teaching efforts in the field. Regarding the first two developments—the fact that the concept will be pursued in the very different venues of the Climate Change Convention, the Biological Diversity Convention, the ICRI, the work on land-based sources of marine pollution, and so on, and the growing convergence of general principles and guidelines for ICM—we urge pursuit of a balance between development of a broadly accepted international model and preservation of flexibility for national governments and local communities to implement ICM according to their own particular needs. It would be counterproductive if various different fora came up with their own models of ICM, forcing national and subnational officials to satisfy several different prescriptions for ICM, or if officials lacked flexibility to tailor the ICM process to address their community or nation's particular requirements.

The third development involves ICM training. All who are interested in seeing widespread application of ICM should welcome the many opportunities being developed—a healthy prospect indeed. It appears, however, that a growing segment of these opportunities are ad hoc in nature, not part of a larger, more coherent whole. We suggest that efforts be made to focus training opportunities at regional centers and to create networks among training institutions. Greater use of joint programs should also be encouraged.

PART III

A PRACTICAL GUIDE TO INTEGRATED COASTAL MANAGEMENT

Chapter 5

Setting the Stage for Integrated Coastal Management

Introduction

In this chapter, the focus shifts from the general concepts and international level to application of ICM at the level of the individual nation. The first part of the chapter deals with the necessity of tailoring the ICM concept to fit the context of the country considering it. Factors important to this tailoring are presented and means of drawing lessons from existing ICM programs are discussed, followed by advice on laying the groundwork for a government decision to develop an ICM program. The importance of framing the proposal so that it is not perceived as a threat to existing coastal management activities is stressed, as is the need to bring into the process as soon as possible the users of the coastal zone and its resources as well as the general public. Ultimately, the decision to put an ICM program in place will depend on the political will of the decision makers.

Tailoring and Lesson Drawing

The nature and structure of an ICM program will depend very much on the physical, socioeconomic, cultural, and political context in which it is to operate. In moving from a general model of integrated coastal management to real-world implementation, one is confronted with the great diversity of national settings. Variables include level of development, concentration of population in coastal areas, types of coastal and marine ecosystems, nature of problems present in coastal/marine areas, sociocultural traditions, and type of political system. For example, problems besetting urban coastal areas in developed countries, such as the Mediterranean coasts of Spain and France, differ dramatically from those of largely unpopulated areas in developing countries, such as the Pacific coast of Colombia. In the former situation, in which little if any coastal land remains available, the main roles of ICM are regulatory and restorative in nature—for example, limiting and regulating new development, maintaining set-back lines, and restoring degraded habitats. In the latter situation, the role of ICM would be primarily developmental—encouraging appropriate settlement; providing incentives for development of particular coastal and marine industries; building roads, water systems, and other coastal infrastructure in environmentally sensitive ways; and the like.

As discussed in chapter 4, development of a synthesis model of ICM and general international guidelines are important in providing a common vocabulary and framework for ICM and facilitating learning through comparison of experiences at the international and national levels. Middle-range models of situations typically encountered in ICM are also useful in illustrating the effectiveness of various approaches and their suitability in different coastal and ocean contexts. A general typology of such contexts can assist the coastal manager in determining which countries' ICM experiences offer the most relevant and applicable information for the situation at hand.

Toward a Typology of ICM Contexts

Although a great many variables could be used in developing a typology of coastal and ocean contexts around the world, we believe that just four variables— socioeconomic, physical, and political—capture the key parameters of such typology: (1) a country's level of development, (2) the concentration of population in its coastal zone, (3) the type of coastal and marine ecosystems it contains, and (4) the type of political system in place. As depicted in figure 5.1, the first three variables establish the context of the problems and opportunities present in the area (the *what*) and the goals and objectives of ICM (the *why*), and the fourth variable addresses the questions of *how* and *by whom* ICM can best be put into effect.

A country's level of development (e.g., developed, middle developing, develop-

Figure 5.1. Major Socioeconomic, Physical, and Political Variables in Coastal and Ocean Contexts

Nation's Level of Development	Concentration of Population in Coastal Area	Type of Coastal and Marine Ecosystems Present	Type of Political System
(Socioeconomic Variables)		(Physical Variables)	(Political Variables)

These variables establish the context for

The *what*:
    — Problems and opportunities
      present in the coastal or ocean area

The *why*:
    — goals and objectives of ICM

This variable addresses

The *how* and by *whom*:
    — *how* can ICM be
      best accomplished?

    — by *whom*?

ing), which can be determined using standardized international data, provides the general context for ICM and helps to define its goals; that is, what is driving the ICM process. For example, as a crude generalization, in many developing countries, the development dimension of ICM—that is, expansion of appropriate coastal and marine enterprises in an environmentally sound manner—will tend to be a major driving force. The *concentration of population in the coastal zone* (e.g., density of settlement, along an urban–rural continuum) further sets the context for ICM and helps to define specific objectives. For example, in a highly urbanized coastal setting, conflict over uses will require that ICM institutions and processes concentrate on conflict resolution and harmonization. The third variable, *type of coastal and marine ecosystems present*, helps to further establish the context for ICM. It is imperative to understand the physical aspects of the coastal zone. The interaction between human populations and the coastal and marine physical and biological systems poses both the problems and the opportunities coastal decision makers must address. Low-lying barrier beach systems, for example, are highly vulnerable to storms and to sea-level rise, often experiencing floods and coastal erosion; physical characteristics such as these determine the type of management approach to be used.

The *type of political system* informs the question of who can develop and carry out integrated coastal management and through what means. Understanding a nation's political system is imperative to answer such basic questions as, Who needs to be convinced to establish ICM? Who will design, implement, monitor, and enforce it? Whose behavior needs to change to make ICM effective? There are many forms of political systems around the world. For example, recent article classifying political systems in developing countries (Berg-Schlosser 1990) found great variety in political structure among the seventy-two countries studied that had stable governments, as shown in table 5.1. The type of political system is especially important when it comes to distributing authority for ICM among national, provincial, and local authorities and among various agencies of the national government.

Table 5.1. Political Systems of Seventy-Two Developing Countries

Type of Political System	Numbers of Countries
Monarchy	12
Old oligarchy	5
New oligarchy	10
Socialist regime	8
Communist regime	5
Semicompetitive system	7
Polyarchy (pluralistic system)	15
Personalistic military regime	5
Corporate military regime	1
Socialist military regime	4

Source: Adapted from Berg-Schlosser 1990.

Table 5.2. Hypothetical Settings for ICM

	Type 1	Type 2	Type *n*
Level of development	Developed nation	Developing nation	. . .
Population concentration	Urban	Rural	. . .
Predominant type of coastal or marine ecosystems	Low-lying barrier beach system	Coral reefs and mangrove forest	. . .
Type of political system	Pluralistic, federal	Monarchy	. . .

Source: Cicin-Sain 1993.

National political systems have myriad features, but in our view two stand out as most important to understand for ICM purposes: (1) the degree of concentration or diffusion of power and authority among national-level government institutions (for example, is it a single-party or a multiple-party system, and how is authority distributed between the executive and legislative branches?) and (2) the division of authority between national and subnational (provincial, local) levels of government (for example, is power concentrated at the national level, as in a unitary system or are power and authority shared by different levels of government, as in a federal system?) (Lijphart 1990).

Combining these four variables in different ways yields a number of "typical" contexts for integrated coastal management, as illustrated by table 5.2. General international guidelines for integrated coastal management can be tailored to fit a range of these typical contexts for ICM. Similarly, such "ICM types" can be useful to coastal managers in finding the most suitable country comparisons for lesson drawing.

Comparative Assessments and Lesson Drawing

Systematic comparative assessments of ICM efforts in typical contexts can provide very useful information. J. C. Sorensen (1993) provides an initial list of variables for use in such assessments. Much can be learned, we believe, from these assessments, in which a number of variables are kept constant while the relative importance of target variables is examined. For example, if one were interested in improving the management of offshore oil resources in the United States (where this issue has posed particularly thorny problems between the national government and the states for many years), much could be learned by examining how other federal systems, such as Australia and Canada, have addressed this problem. On the other hand, one should not ignore ICM applications in dissimilar contexts; such cases can suggest ICM methods that could be adapted to a particular country's setting and needs.

The final step in the tailoring strategy is lesson drawing on the basis of the

Table 5.3. Methods of Lesson Drawing

Copying	More or less intact adoption of a program in effect elsewhere
Emulation	Adoption with adjustments for different circumstances
Hybridization	Combining of elements from two different programs
Synthesis	Combining of similar elements from three or more programs
Inspiration	Use of other programs for intellectual stimulus in developing a unique program without an analog elsewhere

Source: Adapted from Rose 1992, 22.

comparative assessments. Richard Rose writes that lesson drawing addresses the question, "Under what circumstances and to what extent can a program that is effective in one place be transferred to another?" A lesson, he writes, "is more than an evaluation of a program in its own context; it also implies a judgment about doing the same elsewhere. . . . To draw a lesson properly, it is necessary to devote as much care to examining the probability or improbability of transfer as it is to evaluating its initial effect" (Rose 1992, 7). There are various ways of drawing lessons, according to Rose, as noted in table 5.3.

Since World War II, particularly right before and after the great period of decolonization in the 1960s, new nations have routinely engaged in copying, emulation, and hybridization practices. In fact, many political institutions, constitutions, and the like in newly developed countries in Africa, for example, were patterned after those of "metropolitan" colonial countries such as the United Kingdom and France. Similarly, in postwar Japan, institutional reconstruction engineered by American occupying forces followed patterns found in the United States.

The failure of a number of these "copied" institutions over the past twenty years has emphasized the need for careful tailoring and adaptation of institutional and management approaches to a country's specific historical, cultural, socioeconomic, and political context. Thus, coastal decision makers today are more likely to engage in the last two methods of lesson drawing listed in table 5.3, synthesizing and seeking inspiration from the experiences of various countries, particularly those most closely fitting their type of coastal context.

Undertaking ICM: Setting the Stage and Developing the Political Will

The very earliest steps in embarking on the development of an ICM program could be critical to its eventual success. Important early actions are discussed below.

Setting the Stage

Perhaps the biggest hurdle in putting an ICM program in place is obtaining the government's decision to undertake an ICM initiative. A number of factors impinge on such a decision, some of which relate to preconceptions about ICM. It is important, therefore, to provide a clear and accurate picture early in the discussions of what ICM is and what it is designed to do.

For example, ICM should be described from the beginning as a process that for the most part relies on existing programs, improving their outcomes relative to the management of interrelated activities in the coastal zone, through the use of more formalized coordination and harmonization mechanisms. ICM need not be seen as threatening existing sectoral programs. Indeed, it can be presented as an opportunity for existing specialized management programs to improve their efficiency and effectiveness. Also, ICM should not be presented as an elaborate, complex methodology requiring simultaneous full-scale implementation in all coastal areas for the full range of coastal issues. Rather, it should be made clear that ICM can be implemented incrementally—first in those locations most in need of integrated management or for those issues of highest priority. In most cases, ICM can be built on existing management efforts, strengthening rather than replacing them.

ICM must be presented not as a "one size fits all" solution but as a concept that will be tailored to the country's unique circumstances. Similarly, it is important that the need for improved management of coastal resources is made clear and that ICM's role in strengthening and improving the existing management framework is explained in a compelling manner.

Ultimately, however, the decision to develop an ICM program is a political one—and one that generally requires a measure of political will. (We use the term *political will* to refer to the resolve or determination needed by a decision maker to take a course of action that involves expenditure of political capital or comes with some political cost. Noncontroversial decisions do not require political will.) In general, if existing and potential users of the coastal zone evidence clear support for ICM, government leaders will find it much less controversial, and therefore less costly in a political sense, to decide to undertake an ICM program. Hence, the remainder of this chapter describes the initial steps in establishing an ICM program with emphasis on those likely to be most helpful in eliciting the government's consent to proceed.

Assessing the Need for ICM

Early on in the consideration of ICM, the need generally arises for a formal statement of the reasons why ICM (or some version of a more integrated approach to coastal and ocean management) is needed, either for the country as a whole or to address specific problems in a given area. Obviously, it is important that the need for ICM be clearly and accurately stated. Emphasis should be on presenting unambiguous evidence of the problems associated with continuing the existing (sectoral) approach together with a clear and realistic account of the increases in efficiency and effectiveness of management outcomes that can be expected with ICM in place.

Recognition of the need for a more integrated approach to coastal management is likely to arise either from the failure of existing management measures or in connection with new management needs that require a more holistic approach. Conflicts between existing uses, for example, aquaculture and coastal recreation, may increase in frequency and severity, making it clear that the traditional sector-by-sector approach to management is no longer adequate. Emerging needs such as protection of coastal and marine biodiversity or reduction of vulnerability to accelerating sea-level rise clearly require an integrated approach to management and thus could generate the momentum for adopting an ICM program. Also, dealing effectively with the full spectrum of land-based activities that can adversely affect the marine environment is now seen (correctly, in our view) as requiring an integrated management strategy (Kimball 1995). Similarly, the potential for new economic activities on the coast may well require an integrated approach to harmonize new and existing uses.

It is not difficult to tell when existing sectoral management approaches to coastal management are no longer adequate. Two symptoms are often present. First, conflicts between competing or conflicting uses may become more and more prevalent, to the point that an impasse brings decision making to a virtual halt. A second sign of a failing management system involves increasing reliance on the judicial system. In political systems in which such avenues are open, parties who have lost confidence in the integrity of a management scheme typically attempt to achieve their goals through legal challenges. This, too, often indicates that the existing management system is no longer able to incorporate all interests involved in these increasingly complex situations.

Framing the Proposal

As mentioned earlier, where some coastal and ocean management is already taking place, ICM should be presented as a means to strengthen and enhance the existing management efforts rather than as a replacement for them. ICM should be proposed as a complement to any existing sector-based programs, which will in general continue and, indeed, be integral parts of the new effort. It should be stressed that ICM is expected to improve the management of individual sectors, such as fisheries and oil and gas development, as well as help achieve overall national goals.

ICM should also be proposed as a method for better coordinating existing coastal and ocean programs—both regulatory and development oriented—as well as for dealing with new issues. It should be presented as a way to increase efficiency and get more effective government action. Moreover, it can be legitimately described as a way to increase government accountability and increase the likelihood that concrete and measurable outcomes will be achieved from coastal and ocean management programs.

Again, it is useful to prepare an ICM proposal in written form relatively early in the discussions. At this stage, emphasis should be placed on three aspects: (1) a description of the limitations of the existing management approach (probably sector based), with concrete examples of the problems caused by these limitations; (2) a description of the ICM concept, how it would function in the government setting in question, and how it would achieve a higher level of integration and performance; and (3) the steps needed to develop and implement an ICM program, including a timetable, an estimate of the costs involved, and professional staff requirements. It is best to delay politically sensitive topics, such as designation of a lead agency and the nature of the coordination and harmonization mechanism. Such issues are better dealt with at a later stage, when the scope of the program is better known and its goals and objectives are more defined. Early debate over sensitive issues can be divisive and can delay work on more productive aspects.

Overcoming Potential Barriers to ICM

Any proposed initiative in the public sector will probably encounter barriers to its acceptance and implementation. As shown in table 5.4, at least four kinds of resistance can occur, including bureaucratic inertia, turf protection, ideological opposition, and opposition from economic interests.

Since putting an ICM program in place may be seen by some government agencies as altering the existing balance of power among government units, some opposition is almost certain to develop. It is therefore important to place the proposed ICM program in the proper context at the earliest possible moment. Again, a well-written and well-reasoned document that clearly indicates what ICM *is* and, equally important, what it *is not* will help quiet certain kinds of uninformed opposition. Providing full information about the proposal to all potentially affect-

Table 5.4. Barriers to Initiation of ICM

Bureacratic Inertia
    Resistance to change of any sort

Turf
    Opposition to changes seen as competing with or threatening an agency's mission or resource base

Ideological Opposition
    Based on fundamental differences, such as opposition to larger government

Opposition from Economic Interests
    Tied to existing patterns of ocean or coastal use that are benefiting from the status quo

ed organizations early in the debate is useful in this regard. Soliciting comments and suggestions on early drafts of the proposal is also a positive step.

Early efforts should be devoted to assuring agencies with important sectoral management responsibilities in the coastal and ocean area (fisheries, offshore oil and gas development, wetlands protection, coastal water quality, shoreland zoning and land use, etc.) that their participation is vital to the program's success. Indeed, it is only with the active and positive involvement of the sectoral management agencies that ICM can function satisfactorily. As a rule, it is the sectoral agencies that possess the specialized expertise and data concerning key coastal and ocean activities such as fisheries management, coastal erosion, and wetlands management that are of critical importance in any ICM process.

Furthermore, from the earliest discussions it should be made clear that once a decision to develop an ICM program is made, the ICM plan itself will be formulated by a team brought together for this purpose, one that includes representatives from *all* involved or affected agencies. The composition of this team and the work it will undertake should be a prominent part of the proposal document. The proposal should also make clear that the team will consider and make recommendations on institutional aspects of the ICM program, including the nature and operation of the coordination mechanism.

Developing the Political Will to Undertake ICM

An adequate measure of political will is generally needed by decision makers at the executive and legislative levels to commit the resources (funds, staff, etc.) necessary to undertake an initiative such as ICM. Decisions of this nature do not tend to occur spontaneously or readily. Typically, the way must be prepared by providing decision makers with timely information on program benefits and costs and the relevant experiences of other similarly situated countries. Ideally, this information should be included in the proposal document.

Making a decision that involves some political cost—which may sometimes be the case in undertaking an ICM program—is far easier if the decision maker perceives that the proposed action has the support of the groups or interests most likely to be affected by it. In the case of ICM, this means the *users* of the coastal and ocean zones and the concerned public. Therefore, these groups should be brought into the discussions early and provided with sufficient detail that both their input and suggestions and their general support can be obtained and retained throughout the ICM formulation process.

The Importance of Public Participation and Consensus Building

An overarching objective of integrated coastal management is to ensure that coastal and ocean areas serve users and the public in an equitable and sustainable fashion indefinitely, at the least possible cost, and in a way that does not foreclose options for future generations. Thus, the coastal users and the public must be brought into the ICM process at the earliest possible stage, for at least three reasons:

1. The input of those who use and rely on the coastal zone is of great importance in the design of the ICM process. Users have valuable insights regarding both the management needs and the economic opportunities an ICM program should address.

2. Support of the users for development and implementation of an ICM program is crucial to its success.

3. Increasingly, governments are finding it necessary to develop private-public partnerships to fully accomplish the goals of their resource management programs. This is especially true with economic development initiatives related to coastal and ocean resources such as those that might be included in an ICM program.

Creating a Coastal Users' Group

In considering just who are the users of the coast, care should be taken to employ a broad, inclusive approach. It is far better for a group to ask to be removed from the effort than for it to learn of the program later and to demand a place at the table at that point. For example, in the fisheries sector, a coastal users' group should include not only representatives of commercial fishing organizations but also recreational and artisanal fishers, and if indigenous people are involved in fishing activities, they, too, should be invited to participate. Similarly, active outreach activities will be required to ensure that all environmental and conservation groups have an opportunity to be involved. In developed countries, in particular, such groups can encompass a wide range of orientations and goals. Both resource use–oriented conservationists and wilderness-type preservationists may need to be represented. In the latter category will be people who value the coasts and their resources simply because they exist and are there to be appreciated and enjoyed.

Tourism and recreation are growing rapidly in many coastal areas of the world. It is especially important, therefore, that these interests be fully represented in any coalition of coastal users. Depending on the nature of the coastal zone in question and its resource base, any of the following interests, among others, might be represented in a coastal users' group: coastal hotel and resort operators and developers, tour boat operators, dive shop owners, fishing interests, aquaculturists, coastal farmers, real estate agencies, energy companies, port and harbor operators, indigenous peoples, local governments, community and women's groups, environmentalists, and conservationists.

A successful coastal users group will meet these criteria:

• Representatives of all groups that believe they have a stake in the coastal and ocean zone and its sustainable use, health, and well-being are included.

• Members of the group perceive that they are fully informed about the status of ICM planning in their area.

• Members believe that their suggestions and input to the ICM development process will be seriously considered and can affect (improve) the final plan.

• Members do not see themselves as merely a "rubber stamp" to approve work already done by the government staff.

Obtaining Early NGO Support

Nongovernmental organizations (NGOs) are becoming increasingly active in many countries. In many respects, the United Nations Conference on Environment and Development (UNCED), held in Rio de Janeiro in June 1992, served to catalyze the NGO movement globally and nationally. This was especially true in developing countries, where the UNCED preparatory process encouraged NGO participation and the UNCED Secretariat provided significant support. In any event, NGOs now can and do play a role at the national and subnational levels. This means that they can often facilitate the adoption of initiatives such as ICM programs. It also means that they can oppose such initiatives if they have not been convinced of their merit.

It is a fact of life that the agendas of most governments generally contain many more items than can be addressed or successfully acted on at any given time. Lobbying is ongoing with respect to the priorities to be accorded to various issues; thus, a concerted effort is usually needed to move an issue to the top of the agenda. Typically, NGOs help one another promote issues that are mutually acceptable. Therefore, effort should be devoted early on to gaining the support of such groups.

Early interaction with interest groups serves an additional purpose. As the ICM proposal takes form, its political acceptability to various interests, both environmental and development-oriented, can be tested. Changes can be negotiated that make the proposal more acceptable to particular interests without jeopardizing its integrity. To the extent that an informal process such as this addresses a proposal's potentially controversial points, it serves to smooth the way through the formal legislative process.

It is clear, then, that the early support of an existing NGO community can be of considerable importance in securing approval for an ICM proposal. Not only will such assistance broaden the base of support for ICM but it can also help in the implementation and operational stages, when support may be needed for new regulatory authority and additional staff and funding.

Strengthening the Support of the National Government

In the earliest stages of an ICM program, the activities will probably be confined to a single office in a particular ministry or department of the national government. This may be the office that relates to the local government or community that may be providing the impetus for ICM, the office that follows up on the obligations the nation assumes when it embraces specific international agreements, or the office responsible for a sector of the coastal zone that is experiencing urgent problems. In any event, since several different levels of government will ultimately be involved, awareness of ICM and its benefits should generally be extended through much of the structure of the national government.

It is important to disseminate information regarding ICM in a way that sets the stage for constructive dialogue about the program and its potential. Again, care must be taken to describe the ICM process and any institutional mechanisms likely to be associated with it in such a way that it is not viewed as a threat to existing sectoral programs. It would be unfortunate (and unnecessary, in our view) if early negative reactions to the ICM concept were to develop based on incomplete or inaccurate information. To guard against this, the job of building awareness of ICM should be shifted as soon as possible to a more widely representative interagency group. This would also help ensure that no single agency or office is seen as "preordained" to inherit the program at a later date.

In summary, an ICM program will be more likely to succeed at the implementation and operational stage if a broad base of support in the national government can be engendered early on. In our view, the chances for success are increased when the conditions outlined in table 5.5 prevail.

Community-Level Participation and Support

Virtually all ICM programs require active and positive participation at the local or community level; in countries where extensive decentralization has occurred, such as the Philippines, the local community may in fact be the main initiator of ICM. The sole exception might occur where a coastal nation has a unitary form of government—that is, only a national government—with all authority over land and water use authority vested in that central government. However, few nations with this type of government structure exist. A major challenge facing most national governments seeking to initiate an ICM program is how to secure and maintain the positive interest of the local or community level of government.

In addition to exercising certain controls over land and water use, local governments and local communities can contribute much to the ICM process. Cer-

Table 5.5. Conditions Facilitating Successful Initiation of ICM at the National Level

1. The decision to form an interagency group to initiate an ICM program is seen as having been made at the highest level of government (president, prime minister).

2. Positive support from leaders of the relevant ministries or departments is evident.

3. The interagency group is led by a senior policy-level official from a unit above the level of the line ministries—i.e., a national policy office, a national budget office, or the president or prime minister's office.

4. The interagency group includes representatives from all relevant ministries or departments (oceans, fisheries, environment, natural resources, coast guard, etc.) as well as representatives from involved provinces, local governments, and NGOs.

5. Meetings of the interagency group are fully transparent and open to the public.

6. The interagency group operates with written terms of reference and a firm timetable for producing a report and recommendations. In our view, formulation of a proposal to develop an ICM plan (not development of the plan itself) should generally be accomplished in six months or less.

7. The report and its recommendations are presented at formal public hearings before being sent, along with the outcome of such public hearings, to the decision makers for action.

tainly, officials of local governments best understand the social, economic, and environmental issues facing their citizens. Whether the issues in question are declining fish stocks, degraded wetlands, or eroding beaches, local citizens are closest to them and hence are in the best position to provide information and insights. Community members are also likely to have a sense of whether certain solutions will or will not work in the local setting.

Beyond providing needed information, it is also essential that community members support the development and operation of the ICM process in their coastal zone. Such support would seemingly be readily obtained, since in most cases the benefactors of integrated coastal management will be the local users of the coastal zone. Yet this is not always the case. Indeed, sometimes factors such as distrust of the national government or of government programs cause the local community to oppose new initiatives such as ICM. In these cases, showing the local community the benefits that similarly situated communities have received from ICM is a helpful strategy. Local opinion leaders may also be included in special task forces or advisory groups aimed at revising the intergovernmental aspects of the ICM proposal to make it satisfactory to the local community.

Ideally, support at the community level should come from all quarters of society—coastal and ocean users, environmental and conservation groups, the private sector, educational and academic institutions, agricultural interests, aquacultural groups (if present), and the like. An active outreach program specifically targeted to these groups is usually needed. An early task in this regard is development of a program of regular newsletters, public meetings, and other events to keep everyone informed about the ICM effort, its current status, issues facing the program, opportunities to provide input, ways of getting additional information, and so forth. Openness and transparency will help lay the foundation of community trust needed for a successful ICM program.

Options for Early Leadership— Top-Down or Bottom-Up

Recognition of the need for ICM and the first steps in forming a program can take place at the community level or at the provincial, state, or national government level. In the United States, for example, the earliest efforts at coastal management—those in the San Francisco Bay Area—started at the local level, where the problems (associated with the filling of San Francisco Bay for development) and the need for action were most obvious. Shortly thereafter, the state of California initiated a statewide effort based on the San Francisco experience, and within a year or so, several other states had started coastal management programs. These early efforts were soon complemented by legislation at the federal level, one goal of which was to strengthen and support the ongoing state and local efforts. In its present form, the U.S. effort involves the national and state governments, with state governments often including local governments as partners in joint efforts.

Of course, the campaign to put an improved or strengthened form of coastal management in place does not have to originate with a government entity—it can just as well be launched by a coastal users' group or local community concerned with a particular problem. For example, a fishermen's organization concerned about loss of prime fish habitat due to inappropriate coastal development projects may seek measures to better regulate such development. Alternatively or in addition, the fishermen could seek to have the remaining habitat declared a marine protected area, off limits to further development. At this stage, they may or may not recognize that these remedies could be part of a more comprehensive coastal management program. A reasonable first step, however, might be to identify other groups that would benefit from improved planning and management. These might include environmental, conservation, and recreational groups and others interested in protecting coastal and marine ecosystems and preserving natural areas and coastal open space. Such a coalition of interests could approach government officials at both the local and national levels and request action to address their problems. Ultimately, it would probably be recognized that the loss of fish habitat was just one of a number of interrelated coastal issues and that a broader undertaking (ICM) was needed.

It is equally possible, however, for the stimulus to come from the national level of government. The national government might undertake a more integrated form of coastal and ocean management as part of an agreement with a donor organization or a regional development bank or to fulfill the terms of an international agreement. As discussed in chapters 3 and 4, a growing number of international agreements call for an ICM approach, including the Convention on Biological Diversity, the Framework Convention on Climate Change, UNCED's Agenda 21, and, in certain respects, the 1982 Convention on the Law of the Sea.

In most countries, the national government cannot implement an ICM program entirely on its own. Certain land and water use authority is exercised at the provincial and local levels of government; thus, the active participation of three levels of government is needed for an ICM program to succeed. This being the case, the national government's first action may be to arrange a series of discussions with other involved levels of government about the ICM process and the benefits associated with it. Certain incentives might be offered to encourage participation. Financial support is one such incentive; another might be a promise of giving the lower levels of government additional legal leverage over the actions of the national government that have the potential for adversely affecting coastal or ocean resources under provincial or local jurisdiction. This requirement for consistency in the actions of all levels of government participating in an ICM program has been of key importance in some countries, such as the United States, in ensuring harmonization of policy.

Starting Small If Necessary

ICM does not have to be implemented all at once for a country's entire coastal zone. Indeed, in many circumstances, a phased implementation, beginning with the coastal and ocean areas most in need of better integrated management, is a better way to begin. Table 5.6 outlines a number of factors to consider in deciding whether to proceed with a phased approach and in structuring such an approach.

A danger of using a phased or incremental approach, of course, is that the momentum behind ICM can be lost if the effort extends over too long a period. It is best, therefore, to reach agreement as soon as possible, preferably during the first phase, on a schedule for implementing all remaining increments of the program. Although it is difficult to be concrete about this, no more than five additional years should be needed to complete implementation of an ICM program once an initial increment has been put into place.

Summary

This chapter addressed the activities important early in the life of an ICM program. In our view, it is imperative that the stage be carefully set within government for the initial ICM discussion; equally important is the way the ICM concept is described and presented. Any misconceptions regarding ICM must be

Table 5.6. A Phased Approach to ICM: When Is It Appropriate? How Should It Be Structured?

Circumstances Warranting a Phased Approach

Urgent problems are more prevalent in one part of the coastal zone than in other parts.

One coastal area is institutionally, technologically, or politically ready for ICM and others are not.

Funding and staffing limitations dictate an incremental approach.

It seems desirable to conduct a pilot or demonstration ICM effort in a specific coastal region before applying it generally.

Considerations in Structuring a Phased Approach

A concerted effort must be made to learn from the initial ICM program in order to benefit later increments.

Even though the initial coastal area to come under an ICM process may be small, care must be taken to ensure that intergovernmental and intersectoral coordination mechanisms are adequately comprehensive.

The institutional, technical, and political changes made in the initial ICM effort must be capable of being scaled up to handle the entire coastal zone.

If the initial effort is conducted as a pilot or demonstration effort, the program must include tests to judge its success.

dispelled early or they may lead to polarization and mistrust. Recognition of the need for an ICM program can develop in a number of venues—the national government, the local community, the NGO community, a donor agency or development bank—but wherever it first emerges, it must find its way to and be accepted by both the national and local levels of government, since both generally have important roles to play in its implementation.

Suggestions were presented to help ensure that ICM is seen in a positive, nonthreatening way by the sectoral agencies that must cooperate in its implementation. Underscored was the importance of making available early in the process a short but compelling document that outlines the basics: the problems giving rise to the need for ICM, how ICM is going to address the problems, how a better integrated approach will produce public benefits, and how development and implementation of an ICM program might proceed.

The importance of participation and support from local and national levels of government and the coastal users was stressed, and approaches for securing and maintaining broad-based participation beginning at the earliest stages of the ICM program were reviewed. ICM does not have to be applied to a country's entire coastal zone simultaneously. It can be implemented first where it is needed first—in those coastal areas having the most urgent problems and needs. As long as care is taken to construct the program in such a way that its extension to the remainder of the coastal zone is relatively straightforward and attention is paid to ensuring that the integrating mechanisms are fully operational beginning with the first increment of the program, such an approach can be appropriate in a number of circumstances. However, a loss of momentum is possible if the process extends over too long a period of time.

....

Chapter 6

Intergovernmental, Institutional, Legal, and Financial Considerations

Introduction

This chapter discusses four key dimensions in the design of an ICM program: intergovernmental, institutional, legal, and financial considerations. All four dimensions are of fundamental importance to successful integrated resource management. It is important to ensure that the various levels of government (national, provincial or state, and local) have roles in the ICM program consistent with and supportive of their roles in land and water use management. Similarly, existing institutions, be they sectoral (e.g., fisheries, water quality management) or functional (planning, permitting) in nature, must also be factored into the ICM process at appropriate points if ICM is to become an integral part of government functions.

On the legal side, it is essential that an ICM program have the legal authority necessary to undergird the regulatory measures it contains. Because in the first instance this often means relying on existing legislation, a thorough understanding of the legal authorities on which present regulatory and management activities are based is of critical importance. Financial considerations—how the program is to be funded on a continuing basis—are of paramount concern as well. Factors relevant to each of these dimensions in the design of an ICM system are discussed.

Intergovernmental Considerations

Responsibility for the management of coastal and ocean resources rarely falls exclusively on one level of government.[1] Typically, the landward parts of the coastal area are under the jurisdiction of a local or provincial government or both, whereas activities in much, if not all, of the adjacent water area are controlled by

---

1. This discussion relies, in part, on Knecht 1993.

the national (central) government. Given that the natural systems involved are continuous between the coastal lands and the coastal waters, it is essential that policy and decision making in various parts of the coastal zone be internally harmonious and consistent. An important early step in the design of an ICM program, therefore, is to develop a comprehensive understanding of the roles of various levels of government in the protection, management, and exploitation of coastal and ocean resources.

The discussion that follows provides a perspective on the roles of two levels of government almost always involved in coastal zone management—the provincial or local level and the national level. Some nations, such as the United States, Canada, and Australia, have a three-level system with a relatively strong state or provincial level between the national government and the various local governments. As another example, Indonesia has a four-tiered system of government with 27 provinces, 292 municipalities or districts, 3,500 subdistricts, and 66,400 rural or urban villages (Dahuri 1996a). For simplicity, this discussion is cast in terms of a two-level system of government, although much of what is said applies to multi-tiered systems as well.

In this section, we examine the interests and probable motivations of the two levels of government and suggest ways of increasing understanding and cooperation between them, with the goal of developing an effective working partnership in the conduct of an integrated coastal management program. The discussion addresses: (1) the nature of the problem, (2) perspectives of the local level of government (*local* is used generically to denote subnational levels of government), (3) perspectives of the national government, (4) ways to build understanding and cooperation, and (5) ways to create a mutually beneficial partnership.

Nature of the Problem: Harmonizing Local and National Levels of Government

If an ICM program could be operated by either the local or the national level of government alone, a problem would not exist. But in general, it cannot. An effective ICM program, especially one that is integrated in scope and process, demands that both the national and local governments be involved. Typically, both levels have responsibilities for regulating land and water use. Both generally have resource development and management roles as well, and both usually possess critical technical expertise and data. Hence, an effective ICM program clearly requires the positive participation and support of both levels of government.

But the interests of the two levels of government do not necessarily coincide; indeed, sometimes they are in conflict. Differences arise because the national government and the local government have different responsibilities and different legal authorities. Their goals and objectives, though usually broadly the same, often differ in detail and approach. Moreover, the two levels of government serve different constituencies, which are sometimes perceived as having different needs and desires. The national government is often preoccupied with international and national concerns, whereas the local government focuses on local issues—jobs, public health, infrastructural needs, environmental problems, and the like.

A move to institute a coastal management program can arise from either level of government or from a coastal users' group. Sometimes the local community will first see the need for improved planning and management of the coastal area, as, for example, when an important fisheries stock goes into serious decline because of destruction of its habitat or the coastal tourism industry begins to suffer the effects of eroding beaches or contaminated coastal waters. In other cases, it is the national government, perhaps the Ministry of Natural Resources or an environmental unit, that first sees the need for ICM. As discussed in chapter 2, in response to our survey question asking respondents in various countries which level of government was primarily responsible for conducting ICM in their country, 61 percent specified the national level and 16 percent indicated the state level. An additional 20 percent of respondents designated no single level of government (i.e., a mix of levels was involved), and 2 percent were uncertain which, if any, level of government was responsible.

Once the need for improved coastal zone management is recognized, however, all affected interests—government and private—should be brought into the process. Typically, the six stages of an ICM process presented in chapter 2 can be grouped into two phases: (1) the development phase and (2) the implementation and operation phase.

Development Phase

During the development phase, after relevant data on conditions (physical, economic, social) in the coastal zone are collected, the problems and opportunities existing in the area are analyzed (see chapter 7) and options for new policies and improved management are developed. A strategic assessment may be undertaken, and a long-term strategy for the sustainable use of the coastal zone may be formulated. If appropriate, new regulatory and resource management programs are designed. Existing institutions with management, regulatory, or development roles in the coastal zone are analyzed with regard to the adequacy of their legal authorities, technical expertise, funding levels, and so forth. New institutional arrangements may be designed, as needed, to harmonize the various sectoral activities that affect the coastal zone and its resources.

ICM Implementation and Operation Phase

After the necessary legislative and executive approvals and funding are secured, the proposed ICM program moves into the implementation and operation phase. New or strengthened regulatory and resource management programs are put into place. New institutional entities, such as a coastal coordinating council at the ministerial level, go into operation. Monitoring activities begin so that the program's performance can be periodically assessed and midcourse adjustments can be made.

Clearly, both the development phase and the implementation and operational phase of an ICM program require the participation—indeed, the active cooperation and leadership—of both local and national levels of government. Unfortunately, however, this active cooperation is not always easily obtained. In some cases, when the national government has taken the initiative, local government officials may see the ICM program as being forced on them "from above." The normal bureaucratic reaction is to resist this kind of pressure, and all governments have a wide array of techniques for successfully generating such resistance. Conversely, if the local government (or a local user group) has initiated the move for improved coastal management, officials of the national government may be reluctant to get involved, fearing the costs that might be incurred, concerned about the ultimate effect on existing national programs, or perhaps believing that such efforts should be undertaken on a uniform, nationwide basis rather than selectively in a particular coastal locality.

One way to overcome these potential difficulties is to develop cooperative activities involving both levels of government early in an ICM effort. A later section of this chapter offers some suggestions as to how this might be accomplished. First, however, we discuss the different perspectives the two levels of government are likely to bring to an ICM process.

Perspectives of the Local Government

Typically, the local government believes that it is responsible for the most important functions government provides—water supply, sewer systems, local roads, parks and recreation, schools and libraries, public health and welfare, and the like.[2] It believes that as the level of government closest to the people, it knows their wants and needs best and can meet them best. Furthermore, the local level is generally the most involved with local economic activity and most concerned about jobs and the long-term economic health of the region.

Increasingly in recent years, local governments have become concerned about the quality of life in their jurisdictions. *Quality of life* in this context denotes a host of factors that affect the way people in the community live, from the availability of functioning water and sewer systems and the cleanliness of air and water resources to the level of education and quality of health care available. Striking the right balance between environmental quality, economic well-being, and other factors generally requires trade-offs—between higher taxes and better conditions on the one hand and lower taxes and poorer conditions on the other. Local governments strongly believe that they are in the best position to make such decisions. Anything that weakens the power of local authorities over these key aspects

2. In this discussion, the term "local" government refers generically to subnational levels of government (provincial or state, and local or community) and their staffs.

of life in their communities will be resisted, and understandably so. In particular, local governments believe that central governments do not have the information, the sensitivity to local needs and conditions, or the "legitimacy" to make these kinds of decisions. Yet the financial realities are such that local governments usually must look to the national government for substantial funding in order to meet even the most basic needs of their citizens.

Local governments tend to be wary of new programs initiated at the national level. The goals and objectives of such programs may not coincide with their understanding of what is needed locally. They worry about the costs and the distractions involved and the extent to which they may lose control over locally important resource use or allocation decisions. Ultimately, local governments may decide to go along with initiatives of the national government (there may be little real choice) with the expectation of using the new program, and especially any new staff or funding resources, to help them meet certain of their most pressing needs. With this in mind, local governments will generally seek to have as much flexibility and discretion as possible built into such programs.

In political systems having three tiers of government, the national government will sometimes seek to use the intermediate (provincial or state) level of government as an intermediary or overseer of a new program even though much of the program will require implementation at the local level. Local governments will, of course, resist such an approach, preferring that the new program and any associated funding come directly to them.

One of the primary management controls local governments possess involves the use of the land within their jurisdictions. Land use and zoning controls are the key levers available to most local officials in guiding the growth and economic prosperity of their community as well as ensuring the environmental health and well-being of the citizens. Local officials are also concerned about the revenues flowing into local government treasuries and, hence, other things being equal, tend to encourage new taxpaying development projects, especially if they also create new employment opportunities. Thus, to the extent that local governments see new national government programs as weakening their ability to guide the development of their communities as they see fit, they will naturally oppose them. However, programs that are seen as addressing important problems and that also provide additional leverage or funding to solve those problems are obviously more likely to attract a favorable response.

In summary, the strongest local government resistance is likely to be met by national government proposals perceived as:

• Shifting power or authority away from the local level of government.

• Reducing or constraining the discretion available to local government.

• Adding costs or other burdens to the local level without commensurate benefits.

Conversely, the strongest local government support can be expected for new programs perceived as:

- Providing new tools and resources to address important problems facing local communities.

- Flexible and adaptable to the varying situations in different localities.

- Making maximum use of the talent, expertise, and experience of the local community.

Finally, local government officials tend to have a proprietary attitude about their community. After all, it is their place. They understand the locality and the people and their special needs and aspirations as no one else can, and they generally care deeply about the community's present and future well-being. These strongly held feelings must be taken into account if successful intergovernmental partnerships are to be achieved.

Perspectives of the National Government

National governments believe themselves to be at the pinnacle of power and prestige in the broad spectrum of domestic and international governance arrangements. They believe that it is they (individual national governments acting collectively) that create international or global institutions—it is they that create and empower their domestic provincial and local governments, levy taxes, print money, and raise and deploy armies. Although some observers see signs that the concept of the all-powerful, sovereign nation-state is beginning to lose ground as international institutions become stronger and as some regional groupings of nations (such as the European Union) move closer to political unity, nonetheless, for the foreseeable future, the nation remains the primary government actor.

Hence, it is not surprising that an air of superiority is sometimes perceived within national governments, especially by those at the "lower" levels of government. Clearly, the presence of such an attitude makes more difficult the creation of intergovernmental partnerships of the kind required by coastal management programs.

Attitude aside, what are the goals and objectives of national governments and how do they relate to coastal zones and the resources they contain? In an earlier paper, Robert W. Knecht (1986) listed six interests or roles of the national government in the coastal and ocean arena:

- Ensure the nation's security
- Conduct foreign relations

- Minimize transboundary impacts of permitted development
- Ensure a fair return from the use of public resources
- Maintain order and resolve conflicts
- Protect interstate commerce

Clearly, most national governments consider national security and national economic prosperity as their primary goals. Thus, to the extent that either of these is involved in a particular area of the coastal zone, the national government can be expected to exert a strong influence over relevant coastal decision making. In the United States, for example, the national government has been a strong proponent of offshore oil and gas drilling (a big revenue raiser for the national treasury) and has occasionally taken a strong position on the continuing need for military facilities of one type or another in particular coastal areas.

National governments are also interested in seeing their international treaty obligations carried out, some of which may have been entered into without the full involvement of concerned local interests. For example, more than 100 nations have ratified the Convention on International Trade in Endangered Species of Wild Fauna and Flora (CITES) and hence are obligated to prevent and/or regulate such trade. National governments, as direct participants in the CITES program, will generally be more interested than local governments in monitoring and enforcing these activities.

With regard to coastal and ocean management, national governments will be inclined to look for approaches that can be applied more or less uniformly throughout all the nation's coastal zones. The national government may see ICM as a way of addressing nationally important needs such as the siting of new energy facilities or a new port or the adoption of a new system of protected areas in the coastal zone. Alternatively, from the national point of view, ICM may seem attractive as a way of facilitating investment in coastal development such as resorts or other tourism-related activities and thus as a way of increasing foreign currency earnings.

Neither the executive nor the legislative branch of a national government is very well equipped to deal with the differences that usually exist among various parts of a country. Hence, as a general rule, programs are adopted that apply uniformly throughout the country. A national government thus may be inclined to adopt a set of standards for ICM that apply in all coastal zones and to set similar timetables for completion of ICM programs without giving due consideration to the different amounts of work programs in different parts of the country will often require.

The nature of the national government's interest in improved management of coastal areas will often depend on the issues that have given rise to the perceived need for improved management. If pollution of coastal waters is the triggering concern, it can be expected that the national environmental agency (or water pollution agency) will head the national effort and the emphasis, initially at least, will be on pollution issues. Similarly, if coastal erosion is seen as the major problem, the national agency with those responsibilities will be the most involved.

Given the specialized technical expertise that often exists in sectoral ministries or departments at the national government level, a national ICM effort can take on a technocratic orientation that could eventually come in conflict with the more people-oriented process sometimes desired by local interests. A technocratic approach might emphasize such activities as resource inventories, mapping, and development of geographic information systems, all of which are legitimate activities but do not constitute a complete or balanced effort.

In summary, in contrast to the local government, the national government will almost certainly have ongoing resource management and/or environmental concern in place that apply to the coastal zone. As seen from the national level, therefore, the problem is usually not one of creating institutional machinery to address new problems but rather one of adjusting, strengthening, and better harmonizing existing programs to focus them on new problems and opportunities in the coastal zone.

Building Understanding and Cooperation

Given that national and local governments may have rather different views of coastal management and how it should be undertaken, it is not surprising that strained relations between them can sometimes slow the implementation of ICM programs. Hence, in this section we offer several suggestions for creating a positive working relationship early in the development of an ICM program.

Identification and Pursuit of Common Interests. Rather than confronting divisive issues, such as those involving jurisdiction and control, early on, efforts can be made to identify problems, issues, and opportunities that are common to both levels of government. For example, protection of lives and property from coastal hazards such as hurricanes and severe storms may well be a shared concern that both would wish to pursue. Similarly, the protection or restoration of particularly important coastal habitat areas might be of interest to both levels of government. Once one or more common interests have been identified, collaborative or joint programs can be initiated. A working group involving members from both levels of government might be established to address the coastal hazards problem, for example. Such a group could assemble the data and information needed, study the experiences of other countries in handling such problems, develop a set of options for new policies and programs to address the issue, and so forth. In this way, representatives of the two government levels would begin to establish a history of working together and gain an improved understanding of each other's perspectives, concerns, and values. Over time, a level of trust could be established that could then be extended on a broader basis.

Identification and Use of Unique Expertise, Experience, Talent, and Data from Both Levels. Issues can be identified that require specialized expertise of one type or another. These might involve problems as varied as coastal erosion, degraded habitats, or the need to design a more effective public participation process. Combining the expertise found in the national government and that existing locally in a single group of experts to address the problem and recommend solutions would also help establish a pattern of successful work between the two levels of government. Periodic briefings by such working groups of experts given to joint meetings of policy-level officials from the two governments could help convey the positive, cooperative climate to the policy level as well.

Deferral of Difficult Issues. Issues likely to be divisive, such as those involving jurisdiction and the distribution of management responsibilities and revenues, need not be confronted during the early stages of design of the ICM program. Indeed, it can be argued that such questions are better faced in the later stages, when a better understanding of management requirements is in hand. Agreeing at the outset to defer such issues allows work to proceed on more straightforward and less controversial tasks. Once a history of working together harmoniously is established and a certain amount of mutual understanding exists, it should be easier to agree on a way to approach harder questions.

Use of Respected Outside Expertise on Difficult Issues. One way to deal with controversial issues is to make selective use of outside expertise, for example, from the academic community, from a respected international organization, or from the private sector. On basic issues such as the relative power of the national and local levels of government in decisions involving the use of coastal land and water, it is natural that different positions will be held. Outside expertise could be employed to analyze such issues and to present a series of options that merit serious consideration. In this approach, it is important that (1) both levels of government are involved and agree on the details of the request to the outside group and (2) the outside group is seen as fully objective, not partial to either level of government and "above" the politics of the issue.

Creating a Mutually Beneficial Partnership

There are, of course, countless ways to divide ICM responsibilities and duties between the central government and the local government, but certain arrangements seem to be favored, probably because they parallel some existing roles of the two levels of level of government. In this connection, it is useful to review what each government can best contribute to an ICM enterprise, as illustrated in table 6.1. It would seem that a successful working partnership would result if each level of government believed that the arrangements were equitable—that its inter-

Table 6.1. Relative Contributions of Local and National Levels of Government to ICM

What Can the Local Government Contribute to the ICM Enterprise?

• The most detailed understanding of the on-the-ground problems and needs in the local coastal zone.

• The best understanding of the constraints and limitations that will affect the choice of solutions.

• The best data and information concerning the local coastal zone.

• Support of coastal users' groups and the local community.

What Can the National Government Contribute to the ICM Enterprise?

• Specialized data and expertise concerning various sectors of coastal activity (fisheries, erosion, wetlands, water quality, etc.).

• Capacity to harmonize various sectoral activities that affect coastal and ocean areas through the establishment of a coordinating mechanism.

• Funding assistance (in some cases).

• Ties to relevant global and regional coastal and ocean programs.

ests were being fully protected and advanced and that its particular capabilities and expertise were being used in the ICM process.

One obvious division of responsibility suggests itself: the national government, with the advice of local governments and coastal users' groups, would formulate the broad coastal policies and goals for the nation by means of legislation, and local governments would develop plans for their coastal zones that are consistent with and incorporate the national coastal policies. Local governments would then operate regulatory systems (permits, etc.) consistent with their local coastal plans. In this approach, coastal or ocean development projects exceeding some threshold of size or impact could require approval by both the local and the national levels of government. The national government could also hear appeals of coastal permit decisions made by the local government. In addition, and very important, the national government could establish a mechanism to ensure that the programs of its various ministries and departments are also harmonized with the agreed national coastal policies and goals. This concept of a simultaneous top-down and bottom-up approach—national policy setting and sectoral harmonization with local implementation—has much to recommend it, as summarized in table 6.2.

One way to achieve the necessary balance in this kind of intergovernmental arrangement is to have the national government certify that the local coastal program is consistent with national coastal policies and goals and then to require that the actions of the national government, as well as those of the local government and the private sector, be consistent with the approved coastal plan. In the United

**Table 6.2. Benefits of a Simultaneous Top-Down
and Bottom-Up Approach to ICM**

- Uniform national coastal policies can apply throughout the country.

- National policies can be adjusted to fit various local conditions through the local government's implementation process.

- The national government is in the best position to operate a mechanism to harmonize various sectoral programs that affect the coastal zone and its resources.

- Local governments are usually in the best position to integrate a coastal regulatory system with existing regulatory systems.

States this consistency requirement has been found to be an important part of the ICM process, providing a prime incentive for local and state governments to participate in the programs (Eichenberg and Archer 1987).

An arrangement such as this would seem to be a win-win situation for all levels of government involved. The national government has in place a set of policies and goals for the country's coastal zones that serve the interests of the nation as a whole. Local governments are able to reflect needs and realities in their coastal plans and can expect national programs to be consistent with the plans. Moreover, local governments may also receive technical and financial assistance in their ICM activities from the national government which usually has access to more expertise and funding. Hence, approached carefully and with sensitivity to the various interests involved, integrated coastal management can be a unifying factor in intergovernmental relations.

Institutional Considerations

Institutions are, of course, the main actors in the coastal management process. To the extent that the various levels of government perform services and operate programs, each level will be made up of a number of different organizations or institutions, each specialized in a particular function or field. It is through these organizational units, consisting of staff, space, facilities, budgets, and procedures, that governments operate. One of the challenges of ICM is to fashion ways to ensure that the actions of the coastal and ocean institutions at each level of government are harmonized with one another and are consistent with agreed coastal goals and policies.

Complicating the situation is the fact that different institutions are generally guided by different mandates and laws, often containing different goals, objectives, and policies. Agencies may also differ in outlook, type of personnel and their training, and external constituencies. Hence, institutions, even if predisposed to collaborate with sister organizations, may interpret their underlying legislation as restraining or even preventing such cooperation. We can identify at least three factors tending to complicate institutional cooperation in coastal resource management:

1. Organizations in different sectors will probably have different orientations and goals. A fisheries agency's policy regarding the destruction of a wetland will differ from that of a port or harbor authority.

2. Different levels of government will probably have different goals and objectives. The national government and its organizational units, for example, may give higher priority to offshore oil and gas development than does the adjacent state or community.

3. Policy-level leaders may set different policy goals for the units that report to them. Hence, a development-oriented leader will probably set different goals from those of a conservation-oriented leader.

Hence, the challenge is to devise institutional or procedural mechanisms that can coordinate the myriad organizations and agencies often involved in coastal and ocean issues. Indeed, integration is at the heart of the concept of integrated coastal management. Therefore, it is worth investing some time on the topic of integration and how it should be understood in the context of ICM.

The Meaning of Integrated Policy

As succinctly put by E. L. Miles (1992), *policy* refers to a purposive course of action followed by government or nongovernmental actors in response to some set of perceived problems; *implementation* is the process of transferring policy decisions into action; and *management* refers to the actual control exerted over people, activities, and resources. In order to achieve integrated management, then, integration in policy—as the guide to management—must first occur. In the remainder of this section, the concept of integrated policy is explained, methods of achieving policy integration are set forth, and various dimensions of integration in a coastal and marine context are presented.[3]

An excellent discussion of integrated policy was presented by Aril Underdahl in a 1980 article (Underdahl 1980). According to Underdahl, to integrate means to unify—to put parts together into a whole. Thus, integrated policy is policy in which the constituent elements are brought together and made subject to a single unifying conception. To qualify as integrated, a policy must achieve (1) comprehensiveness, (2) aggregation, and (3) consistency at three successive stages in the

3. This discussion relies on Cicin-Sain 1993.

Table 6.3. Aspects of Policy Integration

Stages in the Policy Process		
Inputs	Processing of Inputs	Consistency of Outputs
Comprehensiveness	Aggregation	Consistency
Time Long-range perspective	Extent to which policy alternatives are evaluated from an overall perspective rather than from the perspective of each actor, sector, etc.; i.e., extent to which decisions are based on some aggregate evaluation of policy	Extent to which different policy components accord with one another
Space Extent of geographical area for which consequences of policy are recognized as relevant		**Vertical Dimension (Consistency among Policy Levels)** Extent to which specific implementary measures conform to more general guidelines and policy goals
Actors Extent to which relevant interests are incorporated		**Horizontal Dimension** For any given issue and policy level, extent to which only one policy is pursued at a time by all executive agencies involved
Issues Extent to which interconnected issues are incorporated		

Source: Adapted from Underdahl 1980.

policy process. This notion may best be understood by reference to table 6.3. Thus, according to Underdahl, integrated policy (1) recognizes its consequences as decision premises, (2) aggregates the consequences into an overall evaluation, and (3) penetrates all policy levels and government agencies involved in its execution. Underdahl stresses that this is only an ideal model and that perfect integration will rarely, if ever, be a sensible goal. When costs are taken into consideration, some imperfection usually appears inevitable. We will return to this point later.

How can policy integration be achieved? Underdahl outlines two generic ways of achieving integration: (1) direct methods, that is, specific policy directives for all government agencies to follow, and (2) indirect methods, such as "intellectual" strategies, which seek policy integration through "initiating research, training and socialization aimed at developing a more comprehensive and holistic perspective on the part of decision makers" (Underdahl 1980, p. 167), or "institutional" strategies, which involve some type of organizational change. Institutional strategies are such measures as the naming of a lead agency, the creation of an interministerial coordinating committee, and creation of a new agency (see, e.g., Levy 1988).

As discussed in chapter 2, in the coastal/marine context, several dimensions of integration need to be addressed as a part of an ICM process:

1. *Intersectoral integration*, or integration among different sectors (e.g., oil and gas development, fisheries, coastal tourism, marine mammal protection, port development) as well as integration among coastal and marine sectors and land-based sectors that affect the coastal and ocean environment, such as agriculture, forestry, and mining.

2. *Intergovernmental integration*, or integration among different levels of government (national, state or provincial, and local).

3. *Spatial integration*, or integration between the land and ocean sides of the coastal zone, given the strong connection between land-based activities and their effects—on water quality, fisheries productivity, and the like.

4. *Science-management integration*, or integration among the different disciplines that are important in coastal and ocean management (the natural sciences, the social sciences, and engineering) and the management entities.

5. *International integration,* when countries border enclosed or semienclosed seas or there are international disputes over fishing activities, migratory species, transboundary pollution, establishment of maritime boundaries, passage of ships, or other issues.

Findings on Various Dimensions of Integration in the Cross-National Survey

The severity of problems in intersectoral, intergovernmental, spatial, science-management, and international integration will, of course, vary substantially from nation to nation, depending on the patterns of cooperation traditionally found in each of these areas. Interestingly, the first problem listed—that of intersectoral integration—was the one most often emphasized by respondents to our 1996 cross-national survey, in both close-ended and open-ended responses. For example, in an open-ended question, asking respondents to reflect on "particular 'difficulties' (or obstacles) the ICM effort has faced which might be of interest to ICM managers in other countries," a majority of respondents referred to problems of intersectoral coordination, as reflected in the following responses:

The very few ICM initiatives have had to face the strong inertia of the sectoral administration as well as the lack of tradition and "culture" of integrated management.

—European country respondent

The sectoral system of governmental administration does not facilitate integrated management. Much greater effort is required to coordinate CZM in this situation. Awareness and training in CZM should target key sectors such as fisheries managers who do not consider themselves coastal managers.

—Asia-Pacific country respondent

We also asked respondents to comment on the extent of integration with respect to various dimensions (spatial, intersectoral, intergovernmental, spatial, and science and policy) in their countries' ICM efforts. Table 6.4 compares their responses, listing first responses from all countries and then responses from developed, middle developing, and developing countries.

Some of the findings in table 6.4 are particularly noteworthy. Respondents as a whole described the various dimensions of integration as moderately successful, with the highest marks assigned to intersectoral integration at the state level and the lowest to integration of science and policy. Interestingly, the majority of respondents from developed countries rated only two of the six dimensions (intersectoral integration at the state and local levels) as moderately successful, calling the other four moderately unsuccessful. Most respondents from developing and middle developing countries, on the other hand, placed all six dimensions in the moderately successful category. This finding suggests that in general, integration may become more difficult as a nation develops, as coastal uses (and conflicts) increase, and as government agencies become more specialized and hence more fragmented.

Before proceeding to a more detailed discussion of survey findings on the dimensions of integration in table 6.4, we first explore some caveats on policy integration, or the need to keep issues of policy integration in perspective.

Caveats on Policy Integration

Operationalizing the concept of integration is sometimes difficult. It is important that improved policy integration be informed by the precise needs and circumstances of a particular context. Good information must be available on what problems exist among sectors, what implications they pose, and how much they will cost to resolve (Cicin-Sain, ed. 1992). Hence, although we agree with the Agenda 21 emphasis on integration (e.g., identifying existing and projected uses and their interactions, promoting compatibility and balance of uses, creating coordinating mechanisms for integrated management, etc.), we also think that policy

Table 6.4. Extent and Dimensions of Integration in ICM
Efforts in Selected Countries

Dimension of Integration	Highly Successful (%)	Moderately Successful (%)	Moderately Unsuccessful (%)	Highly Unsuccessful (%)	Undecided (%)
All Countries					
Spatial (land–sea) ($N = 47$)	2	49	34	9	6
Intersectoral					
National level ($N = 47$)	9	48	25	9	9
State level ($N = 47$)	4	64	19	2	11
Local level ($N = 46$)	7	50	13	4	26
Intergovernmental ($N = 48$)	5	51	24	10	10
Science–policy ($N = 48$)	6	43	28	10	13
Developed Countries					
Spatial (land–sea) ($N = 13$)	7	22	64	0	7
Intersectoral					
National level ($N = 13$)	15	23	46	8	8
State level ($N = 13$)	0	54	46	0	0
Local level ($N = 13$)	0	40	30	0	30
Intergovernmental ($N = 13$)	0	31	39	15	15
Science–policy ($N = 13$)	7	32	47	7	7
Middle Developing Countries					
Spatial (land–sea) ($N = 14$)	0	65	14	14	7
Intersectoral					
National level ($N = 14$)	0	65	14	7	14
State level ($N = 14$)	0	72	7	0	21
Local level ($N = 13$)	0	61	0	8	31
Intergovernmental ($N = 15$)	0	55	35	0	10
Science–policy ($N = 15$)	0	49	22	15	14
Developing Countries					
Spatial (land–sea) ($N=20$)	0	55	30	10	5
Intersectoral					
National level ($N = 20$)	10	55	20	10	5
State level ($N = 20$)	10	65	10	5	10
Local level ($N = 20$)	15	50	10	5	20
Intergovernmental ($N = 20$)	10	60	5	15	10
Science–policy ($N = 20$)	10	45	20	10	15

Source: Authors' 1996 cross-national survey.

integration should not be viewed as some kind of quixotic quest. It is an ideal model that suggests appropriate questions to ask and problems to attempt to resolve. Thus, coastal managers should keep in mind the caveats on policy integration presented in table 6.5. Elaborating on caveat (4), we believe that policy integration should be viewed not as an absolute but as a continuum, as depicted in figure 6.1.

Table 6.5. Caveats on Policy Integration

1. *Not every interaction between sectors is problematic and therefore in need of management.* Hence, adequate study of interactions among sectors and uses is imperative in order to understand the extent to which such interactions are harmful or beneficial to one or more sectors in effect or neutral.

2. *Integrated management generally does not replace sectoral management but supplements it.* Generally, different individuals and institutions will be involved in sectoral management and in policy integration (Cicin-Sain et al. 1990).

3. *Policy integration is often best performed at a higher bureaucratic level than sectoral management,* to ensure that an overall, rather than a fragmentary, perspective is pursued.

4. *The costs of policy integration should be kept in mind.* Since policy integration will often be difficult and costly to put into effect, sometimes the costs of integration will outweigh the benefits—such a prospect, of course, should be guarded against.

Incentives for Policy Integration

A central question in the design of ICM programs is how to obtain the necessary level of policy integration. In the preceding section, we suggested that policy integration (or at least some movement toward policy coordination and harmonization) is a requirement for a successful ICM process. But how can this policy integration be obtained? Must it be legislated or otherwise mandated, or are there other ways to encourage or facilitate interagency cooperation?

In fact, there are ways to increase the possibilities that such cooperation will evolve. In a review of the scholarly literature on this question, a study by J. A. Weiss (1987), for example, suggests a number of factors that can be work as incentives in achieving cooperation. After making the case that agency cooperation does not occur naturally or spontaneously, she suggests that at least six kinds

---

Figure 6.1. Continuum of Policy Integration

Less Integrated   More Integrated

◀—————————————————————————————————————▶

Fragmented Approach   Communication   Coordination   Harmonization   Integration

Going from left to right on the continuum—from a situation of less integration toward more integration, one can characterize the following situations:

1. *fragmented approach*—a situation characterized by the presence of independent units with little communication among them;
2. *communication*—a forum or mechanism exists for periodic communication/meetings among independent units;
3. *coordination*—independent units take some actions to synchronize their work;
4. *harmonization*—independent units take actions to synchronize their work guided by a set of explicit policy goals and directions, generally set at a higher level;
5. *integration*—formal mechanisms exist to synchronize the work of various units who lose at least part of their independence as they must respond to explicit policy goals and directions (often involves institutional reorganization).

Table 6.6. Incentives for Achieving Interagency Cooperation

Financial incentives. If funding is tied to interagency cooperation, that cooperation usually occurs, although it can disappear when the funding disappears.

Perception of a shared problem. If the problem being considered can be seen as a problem shared by a number of agencies and not solely the responsibility of one, cooperation clearly becomes easier.

Shared professional values. To the extent that issues can be expressed in professional or technical terms and not in terms of agency missions, cooperative action is facilitated.

Perception of political advantage. If policy-level leaders above the agencies in question make it clear that the issue is also important to them and to the higher levels of government generally, cooperation is likely to result.

Availability of fora for cooperation. Regular opportunities for discussion, accommodation, and, eventually, cooperation—preferably on neutral ground—can be very helpful.

Desire to reduce uncertainties. Virtually all agencies seek to reduce uncertainty in their environments; cooperating with other agencies is one way to do this.

Source: Adapted from Weiss 1987.

of incentives exist or can be developed to enhance the likelihood of interagency cooperation: financial incentives, perception of a shared problem, shared professional values, perception of political advantage, availability of fora for cooperation, and a desire to reduce uncertainties, as elaborated in table 6.6.

Options for Achieving Intersectoral Integration

Agenda 21 calls for states to consider establishing (or strengthening) appropriate coordinating mechanisms (such as a high-level policy-planning body) for integrated management and sustainable development of marine and coastal areas at both the local and national levels and to consider strengthening (or establishing) national oceanographic commissions to catalyze and coordinate the needed research.

Some options for establishing coordinating mechanisms include creating an interagency committee, naming a lead agency, creating a new agency, and training agency personnel to instill an integrated rather than a sectoral perspective. Some specific examples at the national level would be the following:

- *Creation of a special interministerial coastal coordinating council or commission.* This is often a good approach, provided the political will can be generated to create a new government entity of this kind. Important questions of composition, staffing, and funding of a new office such as this are discussed in chapter 8.

- *Assignment to an existing planning, budget, or coordination office.* This can be a satisfactory approach in certain government settings, provided the designated office is at a level above that of the line ministries or departments. The high-level location usually provides adequate legal authority and appropriate legitimacy, but such an office is likely to have other goals (e.g., monitoring or controlling the national budget) that could interfere with timely performance of ICM functions.

• *Designation of an existing line ministry to act as lead ministry.* The challenge in this type of arrangement is to vest sufficient authority in the lead ministry that it has effective control over the activities of sister ministries with regard to ICM decision making.

Again, three conditions, if fulfilled, enhance the effectiveness of the integrated coastal management process: the coastal management entity and process should be at a higher bureaucratic level than those of the sectoral agencies so it has the necessary power to harmonize sectoral actions; the effort should be adequately financed and separately staffed; and the planning aspect of integrated coastal management should be integrated into national development planning.

We now turn to a discussion of our survey findings on various dimensions of policy integration in the context of oceans and coasts.

Findings on Intersectoral Integration in the Cross-National Survey

On the question of intersectoral integration at the national level, respondents were first asked whether one or several institutions were involved in ICM. The results, shown in table 6.7, clearly indicate that responsibilities for ICM tend to be divided among several institutions. Overall, only one-fifth of the responses indicated that a single institution was primarily responsible. However, the results also suggest that the less developed a country is, the more likely it is that responsibility for ICM at the national level resides in one main institution.

Table 6.8 shows survey results pertaining to the nature of relationships among institutions involved in ICM efforts at the national level. The findings suggest that the nature of the relationships varies according to the particular issue involved and

Table 6.7. Institutional Responsibility for ICM at the National Level in Selected Countries

	All (%) (N = 48)	Developed (%) (N = 14)	Middle Developing (%) (N = 15)	Developing (%) (N = 19)
One main institution	19	14	20	21
Several institutions involved	81	86	80	79

Source: Authors' 1996 cross-national survey.

Table 6.8. Nature of Relationships among Institutions Involved in ICM Efforts at the National Level

Nature of Relationship	All (%) (N = 39)	Developed (%) (N = 12)	Middle Developing (%) (N = 11)	Developing (%) (N = 14)
Generally cooperative	30	17	45	29
Have little to do with one another	3	0	0	7
Generally competitive	8	17	0	7
Relationships vary depending on the issue	59	67	54	57
Other	0	0	0	0
Uncertain	0	0	0	0

Source: Authors' 1996 cross-national survey.

that national-level agencies in developed countries tend to be less cooperative and more competitive than those in developing countries.

We next examine the types of national-level coordination mechanisms set up for ICM management in the countries surveyed (see table 6.9). The results suggest that interagency or interministrial commissions are the most frequently used coordination mechanism at the national level with special coordinating bodies also used. Almost half of the respondents reported that lead agencies had also been designated. These findings appeared to be independent of level of development.

Table 6.9. Types of National-Level Coordinating Mechanisms for ICM Programs in Selected Countries

	All (%) (N = 44)	Developed (%) (N = 13)	Middle Developing (%) (N = 13)	Developing (%) (N = 18)
Interagency or interministerial commission	55	54	54	56
Special coordinating commission or committee	32	31	31	33
Lead agency	48	54	46	44
Prime minister's office	11	8	23	6
No national coordination mechanism	16	23	0	22
Other	7	8	0	11

Source: Authors' 1996 cross-national survey.
Note: Percentages do not add up to 100% because respondents were allowed to check more than one category.

Table 6.10. Extent to Which Creation of ICM Coordinating Mechanisms in Selected Countries Was Related to Agenda 21 Prescriptions

Extent of Relationship	All (%) (N = 43)	Developed (%) (N = 13)	Middle Developing (%) (N = 14)	Developing (%) (N = 16)
Highly related	23	8	22	37
Somewhat related	35	31	57	19
Little related	7	8	7	6
Not at all related	23	38	14	19
Uncertain	12	15	0	19

Source: Authors' 1996 cross-national survey.

Another survey question dealt with the extent to which the creation of an ICM coordination mechanism was stimulated or influenced by UNCED's Agenda 21 prescriptions. As shown in table 6.10, some 58 percent of all countries reported that the creation of an ICM coordination mechanism was highly or somewhat related to Agenda 21 recommendations. Middle developing countries were most likely (78 percent), and developed countries least likely (39 percent), to report having been influenced by UNCED.

The nature of the relationships between national-level ICM institutions and provincial- and local-level ICM institutions is also important to the success of ICM. Table 6.11 shows the survey findings in this regard. The results suggest that in slightly more than one-third of the cases, a generally cooperative relationship

Table 6.11. Nature of Relationships between National-Level ICM Institutions and Provincial and Local-Level ICM Institutions in Selected Countries

Nature of Relationship	All (%) (N = 46)	Developed (%) (N = 14)	Middle Developing (%) (N = 15)	Developing (%) (N = 17)
Generally cooperative	37	29	39	41
Have little to do with one another	11	0	7	24
Generally competitive	9	14	7	6
Relationships vary, depending on the issue	41	57	47	24
Other	0	0	0	0
Uncertain	2	0	0	6

Source: Authors' 1996 cross-national survey.

was thought to exist. In another 41 percent of cases, respondents indicated that relationships varied according to issues. Competitive or hands-off relationships were reported in 20 percent of cases. Among developing countries, however, 30 percent reported that national-level ICM institutions had little to do with or were generally competitive with state- and local-level ICM institutions.

Finally, we examined the institutional and policy relationships between coastal management and ocean management in the selected countries. The findings are shown in table 6.12. First, in more than three-fourths of cases, ocean management and coastal management were not administered by the same organizational unit. Furthermore, 71 percent of respondents reported that the programs operated with different policy mandates, program priorities, and resource levels. Although the majority of respondents believed that both programs represented an effort at employing an integrated approach, most also saw them as inadequately coordinated. We comment on the implications of these findings in chapter 10.

Legal Considerations

In an activity like ICM, it is sometimes difficult to differentiate between plans and programs since the terms are often used interchangeably. In our definition, a plan is a set of goals and objectives usually referring to the future use of a specific geographical area. A plan becomes a program when combined with legal devices or other techniques along with the funding needed to bring about (mandate) the desired changes and improvements embodied in the goals and objectives of the plan.

In the past, as a general rule, for any public resource management program— ICM included—to achieve its policy goals, the government had to act as a "regu-

Table 6.12. Relationship of Coastal Management and Ocean Management in Selected Countries

All Countries (N = 48)			
Relationship of Programs	Yes(%)	No(%)	Uncertain(%)
Administered by the same organizational unit	19	77	4
Administered separately			
Programs operate with similar policy mandates, program priorities, and resource levels?	21	71	8
Both programs represent an effort at employing an integrated approach	47	34	19
Programs are adequately coordinated	20	67	13

Source: Authors' 1996 cross-national survey.

lator" and for that regulatory role to be exercised, the responsible government agency had to possess the appropriate legal authority. Assignment of that legal authority was typically made through the action of a duly elected legislature, that is, a congress or parliament, or by an action (decree) of the chief executive if the nation's constitution assigns such power to that office. In recent years, another approach—largely employing economic incentives—has joined the regulatory approach as a potentially important tool in management of natural resources and the environment, including coasts and oceans. This section focuses primarily on the legal aspects associated with the regulatory approach—the distribution of legal jurisdiction in coastal and ocean areas, the types of legal authority usually needed in an ICM program, how such authority is obtained, and limits and constraints on the use of regulatory power. More information on the use of economic incentives and other nonregulatory devices to achieve the goals of an ICM program is provided at the end of this section.

Legal Jurisdiction in the Coastal Zone

In most governance systems, the national government has primary, if not exclusive, legal jurisdiction over the ocean areas adjacent to the coast. Having legal jurisdiction over a publicly owned area such as the ocean means having control over virtually all activity in the area consistent, in this case, with international law. As discussed in chapter 3, under the terms of the 1982 Convention on the Law of the Sea, nations are permitted to claim an exclusive economic zone (EEZ) 200 nautical miles in width and a territorial sea as much as 12 nautical miles in width as measured from their shorelines. By the beginning of 1996, the majority of eligible coastal states had declared such EEZs or fishing zones of the same width and had adjusted their territorial seas to the 12-mile limit.

In general, national governments exert direct jurisdiction over their 200-mile zones and often divide or otherwise apportion jurisdiction of their 12-mile territorial seas with the adjacent subnational government, such as a state or province. Within these ocean zones, including the intertidal area and the shoreland immediately landward of the water's edge, the government with jurisdiction over the area needs to have the legal authority for the following management activities:

• Management of beaches and dunes

• Protection of wetlands and other valuable habitats (mangrove forests, etc.)

• Protection of estuarine and coastal water quality

• Acquisition of public access for recreational purposes

- Protection of threatened and endangered species and biological diversity in general

- Management of shorefront space, with emphasis on water dependence

- Management of coastal lands subject to natural hazards (hurricanes, inundation, erosion)

- Management of living marine resources on a sustainable basis

An ICM program will, in general, contain management elements for each of these, consisting of two parts: the policy goals to be achieved and the regulatory or other devices to be used in achieving them.

Although jurisdiction over the intertidal area (that area regularly washed by tides) may also be in the hands of the provincial or state government, in some countries whose legal systems are grounded in English common law (the United Kingdom and the United States are two examples), a principle called the public trust doctrine guarantees citizens certain rights related to such traditional uses of the intertidal zone as navigation, fishing, and recreation) (Archer et al. 1995). The shoreland itself is usually under the jurisdiction and control of the local level of government, but in an increasing number of countries, it may also be subject to a narrow exclusionary zone immediately adjacent to the water's edge, created by the national government. Typically, these exclusionary zones, which may be seen as elements of a rudimentary coastal management scheme, prohibit the building of structures within some fixed distance of the water's edge or tide mark. Farther inland, jurisdiction is held by the local government and, in the absence of a coastal management program or other environmental or zoning controls, private landowners are essentially free to develop their land as they see fit.

Creating an Adequate Legal Framework for ICM

It is in this somewhat complex jurisdictional setting that integrated coastal management must operate. To be successful, an ICM program must be able to operate over the full span of geography relevant to the issue at hand, regardless of the legal jurisdictions involved. For issues related to water quality, the geography from the pollution source in the upland drainage (catchment) basin all the way down to the coastal waters affected by the pollution discharge should be within the ambit of the ICM process. Similarly, for issues related to beach management, the management area should embrace the full beach and dune system, the active portion of the submerged beach, and the littoral cell containing the circulation of the long-shore drift of sediments.

As a general rule, then, the management zone should be legally delimited so as to extend inland far enough to include all lands whose uses can affect the coastal waters and should extend seaward far enough to include all waters whose uses can affect the shoreland. In reality, of course, nations will set their management zone boundaries to embrace all the activities, land based and sea based, their coastal management programs must address if they are to meet policy goals. Thus, within the area, both land and water, included in the management zone, the ICM program must include adequate legal authority to regulate such activities. This does not mean, of course, that new legal authority will necessarily need to be enacted. In many cases, it is likely that the needed authority already exists in legislation vested in an existing coastal or ocean agency. What will ultimately be required is that the institution exercising this regulatory authority and the legal authority itself be incorporated into the ICM process.

As with all legal authorities proposed for inclusion in an ICM process, analyses will be needed to confirm that the legal or regulatory authority in question can, in fact, help achieve the coastal policy goals as well as those for which it was originally enacted. Thus, a key part of the legal analysis is ensuring that the legal authorities in use by sectoral agencies are adequate for broader coastal management purposes as well. To the extent that the existing legal framework is deficient, new authority may have to be sought to fill specific gaps. An excellent analysis of various legal approaches to ICM may be found in Boelaert-Suominen and Cullinan 1994.

Regulatory Power: Its Exercise and Its Limits

Most of the legal authority just discussed could be characterized as regulatory power—that is, a government agency is authorized by legislation or presidential decree to control certain activities by setting conditions, constraints, and limitations on the way in which property or resource owners use their land or the agency is authorized to set limits or conditions on access and use of common property resources such as fish stocks or public lands. In some Western legal systems, these powers are called police powers and are placed in the same category as laws to reduce nuisances and to protect the public health and welfare. The system of zoning of urban land used by many cities is also based on local government use of the police power. Most environmental and resource-based laws and regulations also rely on this legal principle.

Recently, however, the judicial branches of government in some countries, notably the United States, have begun to look more carefully at the use of the police power in the context of environmental and resource management in response to protests from property owners that through the application of environmental regulations, their property was being taken by the government without due compensation (something the United States Constitution expressly forbids). Thus, in certain circumstances, the "takings issue" is beginning to cast doubt on the appropriateness of the police power as legal authority for environmental controls. Although the precise direction in which the law will evolve is still unclear (even in the United States), one useful if somewhat oversimplified rule may be helpful: if a harm to the environment or to humanity is prevented by a government regulatory action, a taking has not occurred—on the other hand, if a public good is secured by a government regulatory action, a taking may have occurred and compensation may be required. Students of this issue assert that governments may draft and enact environmental regulations based on the police power provided close attention is paid to the nexus between the damage or harm to the environment or natural resources being protected and any exaction required of the applicant. Thus, they believe, as long as there is a logical relationship between the damage done and the requirement placed on the landowner (for example, donation of a public easement if the applicant's project reduces public access), the courts will generally continue to allow this use of the police power.

One type of regulatory power does not depend on use of the police power—the public trust doctrine. As mentioned earlier, the doctrine derives from the common law principle that gives citizens a permanent, virtually inextinguishable right to use intertidal areas of the oceans, as well as the banks of rivers and streams, for such traditional purposes as navigation, fishing, and recreation. Largely built up by means of court decisions, a substantial legal framework based on the public trust doctrine now exists in some nations. Additional information on the doctrine and its application to coastal zone management can be found in Archer et al. 1995). In countries with a common law background embodying this principle, the public trust doctrine could provide the legal basis for certain regulatory activities in the coastal zone, especially if it also served to protect traditional uses of the coastal zone. As yet, the doctrine has not been widely acknowledged as pertaining to the EEZ and its resources, although some legal scholars have suggested that a nation's proclamation of "sovereign rights" over the resources of the EEZ carries a concomitant duty for the nation to ensure that those resources will be as available to tomorrow's generations as they are to us today (Archer and Jarman 1992).

Coordinating the Use of Regulatory Power

As mentioned earlier, several levels of government may have jurisdiction over the same areas or activities in the coastal zone. The local, provincial or state, and national governments, for example, may all have laws referring to the dredging and filling of coastal wetlands. On occasion, all three levels of government can be involved in regulatory activity pertaining to a proposed development. Although some environmental interests applaud this situation as providing multiple opportunities to stop developments they see as adversely affecting the environment or sensitive resources, most observers see such overlaps as counterproductive and time-consuming.

Increasingly, ways are being found to join these regulatory activities into a single "joint review" process, which greatly simplifies matters for applicants and the government agencies and NGOs involved. One of the earliest uses of the joint review process occurred in Santa Barbara County, California, in the United States, in connection with regulation of offshore oil and gas development in the Santa Barbara Channel. Beginning in the mid-1980s, the county government, working cooperatively with state and federal regulatory agencies, created a process wherein all government permits and approvals for a particular development proposal were considered together as a part of a single, coordinated process. Proponents of the joint review process believe that it is more efficient and more thorough than separate reviews and results in better decisions (Hershman, Fluharty, and Powell 1988).

Economic Incentives for Modifying Behavior

Again the use of legally enforceable regulatory power is not the only way to secure the kinds of changes in behavior with respect to resource use and protection that will be necessary if we are to move toward sustainable use of the coasts and oceans. Increasingly, economic measures, sometimes called market-based approaches, are seen as desirable ways to obtain these changes. Such measures have the distinct advantage of not requiring regular and repeated use of the government's police power authority and hence do not suffer from the legal uncertainty discussed earlier. Also, market-based approaches are by no means as difficult to enact through the legislative process, since they often engender the support of both free-market proponents and environmentalists in legislatures. On the negative side, however, is the fact that these measures generally cost money—either in reduced tax revenues to the nation or through increased payments or subsidies as incentives to landowners. Furthermore, the economics have to be correct as well as convincing to those whose behavior is to be changed, and such programs must be built on a foundation of solidly based environmental standards. A wide range of economic measures exist to induce desirable action, including the following from Cicin-Sain et al. 1997:

- Charges applied directly to pollutant discharges (e.g., emissions or effluent charges)

- Charges applied to production inputs or outputs

- Property or rights in open markets (e.g., tradable development rights, emission or effluent permits)

- User fees (e.g., for water supply and sanitation, access to marine protected areas)

- License and permit fees (e.g., for exploitation of renewable and nonrenewable resources, mariculture, discharge of wastes, recreational use)

- Indirect use fees (e.g., tax on boat fuel, motor vehicles, or coastal use; environmental tax)

Unfortunately, however, thus far relatively little experience with most of these approaches has been reported in the context of ICM. For the most part, it appears that application of these concepts to ICM is still largely in the discussion stage. Exceptions include the use of pollution "umbrellas" in air quality regulation and the trading of pollution credits. Individual transferable quotas (ITQs) now being used in certain fisheries also represent the use of a market-based approach in allocation and management of fish stocks.

Financial Considerations

How to pay for an ICM effort is a critically important question. Obviously, adequate financial resources must be available at both the development phase and the implementation and operation phase if the program is to be successful. Funding options for each phase are discussed in this section.

Funding the Development Phase

As discussed earlier, the process of formulating the plan for an ICM program in a given country or coastal region can extend over a one- to two-year period and involve considerable effort. The cost of such an effort can vary a great deal, depending on a number of factors, such as

• The extent to which staff members are already in place or will be seconded (temporarily reassigned from other posts)

• The number of consultants who will be hired

• The amount of new data, especially field data, that must be collected

• The number and types of studies for which outside contracts will be issued

Clearly, costs of this phase can be kept to a minimum by relying principally on in-house staff or staff seconded from sister government units. An added advantage of this approach is that the resulting plan will be more apt to reflect reality if it is prepared largely by local people, who will be well versed in the real problems, opportunities, and constraints of such an effort. Furthermore, assignment of staff members from sister coastal and ocean agencies to the ICM development team will help ensure that a wider range of views and experience is brought to the planning process than might otherwise be the case.

Costs can also be constrained by minimizing the new data that is required. No ICM planning effort should require funding for collection of large amounts of outside data. The rudimentary data and information needed to initiate ICM planning work almost certainly already exists in virtually every coastal country. In geographical areas or sectors where more data are needed, they can often be obtained later, as part of the management oversight process. A comprehensive search may also uncover useful information at universities in the area or elsewhere in the country and in regional and global organizations (for example, the Food and Agriculture Organization of the United Nations for fisheries information and IUCN and UNEP for data on threatened and endangered species).

Even with the cost-limiting approaches just outlined, some new funding is usually needed to embark on an ICM effort. Typically, the national government will provide funds expressly to cover the development phase of an ICM program; these funds may represent a fixed sum of money available for only a specific period of time. In smaller developing coastal countries, however, the national government may be unable to provide such funding or may be able to provide only a portion of what is needed. In these circumstances, outside funding becomes a necessity.

Recently, especially since the Earth Summit, an increasing number of international organizations and donor agencies have been willing to fund ICM-related activities. In chapter 4, we discussed the capacity-building efforts in ICM that have emerged since June 1992. Much of this work focuses on the development of training programs in ICM-related topics and hence provides a relatively low-cost way for governments to get key staff members trained in the skills required to plan and operate an ICM program. Alternatively, some international organizations provide on-site funding and technical assistance to selected countries for the development and implementation of ICM programs in particular regions of their coastal zones. The International Maritime Organization's (IMO's) Regional Programme for the Prevention and Management of Marine Pollution in the East Asian Seas is a case in point. Demonstration ICM programs at three coastal locations in China are now under way as a part of that program.

Perhaps the best way for small developing countries to secure the funds necessary to initiate an ICM effort is through the World Bank, one of the regional development banks, or the Global Environment Facility (GEF), a funding mechanism operated jointly by the World Bank, UNEP, and UNDP. Funding for ICM in the form of loans or, sometimes, grants can be obtained in association with a particular coastal development project or as part of a government-wide effort to strengthen and enhance a nation's environmental and resource management system. Another possible source of ICM funding is, of course, the international aid agencies in countries such as Canada (CIDA), the United States (USAID), and Sweden (SIDA). Furthermore, specialized agencies such as the U.S. Country Studies Program provide funding assistance for ICM to developing countries to help them prepare to respond to the effects of climate change (Huang and Dixon 1995).

Given the present high level of support and encouragement for ICM by these funding sources, it seems likely that any nation willing to commit itself to a sound ICM effort should be able to obtain the needed funds from one source or another. This should be especially true since, as mentioned in chapter 3, most of the key international agreements regarding ocean and coastal areas now endorse ICM as the appropriate framework for implementing activities at the national level.

Funding the Implementation and Operation Phases

Funding the ongoing operations of an ICM program is clearly an important aspect of successful implementation. Ideally, the costs of running an ICM program should come from the recipients of the program's benefits—the users of the improved or better-protected amenities such as beaches and clean swimming waters; those who benefit from the more productive fishery nursery grounds and habitats and the shellfish and fish they produce; those whose property is better protected by healthy sand dunes, coral reefs, mangroves, and so on. Funds equivalent to only a small percentage of these benefits would probably pay the costs of operating an ICM program. But collecting these funds is inherently difficult. Market systems are generally organized in a way that precludes identification and monitoring of the benefits of ICM, except as they are accounted for by use fees at public beaches. Hence, other, less appropriate ways must be found to fund ICM over the long term.

General Fund

One of the usual ways of funding ICM on a continuing basis is by use of the general fund of the national or provincial or state government. This fund, raised and replenished on a regular basis by taxes (on personal and corporate income, sales, tobacco and liquor, etc.), is used to run all parts of the government. In some systems, the national government may provide some of the funding for ICM and require the state or province to match a portion of that share. In the United States, for example, the national government requires that states provide as much as one-third of the funds to operate the CZM program, with the federal government providing two-thirds.

There are difficulties, however, with using the general fund to cover the continuing costs of programs such as ICM. First, since most parts of government are supported by this fund, the competition is keen, and constantly changing priorities can affect the amount of money available for a given program. Second, since the general fund usually depends on taxes for its replenishment, it can vary according to the state of the economy and other factors. During economically depressed periods, funds become short and all programs suffer from smaller budgets. Finally, general funds are usually subject to appropriation by a legislature, congress, or parliament, making them vulnerable to political pressures. If a program becomes controversial because, for example, its decisions are seen to adversely affect a strong economic interest, its funding could be decreased or even eliminated by legislative action.

Dedicated Funds

An alternative to use of the general fund is use of one or more dedicated funds. In this approach, a portion of tax revenue from a specific activity in the private sector (e.g., real estate transactions, hotel taxes) supports a specific government activity. For example, a very small percentage of tax on real estate transactions in the coastal zone would raise sufficient money to support an ICM program. Unfortunately, it is difficult to enact such arrangements in most legislatures; hence, at present few, if any, coastal management efforts are funded in this manner.

Summary

This chapter paid considerable attention to institutional issues in ICM, a question of paramount importance, in our view, in the design and implementation of ICM programs. In the intergovernmental dimension, an effective ICM program generally must involve both national and local levels of government, since both typically have responsibilities for land and water management. But the interests of the two levels of government do not necessarily coincide and in fact may often conflict, as suggested by the responses to our cross-national survey. The discussion in this chapter addressed some strategies for bringing together the different (and valuable) perspectives of each level of government in an effective partnership.

The chapter also considered in some detail the problems of overcoming sectoral fragmentation through ICM, a problem mentioned frequently by respondents to our survey. The concept of policy integration was defined, and the view was put forth that integration should be considered not as an absolute but rather as a continuum. The continuum begins with a fragmented approach wherein institutions function independently and with little communication and progresses through the stages of communication and coordination, in which a forum for communication exists and institutions begin to synchronize their actions; harmonization, in which this synchronization is guided by a set of explicit policy goals and directions, generally set at a higher level; and integration, in which more formal mechanisms to ensure harmonization are put in place. The importance of providing incentives (such as joint funding) in order to achieve lasting cooperation among sectoral entities was highlighted.

Several alternatives for achieving sectoral integration were examined, including creation of coordination commissions or councils and naming of a lead agency. Results of the cross-national survey were discussed, revealing, among other findings, that about half of the nations surveyed had an interagency or interministerial commission as the national-level coordination mechanism for ICM.

A discussion of legal considerations pointed out the need for legal analyses to determine whether existing legal authorities in use by sectoral agencies are adequate for application to broader coastal management purposes. To the extent that existing legal frameworks are found to be deficient, new legal authority may be needed to fill specific gaps.

A final section on financial considerations raised the need to think creatively about various forms of financing alternatives, drawing from government, private, and nongovernmental sectors, to ensure sustainable funding of ICM activities over time.

Chapter 7

Informing the ICM Process:
Building the Science
and Information Base

Introduction

This chapter focuses on the information, science, and tools needed to inform the ICM process and where such information can be found. Perhaps the most fundamental tenet underlying the ICM concept is that ICM decision making is based on the use of the best information and the best science available. This does not mean, of course, that such decisions are based solely on data and information derived from the natural sciences. Indeed, in some situations it is appropriate to give greater weight to social, cultural, and economic factors provided irreversible damage is not done to the ecosystem in the process and unsustainable practices are not put in place. A case in point is the continued taking of minimal numbers of endangered whales by the indigenous Inuit peoples of the Arctic. The social and cultural benefits of this very small amount of whale taking have been judged to outweigh the detrimental effects to the stocks. Thus, both the natural sciences and the social sciences are important in the ICM process. An understanding of the physical functioning of a beach system may not lead to a solution for beach management problems if the behavioral characteristics of the beach users are not also considered.

Indeed, one of the strengths of the ICM process is its reliance on accurate information and data. This is often the only way in which controversial decisions can be implemented and enforced. Without an adequate technical basis for decision making, the special interests that are always present will tend to dominate the process. Of course, there is never enough information or data to answer all questions concerning a proposed activity; hence, an important part of the ICM process involves decision making in the presence of uncertainty. Rules can be fashioned and risks assessed even in the absence of complete information.

In focusing on the data, information, and scientific needs of ICM, this chapter necessarily takes a broad-brush approach. Each nation will have specific data and information needs, depending on its situation—in particular, on the nature and current uses of the coastal zone and on its resources and physical characteristics. Moreover, the specific institutional and legal information needed will depend on the structure of the government system, the distribution of power among levels of government, the resource management programs already in existence, and the effectiveness of these programs Hence, in this chapter, we describe generic categories of data and information needed by all countries as they consider the formulation of an ICM program. Outlining a particular country's ICM-related data and information needs is an important early step in formulating an ICM program and is discussed more fully in chapter 8.

In describing data and information needs, we discuss the special information requirements associated with some areas of concern addressed in most ICM programs. In each case, we outline typical management problems, sketch some relevant scientific questions, and list representative data and information needs. Obviously, all the information cited need not be available at the beginning of an ICM effort, but as time and resources become available, adding this kind of information to the ICM database should improve the performance of the program.

This chapter also contains suggestions for assembling data and information in forms useful to the analyst and the resource manager. The roles of resource inventories, environmental profiles, mapping, and geographic information systems (GIS) are also discussed. Potential applications of remote sensing techniques are reviewed as well. Finally, various analytical tools of potential use in an ICM program are briefly discussed. These extend over the full range from rapid appraisal techniques, environmental impact assessment, and benefit-cost studies to risk assessment techniques, resource valuation procedures, and habitat assessment techniques.

A continuing challenge in the management of coastal resources generally centers on the science-policy interface. Improvements in resource management usually depend on improvements in our understanding of the processes involved, yet obtaining these gains in scientific understanding has proven to be difficult and slow. Managers and policy makers seem to have difficulty motivating the scientific community to carry out the needed research, perhaps because it is perceived as "too applied." Similarly, researchers complain that they do not get clear messages from policy makers as to what is needed and, furthermore, that the managers and policy makers do not seem to use much of the information scientists have already produced. This chapter elaborates on this issue and provides some suggestions for improving the effectiveness of communications across this importance interface. Here, we rely on insights derived from a study on science-policy interactions conducted by the U.S. National Academy of Sciences, which author Biliana Cicin-Sain cochaired (NRC 1995).

An orientation we bring to the topic of scientific information for ICM involves the concept of adaptive management. This management approach is somewhat flexible and is modified from time to time on the basis of new information collected as a part of the management process itself. In effect, the management process "learns" as it adapts to changing circumstances (Lee 1993; Hennessey 1994). Properly designed, such an approach can be used to begin an active management program even before all the critical information is in hand. It is, of course, important that a conservative approach be taken until adequate information has been collected, as we discuss later in the chapter.

Importance of a Solid Scientific Basis for ICM

As succinctly stated in the National Academy of Sciences report on science-policy interactions (NRC 1995, 7), "Scientific information is needed to guide the wise use of coastal resources, to protect the environment, and to improve the quality of life of coastal zone residents. This need is becoming more evident as the complexity of the relationships among the environment, resources, and the economic and social well-being of human populations is fully recognized and as changes and long-term threats are discovered."

Similarly, a recent assessment of scientific requirements for ICM undertaken by the Joint Group of Experts on the Scientific Aspects of Marine Environmental Protection (GESAMP) underscores the need to bring in both the natural and the social sciences to contribute to all phases of the ICM process. As stated in the report's executive summary (GESAMP 1996, pg. iv): It is clear that the management of complex ecosystems subject to significant human pressures cannot occur in the absence of science. The natural sciences are vital to understanding ecosystem function and the social sciences are essential to elucidating the origin of human-induced problems and in finding appropriate solutions. The need to design studies in accordance with clearly stated objectives is particularly important. Scientific techniques and procedures that are particularly useful to ICM include resource surveys, hazard and risk assessments, modeling, economic evaluations, and analyses of legal and institutional arrangements. Scientific support is also needed in the selection of management control measures and in preparing material for public information and education.

Although heightened pressures on coastal environments are posing new challenges for management and the science that supports management, as the 1995 NRC report notes, in many instances science and technology have played important roles in the resolution of coastal problems. Some examples cited in the NRC report include scientific work that led to the formulation of policies and regulations regarding the release of artificial radionuclides from energy facilities into the atmosphere and the ocean; work on the effects of halogenated hydrocarbon biocides such as DDT on nontarget organisms, such as fish-eating birds, which led to legal constraints on their use in pest control (increasing the rate of recovery of threatened species); strategies to clean up coastal systems overloaded with organic and nutrient wastes, such as Lake Erie in the United States; identification of the effects of tributyl tin (a toxic compound found in marine paints) on marine organisms, which led to the banning of this compound by several nations; and imposition of technology-based standards for pretreatment of industrial and municipal discharges, which has resulted in improved coastal water quality in U.S. waters.

In other cases, however, as the NRC report notes, environmental and resource problems have worsened because of inadequate scientific information or because of lack of foresight in anticipating such problems—as in the case of the widespread eutrophication (nutrient overenrichment) of coastal waters or the absence of good models for relating water quality, habitat conditions, and fish productivity (which, in the case of fishery depletion, would allow differentiation between problems caused by environmental factors and problems caused by overfishing). There have been times, too, when scientific information was available and there was scientific consensus, but no appropriate policies were enacted. Such cases include the collapse of fishery stocks in the New England region of the United States (NRC 1995, 8), in spite of substantial scientific evidence on fishery decline, and the unnecessary costs incurred by society in paying for secondary treatment of municipal discharges in some California communities, in spite of scientific evidence of more effective and less expensive alternatives (NRC 1993).

The scientific community often perceives that coastal policy is made more in response to public needs and demands than in response to scientific understanding of the problems involved. This is inevitable in the sense that in democratic societies, it is appropriate and necessary for the public to be influential in policy making. Nevertheless, if good scientific analysis does not underlie the coastal policies enacted, the public good may be damaged in at least two ways: natural resources and environments may suffer irretrievable damage or, alternatively, many expenditures may be incurred in controlling situations that do not need controlling (as in the municipal discharge case cited earlier). To avert these situations, better ways of interjecting scientific information into the policy process and an improved public understanding of science are needed. As noted in the NRC report (NRC 1995, 9):

> New means must be developed to improve interactions among scientists, policymakers, and the public so that policymakers can obtain the information they need about social and natural systems, scientists can determine from policymakers what kinds of scientific questions are relevant to policy, and the public can be introduced as an integral participant in coastal management and policy making.

There are also important political reasons for grounding ICM in good science. Decisions made as part of the ICM process often have important effects on coastal stakeholders. Some decisions, for example, may reduce the profit a private corporation derives from a coastal or ocean activity by increasing the cost of doing business or by decreasing the amount of resource that can be taken. Understand-

ably, such decisions are unpopular with those affected by them, who will sometimes seek ways to overturn them. Decisions are much more easily reversed or set
aside if they are not scientifically defensible—that is, if they are not based on the
best data and scientific information available. It is important, therefore, that the
decision-making process embodied in an ICM program be as objective and as
soundly based as possible. Furthermore, the reasoning underlying the decision
making should be clear to all involved in the process. The science and data on
which decisions are based should be explicitly spelled out and made available for
examination and study by any interested party. As stressed in chapter 9, it is
important that the decision process be made fully transparent so that interested
parties can determine the basis (data, information, science) on which the decisions
are made.

Dealing with Scientific Uncertainty

The coastal zone and adjacent ocean represent a complex and dynamic environment in which a number of physical, biological, geological, and chemical processes take place. Coexisting in the coastal zone are large numbers of people joined
together in a wide variety of social institutions that themselves are made up of
social processes and behaviors. Major uncertainties exist in current understanding
of the behaviors of the physical and social systems making up the coastal zone.

Therefore, an important challenge in ICM is the design of a decision-making
system that can function in the presence of gaps in information and understanding. It is unrealistic to delay utilization of coastal zones and their resources until
all important data have been obtained. In many cases, the pressures are such that
decision making must go forward in the absence of full information. In such circumstances, which are elaborated later in this chapter and in chapter 9, certain
principles become important:

- Do not make decisions that have irreversible consequences.

- Do not make decisions that could seriously threaten the resource base over
  the long term.

- Do not make decisions that could foreclose options for future generations to
  utilize coastal and ocean resources.

Taken together, these principles make up the concept of intergenerational equity and are a fundamental part of the concept of sustainable development. Another important principle that should guide decision making in the presence of scientific uncertainty is the precautionary principle. It says, in effect (Van Dyke
1996, 10):

When full information is not available, the burden of proof should be on those wishing to develop or utilize a resource to demonstrate that their proposed activity will bring no harm to the environment or its resources. If lack of data or information prevent the demonstration of "no harm," then a very conservative regulatory approach should be taken until convincing evidence is obtained.

The foregoing discussion should not obscure the fact that much useful data and information concerning the coastal zone already exist and obviously should be used in ICM programs. The challenge here is to identify, locate, and organize the information in ways that will make it accessible and usable in the ICM decision-making process. In recent years, great strides have been made in procedures for collecting and handling information and data; these methodologies should be incorporated as fully as possible into ICM programs. Additional information on such methodologies is provided later in this chapter.

Data, Information, and Scientific Requirements

As mentioned earlier, a book of this type cannot cover the data and information needs of all the world's coastal areas. Much of the needed information is site specific and will be of interest only to the country containing particular kinds of coastal geography. Thus, we refer the reader to efforts by others to develop methodological approaches to ICM and to specify ICM data requirements, such as Denis et al. 1996.

In this section, we review the types of data and information required by the typical ICM program. This generic approach should provide a useful guide to program managers as they consider the data and information needs of their specific coastal areas and problems. It should be noted that different types of information are needed at different stages in the ICM process. Table 7.1 provides a general perspective on this matter.

It is difficult to provide a broad-brush description of the scientific requirements of an ICM program, given the great differences that exist in coastal management problems from country to country and region to region. However, a few of the more important natural science and social science needs related to ICM are listed here, by way of illustration.

A Sampling of Natural Science Needs

- An understanding of coastal physical, biological, chemical, and geological processes—for example, the forces that affect the long-shore and across-shore transport of sand.

Table 7.1. Types of Data and Information Needed at Various Stages in the ICM Process

Stage in the ICM Process	Types of Data and Information Needed
1. Initiating the effort: scoping, issue identification, and priority setting	1. Information and data on coastal problems triggering the need for ICM (pollution; resource depletion; degraded wetlands, mangrove forests, corals, etc.)
2. Formulation of the ICM plan	2. Information and data on resources, coastal uses, problems and conflicts, social organization, existing management efforts, capacity of existing institutions, and existing legal authorities
3. Formal adoption, implementation, and operation of the ICM program	3. Information on benefits and costs of ICM and on regulatory and management measures, conflict resolution techniques, and relevant analytical methods
4. Monitoring and assessment of program performance	4. Monitoring of biological and physical parameters and socioeconomic variables to ascertain changes in natural and social systems; observations of program outcomes; information on user satisfaction; users' suggestions for new management measures and new products

- An understanding of the concept of coastal ecosystem health—what it is, how to measure it, and how to sustain it.

- An understanding of the management of coastal and ocean biodiversity.

- An understanding of the various functions coastal wetlands perform and a better appreciation of the valuation of each.

- An understanding of climate variability, especially as related to global warming and its effects on the coastal zone (sea-level rise, saltwater intrusion, etc.).

A Sampling of Social Science Needs

- An understanding of the structure and dynamics of coastal settlements and of particular user groups that depend on the coast and ocean.

- Research on the facilitation of interagency cooperation in coastal resource management.

- Development of the concept of adaptive management.

- Development of workable market-based techniques to supplement regulatory approaches in coastal management.

- An understanding of the role of indigenous peoples and traditional knowledge in coastal management.

- Translation of the concept of intergenerational equity into some working principles and policies.

Data and Information Needs for Eight Areas of Concern in ICM

In the previous section, we provided generalized information on the data and information needed in the development and operation of an ICM program. In the list that follows, we provide a more specific treatment of eight issues that many ICM programs are designed to address: beach management and coastal erosion, wetlands protection, coastal hazards (storms, etc.), nonpoint-source pollution, sea-level rise, coastal and estuarine water quality, threatened and endangered species, and coral reef management. For each issue, we briefly describe the typical management problems confronted; some of the scientific questions involved; and some of the more important data and information needs with regard to the issue (see table 7.2). We hope this approach will contribute to an understanding of the kinds of data and information needed to enable policy making and decision making in an ICM program to be conducted in a technically sound fashion. Please note that none of these topics is treated comprehensively; only some examples are listed. Clearly, if these areas of concern become part of a specific ICM program, additional information will be needed, especially regarding the particular management problems to be confronted and the data and information needs associated with those problems.

Coping with Missing or Sparse Information

It is axiomatic that there are never enough data and information to support conscientious management efforts. Lack of funds, lack of time, lack of foresight, and lack of understanding are all responsible in varying degrees for this situation. But often, the start of a management program cannot await the missing data. Management must begin using what is obtained. Shortcut ways to fill information gaps do exist (see, for example, the RACE technique described later in this chapter). But even using these, gaps are still likely to be present. In such cases, as mentioned earlier, a conservative approach is called for until adequate data and understanding are in hand. In the presence of uncertainty, prudence would suggest that no project should be allowed to go forward that is likely to have irreversible con-

Table 7.2. Data and Information Needs for Eight Areas of Concern in ICM

	Typical Management Problems	Some Relevant Scientific Questions	Representative Data and Information Needs
1. Beach Management and Coastal Erosion	• Controlling beach erosion • Designing beach replenishment projects • Regulating the use of hard erosion control structures on the beach	• Is the erosion episodic or continuous? • What are the magnitude and direction of the long-shore drift of sand? • Does a littoral cell of sand circulation exist? • How does the beach profile change in shape with seasonal changes and storms? • How do differing degrees of coastal erosion affect the beach recreational experience?	• Data on properties of the beach (grain size, sorting, slope, profiles) • Data on long-shore transport of sand • Data on wave climate affecting the shoreline • Data on storm frequency and intensity
2. Wetlands Protection	• Maintaining healthy and functioning wetlands • Restoring original values and functions to degraded wetlands • Creating new wetlands from "clean" dredged materials	• How much can the water circulation change without damaging the wetland and its functions? • How much pollution can occur in the wetland without damaging its functions? • Can most or all of a degraded wetland's original functions be restored? • Is it really possible to create new fully functioning wetlands? • How should the costs and benefits associated with wetlands restoration projects be measured?	• Data on the functions the wetland performs • Data on circulation, soils, and vegetation types in the wetland • Data on animal, bird, and fish species present • Data on biodiversity present
3. Coastal Hazards (Storms)	• Establishing set-back lines to control placement of buildings in the coastal zone • Establishing and managing exclusionary zones marking hazardous locations • Establishing and managing storm evacuation plans and procedures	• How can an appropriate erosion rate be established for use in a dynamic definition of the set-back line? • How can a general retreat from the immediate water's edge be motivated? • Will global warming increase the frequency and intensity of coastal storms and hurricanes? • How do coastal users differ in their perceptions of "risk"?	• Frequency of occurrence of coastal storms and hurricanes • Erosion rates at various shoreline locations • Behavioral and demographic information about coastal land owners • Accurate topographical maps of the coastal zone for evacuation planning

(continues)

Table 7.2. (*continued*)

	Typical Management Problems	Some Relevant Scientific Questions	Representative Data and Information Needs
4. Nonpoint Source Pollution	• Controlling runoff from agricultural lands • Controlling runoff from urban areas (streets, parking lots, etc.) • Managing storm water	• What types of pollution do various agricultural activities contribute? • How effective are different types of buffers? • What are the effects of different types of best management practices (BMPs)? • What is the extent of pollution caused by atmospheric deposition?	• What are the attitudes of farmers toward voluntary versus regulatory nonpoint source pollution control measures? • Data on location and extent of agricultural and forestry activities in the watershed • Areal extent and nature of impervious surfaces in urban areas • Rainfall, storm frequency, and other meteorological data
5. Sea-Level Rise	• Establishing an appropriate strategy (defense, armor, or retreat) • Preserving important recreational beaches and wetlands • Preventing increased salt-water intrusion • Managing discharges from commercial and recreational vessels	• What causes sea-level rise (global warming, land subduction, isostatic equilibrium, etc.)? • Has sea-level rise due to global warming been observationally confirmed? • What magnitude of sea-level rise is expected at particular locations? • How will sea-level rise affect fisheries, water quality, and other aspects of the coastal zone? • Under what conditions will policy makers act to reduce the threats posed by remote and distant hazards such as sea-level rise?	• Accurate measurements of relative sea-level at a number of locations • Global measurements of eustatic sea-level rise • Data and information on economic and social values of coastal facilities and coastal resources that may need protection
6. Coastal and Estuarine Water Quality	• Managing point-source pollution entering marine waters • Managing nonpoint-source pollution (agricultural and urban runoff) • Upgrading sewage treatment plants to improve effluent discharges and dealing with combined sewer overflows (CSOs)	• What is the assimilative capacity of the coastal ocean? • Can sewage sludge be be dumped into the ocean safely? • How much and what kind of a land buffer is needed to significantly reduce the nonpoint-source pollution from agricultural activities reaching coastal waters?	• Existing quality of coastal waters in various locations and under various meteorological conditions • Properties of effluents entering the coastal waters from various point and nonpoint sources • Extent and nature of chemicals (fertilizers, pesticides) in the agricultural and

	Typical Management Problems	Some Relevant Scientific Questions	Representative Data and Information Needs
6. Coastal and Estuarine Water Quality (*continued*)		• How important is coastal water quality to the recreational boating experience?	drainage basin and trends in this regard • Population of commercial and pleasure craft using estuarine and coastal waters
7. Threatened and Endangered Species	• Reversing declines in populations of threatened and endangered species • Identifying and protecting the habitats of such species • Developing restoration plans for depleted stocks	• Where do such species fit in various food chains and webs? • Can colonies of threatened or endangered species be successfully moved to better locations? • Do whale-watching boat tours adversely affect whale populations? • How much public support exists for measures aimed at protecting particular endangered species?	• Established populations (and trends) of threatened and endangered species in the area • Locations of existing and potential habitat for such species • Socioeconomic and demographic data related to the use of such species by indigenous or subsistence-based peoples
8. Coral Reef Management	• Developing management programs to prevent sediment from entering coastal waters • Managing the taking of corals and fishes from the reef • Preventing anchor damage from dive boats • Maintaining adequate water quality, circulation, salinity, and temperature for coral health and growth	• What are the most effective techniques for restoring coral health? • Can some kinds of coral be harvested on a sustainable basis? • Should special protective measures be taken during times of coral spawning? • What risks to coral reefs are posed by various types of recreationists (snorklers, scuba divers, swimmers, etc.)?	• Inventories of coral reef locations and conditions (and trends) • Information and trends regarding water quality, temperature, salinity, and clarity in coral reef areas • Management measures currently in effect regarding sedimentation, pollution, and the like • Management measures currently in effect regarding activities in and around coral reefs (dive boats, diver behavior)

sequences on the resources in question. For proposed activities the precautionary principle touched on earlier, endorsed at the United Nations Conference on Environment and Development and now gaining acceptance generally, would prescribe that in the absence of convincing evidence to the contrary, a conservative regulatory and management approach be taken. In such cases, the burden of proving that no adverse effects will occur should fall on the developer. The government, acting on behalf of the public interest, should not have to demonstrate that harm will occur in order to regulate.

Another strategy for coping with sparse or nonexistent data concerning the effects (or lack thereof) of a proposed coastal or ocean development activity is to require that the needed data, or as much data as feasible, be collected as a condition of approval of the initial development activity. For example, data on the abundance of a heretofore unfished stock could be required to be gathered by the person(s) seeking a permit to begin to exploit this resource. Catches could be restricted to conservative levels until sufficient data are in hand to establish technically sound catch limits.

Where adverse environmental effects of a proposed development activity are the concern, it may be possible to design a monitoring program to detect any such effects before serious degradation takes place. The development activity could then be scaled back or terminated, depending on the circumstances. Of course, the design and operation of a sufficiently sensitive monitoring program requires a degree of understanding of the physical processes involved that may or may not be present at this stage.

In smaller-scale developments that might affect sensitive coastal areas, yet another approach could be considered. The applicant developer could be required to post a bond or financial guarantee to be held by the government during the duration of the development or exploitation activity. A comprehensive monitoring program could be required as well. If, during the course of the development activity, unacceptable adverse effects are detected by the monitoring program, the activity could be terminated and all or a portion of the proceeds from the bond could be used to mitigate the damage. Again, the success of such an approach will require sufficient scientific understanding and technical skill to carry out the mitigation effectively.

Useful Methodologies, Technologies, and Analytical Tools

A wide variety of tools are available to help ensure that decisions made in the ICM process are as sound and informed as possible. In this section, we provide brief overviews of nine of these tools and discuss situations in which they might be applied. The discussion is not comprehensive, but sources of additional information are provided.

Resource Inventories and Environmental Profiles

Resource inventories and environmental profiles are collected early in the process of developing the management program. Inventories are needed of important coastal resources (fish stocks, wetlands, beaches, coral reefs, mangrove forests, etc.) and their locations, how much of each type or species is present, the condition of the resources (pristine, overused, degraded, depleted, etc.), and any trends that are evident. Socioeconomic data on the users of the resources (demographics, extent of economic dependence, cultural and social aspects, etc.) should also be obtained. Which resource inventories are needed will depend, of course, on the resources present, the goals and priorities of the ICM program, the nature of the interactions between resource-related activities, and the like. Environmental profiles tend to be nation-based and often are a part of a donor agency's effort to ensure that all environmental effects of a proposed development project are identified and satisfactorily addressed. As such, they provide a reasonably detailed description of the environmental and resource conditions prevailing at the time of the profile's preparation. Hence, environmental profiles describe the environmental setting and provide some early judgments about management problems. Typically included are information and data on air and water quality; waste disposal programs; principal natural resources present and their condition; nature and status of agricultural activities; mining and forestry and other extractive activities; main economic drivers in the country; nature and capacity of the environmental regulatory system; organization of the government in relation to environmental and resource issues; principal environmental issues facing the nation; and so on. Profiles of this kind are often an initial element in the development of a national environmental action plan (NEAP). The development of NEAPs has been encouraged and supported by the World Bank as one of its environmental initiatives following UNCED.

For Additional Information

Chua, T. E., L. M. Chou, and M. S. M. Sadorra, eds. 1987. *The Coastal Environmental Profile of Brunei Darussalem: Resource Assessment and Management Issues.* ICLARM Technical Reports on Coastal Area Management Series No. 18. Manila: International Centre for Living Aquatic Resources Management.

Levitus, S., and R. D. Gelfeld. 1992. *National Oceanographic Data Center Inventory of Physical Oceanographic Profiles: Global Distributions by Year for All Countries.* Silver Spring, Md.: U.S. Department of Commerce, and National Oceanic and Atmospheric Administration, National Environmental Satellite, Data, and Information Service, National Oceanographic Data Center.

MTE (Multidisciplinary Team of Experts). 1996. *The Coastal Environmental Profiles of the Batangas Bay Region.* MPP-EAS Technical Report No. 5.

GEF/UNDP/IMO Regional Programme for the Prevention and Management of Marine Pollution in the East Asian Seas, Manila, Philippines.

White, A., P. Martosubroto, and M. S. M. Sadorra, eds. 1989. *The Coastal Environmental Profile of Segara Anakan-Cilacap, South Java, Indonesia.* ICLARM Technical Reports on Coastal Area Management Series No. 25. Manila: International Centre for Living Aquatic Resources Management.

Mapping and GIS Systems

Like resource inventories, mapping activities are frequently undertaken as a part of early data and information collection activities prior to the onset of active management, although they can also be useful in assessing certain aspects of program performance. Mapping programs are pursued when the spatial locations and relationships of particular resources, activities, or problems are important. The appropriate scales to be used in the mapping depend, of course, on both the potential use to which the maps will be put and the nature of the data being mapped. For example, in most ICM programs it will be necessary to map the locations of coastal wetlands, since property owners will need to know whether their land or a portion of it is affected by management measures. Periodic mapping of sensitive resources is also performed as a way to observe trends and changes. Geographic information systems (GIS), of course, can play a very useful role in this regard. GIS systems, fundamentally, are computer-based tools to aid in the display and analysis of geographically based information. Different types of information can easily be "overlaid" to help identify and assess the effects of human activities on resource systems. GIS can provide information in a form that is useful in instituting zoning schemes and in designating exclusionary areas, high-risk zones, and the like. It can also be used to analyze other kinds of information, such as mapped information and data derived from remote sensing activities.

For Additional Information

Antenucci, J. C., K. Brown, P. L. Croswell, M. J. Kevany, and H. Archer. 1991. *Geographic Information Systems: A Guide to the Technology.* New York: Van Nostrand Reinhold.

Johnson, A. I., C. B. Patterson, and J. L. Fulton, eds. 1992. *Geographic Information Systems and Mapping: Practices and Standards.* Philadelphia: ASTM.

Lyon, J., and J. McCarthy, eds. 1995. *Wetland and Environmental Applications of GIS.* Boca Raton, Fla.: CRC Press.

Remote Sensing

Remote sensing is a technology involving the use of satellites or aircraft to observe the characteristics of broad areas of the earth's surface in relatively short periods of time and in considerable detail. Active modes (using radar signals, for example) and passive modes (using emitted or reflected radiation) are employed across the electromagnetic spectrum (visible, infrared, radar) to obtain wavelength-dependent information on a periodic basis, depending on satellite trajectories and aircraft schedules. This information allows the discrimination of various types of coastal zone vegetation, the health of wetlands, the amount of sediments entering coastal waters, waves and currents, chlorophyll concentrations, and other information of potential use to coastal managers. Remotely sensed data allow information obtained by more conventional means (for example, in situ observations) to be extrapolated over wide areas with relative ease. Remote sensing can play a role both in the initial mapping of resources and their conditions and in the provision of data showing trends over time. For example, one could monitor the recovery of a degraded wetland by periodically obtaining remotely sensed data on the types of vegetation present and the relative abundance and condition of such plants. Satellite data can cover extensive areas of the earth's surface (tens of thousands of square kilometers) but at a relatively coarse resolution (typically tens of meters), whereas aircraft platforms can provide coverage of smaller areas (tens or hundreds of square kilometers) but at a higher resolution (typically a few meters). Although remote sensing technology is still evolving and acquisition and interpretation of the data are still somewhat expensive, it is likely that costs will come down in coming years as greater use is made of computers and GIS methodologies and as satellite costs continue to decrease.

For Additional Information

Bennett, E. C., and L. F. Curtis. 1992. *Introduction to Environmental Remote Sensing.* 3rd ed. New York: Chapman and Hall.

Englehardt, R., ed. 1994. Remote sensing for marine and coastal environments. *Marine Technology Journal* 28 (2): 83.

Matson, P., and L. Ustin. 1991. The future of remote sensing in ecological studies. *Ecology* 72:1917–1945.

Sample, V. A., ed. 1994. *Remote Sensing and GIS in Ecosystem Management.* Washington, D.C.: Island Press.

Simpson, J. 1994. Remote sensing in fisheries: A tool for better management in the utilization of a renewable resource. *Canadian Journal of Fisheries and Aquatic Sciences* 51:743–771.

Rapid Appraisal Techniques

Michael Pido and Chua Thia-Eng and their colleagues, working in Southeast Asia, have developed a technique for rapidly analyzing the environmental and socio-economic conditions of a particular coastal area. Using techniques developed in the field of agriculture, they focus on the use of key indicators to provide the necessary information. Their approach, called RACE (rapid appraisal of coastal environments), involves the use of informal workshops, secondary data sources, direct field observations, "stories and portraits" by community members, and the like. According to its developers, RACE can be a catalyst for coastal zone planning and management and is a useful tool for "prognosis and diagnosis of coastal zone problems." Pido and Chua assert that use of the RACE technique can reduce the time needed to appraise and diagnose the situation in a given coastal area from several years to about one year.

For Additional Information

Pido, M. D., and T. E. Chua. 1992. A framework for rapid appraisal of coastal environments. In *Integrative Framework and Methods for Coastal Area Management,* ed. T. E. Chua and L. F. Scura, 144–147. ICLARM Conference Proceedings No. 37. Manila: International Centre for Living Aquatic Resources Management.

Environmental Impact Assessment

Environmental impact assessment (EIA) is a tool used in many coastal management contexts. In fact, it is a mainstay of many nations' environmental protection programs. It is a process whereby developers or others proposing a project or activity conduct an analysis to determine its likely effects. The process originated in the United States and was institutionalized in the National Environmental Policy Act of 1969. An important part of the EIA process is the determination of alternative approaches to a project that will reduce or eliminate adverse environmental effects. John Clark (1996, 118) provides a reasonably detailed account of the various aspects of the environmental impact assessment process. He emphasizes two important aspects of the EIA process—"identification of alternative designs and locations and operational precautions to prevent *avoidable* negative impacts" and "identification of measures to reduce effects of *unavoidable* negative impacts, including habitat rehabilitation."

In administering an EIA process, care must be taken that the assessment does not get too detailed or technical. Project proponents should be required to provide succinct, understandable, and clearly written assessments that emphasize the most pressing adverse effects and provide a serious consideration of realistic alternatives. In some legal systems, individuals or organizations can challenge the accuracy or comprehensiveness of an assessment and substantially delay the project. Although the EIA should undoubtedly be a part of an ICM process, it cannot by itself substitute for ICM. It tends to operate on a project-by-project basis and hence is not a good tool for comprehensive, areawide planning or for the assessment of cumulative adverse effects. Moreover, preparation of an EIA statement by no means guarantees that a developer will take the least environmentally adverse alternative. It merely requires that he or she follow the process and prepare an accurate and complete assessment of potential effects.

For Additional Information

Cort, R. P. 1995. A practical guide to environmental impact assessment. *Quarterly Review of Biology* 70 (4): 533.

Gilpin, A. 1995. *Environmental Impact Assessment: Cutting Edge for the Twenty-First Century.* Cambridge: Cambridge University Press.

Jernelov, A. 1988. *EIA—A Practical Approach: Proceedings of the ROPME Workshop on Coastal Area Development.* UNEP Regional Seas Reports and Studies No. 90, 143–169. Nairobi, Kenya: United Nations Environment Programme.

Morris, P., and R. Therivel, eds. 1995. *Methods of Environmental Impact Assessment.* Vancouver, B.C.: UCL Press.

Smith, L. G. 1996. Introduction to environmental impact assessment: Principles and procedures. *Environment and Planning* 28 (2): 373.

Vanclay, F., and D. Bronstein. 1985. *Environmental and Social Impact Assessment.* New York: Wiley.

Benefit-Cost Studies

Virtually any development project carries some environmental risks; clearly, the gains to flow from the project should outweigh these risks. Similarly, if public funds are to be used in a development project, the benefits from the development should exceed the costs. Hence, benefit-cost studies have a distinct place in resource management programs such as ICM. Economics texts, of course, explain the mechanics of the benefit-cost calculation at considerable length. Suffice it to say here that the calculations are often not straightforward and the results are sometimes difficult to interpret. In the environmental field, both costs and benefits are often difficult to calculate. Some of the costs (adverse effects on other activities) are not internalized and hence are ignored; others are nonmonetary in nature and therefore are difficult to handle quantitatively. Similarly, some benefits are difficult to quantify as well. Furthermore, bringing everything "back" to the present involves the use of present-value calculations employing often controversial discount rates. This methodology definitely has a role to play but must be employed with due care and consideration.

For Additional Information

Edwards, S. F. 1987. *An Introduction to Coastal Zone Economics: Concepts, Methods, and Case Studies.* New York: Taylor and Francis.

Field, B. C. 1994. *Environmental Economics.* New York: McGraw-Hill.

Hanley, N. 1993. *Cost-Benefit Analysis and the Environment.* Aldershot, England: Elgar.

Lipton, D. W., and K. F. Wellman, with the collaboration of I. C. and R. F. Weiher. 1995. *Economic Valuation of Natural Resources: A Handbook for Coastal Resource Policymakers.* Silver Spring, Md.: U.S. Department of Commerce, National Oceanic and Atmospheric Administration (NOAA), Coastal Ocean Office.

Risk Assessment

Increasing attention is being paid to the assessment and management of risk. As governments tighten their budgets, greater effort is made to concentrate limited resources where they will do the most good—in managing the "riskiest" activities. Risk in this sense involves both the frequency or likelihood of a negative event occurring and the severity of its consequences. Techniques are being developed to quantify risks and thus to provide some guidance as to which environmental problems need attention first and the appropriate level of public investment in addressing such issues. Risk assessment has most often been employed in situations involving public health, as in risks to health from air and water pollution or from contaminated seafood. In the coastal and marine field, such assessments are regularly used to estimate the likelihood of ship accidents, hazardous spills, oil drilling blowouts, and the like. Risk assessment also has application in dealing with coastal hazards such as hurricanes and other storms. The risk (probability) of a large hurricane striking a particular section of coastline can be calculated on the basis of historical data. Likely losses associated with such a storm can be estimated, and hence the appropriate level of effort and expenditure in dealing with this risk can be determined.

For Additional Information

EPA (Environmental Protection Agency). 1992. *Framework for Ecological Risk Assessment.* EPA/630/R92/001. Washington, D.C.: Environmental Protection Agency.

EPA (Environmental Protection Agency). Science Advisory Board. 1990. *Reducing Risk: Setting Priorities and Strategies for Environmental Protection.* Washington, D.C.: Environmental Protection Agency.

Gregory, R., T. C. Brown, and J. L. Knetsch. 1996. Valuing risks to the environment. *Annals of the American Academy of Political and Social Science* 545: 54–63.

Macilwain, C. 1996. Risk: A suitable case for analysis? *Nature* 380: 10–11.

Mlot, C. 1989. Global risk assessment. *BioScience* 39: 428–430.

Swaney, J. A. 1996. Comparative risk analysis: Limitations and opportunities. *Journal of Economic Issues* 30 (2): 463–473.

Trends and challenges: The new environmental landscape. 1996. *Environmental Science and Technology* (special issue) 30: 24–44.

Valuation of Resources

Increasing attention is being paid to economics in the management of coastal and ocean resources. Understandably, economics drives private decision making with regard to investments in various resource-related activities. As governments tighten their financial belts, economics is playing an ever larger role in public programs as well. Economic justifications are required before new expenditures are made; user fees are becoming commonplace. But a weakness in economics has been the difficulty of assigning monetary value to many of the goods and services provided by natural resources and the ecosystems containing them. This first came to light in connection with efforts to assess the damage to coastal resources caused by oil spills: How much is a seabird worth? How valuable is a pristine beach or a functioning coastal wetland? In another context, can a value be placed on clean air over, say, the Grand Canyon in the United States? Analysts use the terms *option value* to denote the value of keeping options open to use the resource later and existence value for the benefit or value associated with simply knowing that the resource exists and will exist in the future (Clark 1996). An additional complication in efforts to value resources arises when certain benefits of the resource occur off-site, that is, some distance removed from the resource itself. Coastal fish that thrive and grow in the coastal ocean because of nutrients flowing out from coastal wetlands are one example. The wetlands clearly benefit the fish but in ways that are difficult to capture or quantify. Nonetheless, this methodology will surely continue to develop, will become even more important in the future, and will be needed by most ICM programs as they develop.

For Additional Information

Dixon, J., and M. Hufschmidt, eds. 1986. *Economic Valuation Techniques for the Environment: A Case Study Workbook.* Baltimore: Johns Hopkins University Press.

Freeman, M. A. 1993. *The Measurement of Environmental and Resource Values.* Washington, D.C.: Resources for the Future.

Lipton, D. W., and K. F. Wellman, with the collaboration of I. C. and R. F. Weiher. 1995. *Economic Valuation of Natural Resources: A Handbook for Coastal Resource Policymakers.* Silver Spring, Md.: U.S. Department of Commerce, National Oceanic and Atmospheric Administration (NOAA), Coastal Ocean Office.

Habitat Assessment Techniques

Habitats can perform many different functions. Wetlands, for example, can supply nutrients to coastal waters, as previously discussed; act as nursery grounds for juvenile fish; provide food for certain marine species; act as a buffer for floodwater; and remove certain pollutants from the water column. Understanding the full range of functions of various habitat types is important in the valuation efforts mentioned earlier, particularly when the question of damage and replacement arises. If a developer proposes to replace a wetland that is to be destroyed in a development project, the coastal manager would want to compare the functions being lost and their values with those of the replacement wetland. Will all lost functions and values be fully replaced? Clearly, such information is critical to the ultimate success of a wetlands mitigation program. Although they are still in a state of development, acceptable techniques for assessing the values of various kinds of coastal habitats—techniques that will take into account the full range of services (functions) provided by these habitats—will emerge soon. Obviously, tools such as this will be a valuable part of all ICM programs.

For Additional Information

Lipton, D. W., and K. F. Wellman, with the collaboration of I. C. and R. F. Weiher. 1995. *Economic Valuation of Natural Resources: A Handbook for Coastal Resource Policymakers.* Silver Spring, Md.: U.S. Department of Commerce, National Oceanic and Atmospheric Administration (NOAA), Coastal Ocean Office.

Focusing the Science on Management Needs: The Science-Policy Interface

As has been emphasized, ICM should be built on the use of the best science and the best information available. But the coasts and the oceans are dynamic places with complex, interrelated ecosystems. Although our understanding of these systems and the physical, chemical, and biological processes that drive them is continually advancing, there are still large gaps in our knowledge. This is especially true with regard to the detailed understanding generally needed to support informed management in coastal areas. The challenge is to focus more scientific attention on management-related research questions. Scientists, from their side, claim that resource managers do not convey their research needs clearly and effectively and, furthermore, that they do not appear to use the research results that already exist.

The Cultures of Science and Policy

One of the problems, of course, has to do with the different cultures to which the two sides belong. The scientists (both natural and social scientists) and the resource managers (coastal policy makers and coastal managers) are often worlds apart in their value systems and their ways of looking at the world. Table 7.3, taken from the 1995 NRC report, illustrates these differences.

Table 7.3. Behaviors and Points of View Typically Associated with the Cultures of Science and Policy

Factor	Science	Policy
Valued action	Research, scholarship	Legislation, regulations, decisions
Time frame	That needed to gather evidence	Immediate, short-term
Goal	Increase understanding	Manage immediate problems
Basis for decisions	Scientific evidence	Science, values, public opinion, economics
Expectations	Understanding is never complete	Expect clear answers from science
Grain	Focus on details, contradictions	Focus on broad outline
Worldview	Primacy of biological, physical, chemical mechanisms	Primacy of political, social, interpersonal, economic mechanisms

Source: NRC 1995.

Obstacles Arising from Cultural Differences

The cultural differences between science and policy pose a number of problems or obstacles in relationships between scientists and coastal policy makers: (1) lack of understanding, (2) lack of communication, (3) lack of, or misuse of, each other's products, and (4) conflict and competition instead of cooperation.

Lack of Understanding. As the NRC report states (NRC 1995, 34), "Human ego is a powerful thing, and few things offend us and make us react in negative ways as much as the knowledge that another person does not value, respect, or understand what we are as individuals or what we do professionally." Unfortunately, all too often in debates about coastal policy one sees natural and social scientists, fishermen, environmentalists, private property advocates, and policy makers dealing with one another without mutual respect for perspectives and positions. Understanding, as the NRC report points out, "doesn't have to mean admiration or agreement, but simply an acceptance of the fact that the other party has a legitimate status and role in the human ecology of the policy-making process and views that must be understood in the context of that status and role."

Lack of Communication. "Cultural differences, whether they stem from language, occupation, or advocacy position, tend to make communication more difficult. Not only are we less likely to communicate at all with different cultures and subcultures, but communication that does occur tends to be fraught with misinterpretation or lack of understanding" (NRC 1995, 34). This frequently happens in discourses between scientists, managers, and stakeholders; even though they may all be speaking at the same public hearing or meeting, there is all too often very little real communication taking place.

Lack or Misuse of Each Other's Products. As the NRC report notes (1995, 34), "It is often the case that an administrator will not know how to use the contents of a scientific report. It is often the case that a scientist will not understand the genesis or rationale for a particular public policymaking process. Private citizens will often be confused by both a scientific report and a policy process. The unfortunate response is for individuals to disengage—that is, to withdraw from the interaction or process—or simply to ignore the activity or viewpoint of others."

Conflict and Competition Instead of Cooperation. As noted in the NRC report (1995, 35), "All of the above effects lead to conflict and competition in place of cooperation. They are all dimensions of the potentially negative public policy outcomes that can result from cultural differences, when those differences are not recognized, understood, and addressed."

Results of the Cross-National Survey Regarding Science-Policy Integration

It is interesting to note that in our cross-national survey (see table 6.4 in chapter 6), among all nations, integration of science and policy was rated the least successful of the four dimensions of integration examined (spatial, intersectoral, intergovernmental, and science-policy). Respondents from developing countries, however, reported greater success in integrating science and policy, as reiterated in table 7.4. This could be due to a variety of reasons. For example, in developing countries, natural scientists, in particular, tend to be more cognizant and attuned to the social development and management needs of their countries than are scientists in developed countries. Similarly, scientists in developing countries tend to be closely tied to government funding of their research, perhaps, making them more aware of the need for application of their scientific studies.

Ways to Improve Science-Policy Interactions

Some ways to improve science-policy interactions are noted in the NRC (1995) report: (1) improve mechanisms for interaction between scientists and coastal policy makers; (2) enhance communications among scientists, and the public policy makers; (3) build capacity for science-policy interactions; (4) employ integrated and adaptive approaches in coastal policy making and implementation; and (5) deploy resources to support the foregoing objectives. Each of these is described briefly in the paragraphs that follow.

Improve Mechanisms for Interaction between Scientists and Coastal Policy Makers. The fundamental point here is that if there are no established fora or regular mechanisms for interaction between scientists and policy makers, few, if any,

Table 7.4. Extent of Success in Integrating Science and Policy in Selected Countries

	Successful (%)	Not Successful (%)	Uncertain (%)
All countries	48	37	13
Developed countries	38	53	7
Middle developing countries	47	33	13
Developing countries	55	30	15

Source: Authors' 1996 cross-national survey.

interactions will occur. A variety of methods whereby scientists and policy makers may interact are cited in the NRC (1995) report: (1) scientific advice can be provided internally within agencies (agencies hire scientists to advise them in coastal policy making); (2) advisory groups external to agencies can be created, such as scientific advisory committees or groups for the coastal agencies; (3) workshops can be held to bring together coastal policy makers and managers, stakeholders, and natural and social scientists; (4) informal policy advisory groups can bring the published results of scientific research performed outside an agency to the attention of coastal decision makers through such means as electronic mail.

The NRC report also counsels agencies to involve stakeholders in all phases of the ICM process, including the planning and application of policy-relevant scientific research. This is a relatively new practice but one used with increasing frequency. Given that the general public and particular stakeholders will be important in the final decisions made about coastal use, it is useful to include them from the outset in the design and conduct of scientific studies designed to influence the outcomes of coastal decisions. Similarly, governments should encourage the formation of regional problem-solving task forces or groups to address coastal problems that cross subject areas, legal jurisdictions, and policy sectors, using, when appropriate, an ecosystem approach (NRC 1995, 69).

A final point is that scientists also need to be encouraged to reach scientific consensus about important coastal problems—something that happens all too rarely given the typically individualistic nature of the scientific enterprise. In this regard, the NRC report encourages professional scientific associations, groups of scientists, and university research consortia to develop syntheses of the state of knowledge on important coastal issues and plans for strategic research.

Enhance Communication among Scientists, Policy Makers, and the Public. The NRC report provides three recommendations for enhancing communication among the groups involved in ICM efforts. (1) Policy makers and implementers should be encouraged to "clearly identify their short-term and long-term research needs, and to indicate how the information is to be used, what resources are available to support the collection and analysis of information about natural and social systems, and when the information is needed" (NRC 1995, 70). (2) government agencies, with the assistance of universities, NGOs, and others, should ensure that the results of policy-relevant scientific research are summarized in a manner intelligible to the lay public and are widely disseminated to decision makers and the public. One way of accomplishing this would be through requirements imposed on funded research projects. (3) Agencies, scientists, NGOs, and others should help representatives of the print, radio, and television media to understand and disseminate the results of policy-relevant scientific research (71).

Build Capacity for Science-Policy Interactions. The NRC report recommends a variety of avenues for capacity building in science-policy interactions. (1) Agencies that have made innovative efforts to apply scientific expertise in the design and implementation of coastal programs (such as the Chesapeake Bay program) should be encouraged to prepare assessments of effective models of science-policy interactions as a guide for use in other cases. (2) Scientists working within government agencies should be encouraged to maintain their expertise and to stay current with developments in their scientific fields. (3) Universities should be encouraged to develop cross-disciplinary training of natural and social scientists on coastal topics (i.e., enhance the natural science training of social scientists and vice versa). (4) Creation of training programs for "science translators" (people who can work across disciplines and interact with coastal policy makers) should be encouraged. (5) Consortia for strategic research should be created to facilitate regular communication of state-of-the-art science to policy makers (through summer institutes, trips to research sites and laboratories, etc.). (6) The academic reward system should be modified to encourage the involvement of scientists in the policy development and implementation process (a time-consuming activity that is rarely rewarded by academic officials). (7) Government programs should be evaluated, in part, on the basis of their efforts and successes in incorporating science in their decisions.

Employ Integrated and Adaptive Approaches and Deploy Sufficient Resources to Improve Interaction. Two final recommendations in the NRC report are to employ integrated and adaptive management approaches in coastal policy making and implementation (largely the subject of this book) and to allocate and coordinate resources to improve interaction between coastal scientists and policy makers. To accomplish the latter goal, agencies should (1) require that a portion of scientific research budgets be devoted to the translation and dissemination of scientific results; (2) promote, in their requests for proposals for funding, the formation of interdisciplinary teams to carry out policy-relevant scientific research; (3) develop mechanisms for better integration of their own policy and science capabilities through such means as data sharing, colocation of facilities, and establishment of cooperative programs; and (4) facilitate personnel exchanges or staff-sharing arrangements whereby scientists, NGOs, and industry personnel spend time in government agencies and government employees work in universities, NGOs, and corporations on temporary assignments.

As noted in the NRC report, an important avenue for making ICM more scientific is to craft better interdisciplinary teaching and training programs. This point is underscored as well by Gunnar Kullenberg (1995, 41):

> There is a need to re-think education and its content. Real interdisciplinarity cannot be obtained without the educational system providing a base for it. Traditional education and the resulting human capacities are not sufficient for sustainable development. There is a need to bring economics and other social sciences into [marine sciences] education. There is also a need to clarify how to handle the interfaces, for

instance, between producers of data and forecasts and the end users: various sectors or activities of society, such as tourism, shipping, aquaculture, fisheries, and coastal area protection. How do we know that the products are what the different use sectors need and can use?

Summary

This chapter underscored the importance of a strong scientific basis, including both the natural and the social sciences, for ICM. In cases of scientific uncertainty the precautionary principle should be followed in ICM decision making and policy makers must avoid making decisions that have irreversible consequences, threaten the resource base over the long term, or foreclose the options of future generations.

Different types of data and information needed at each stage of the ICM process were illustrated, as were data and information needs associated with eight areas of concern in ICM: beach management and coastal erosion, wetlands protection, coastal hazards, nonpoint source pollution, sea-level rise, coastal and estuarine water quality, threatened and endangered species, and coral reef management.

Also discussed were a number of useful techniques to aid in ICM decision making, such as resource inventories and environmental profiles, mapping and geographic information systems, remote sensing, rapid appraisal techniques, environmental impact assessment, benefit-cost studies, risk assessment, resource valuation studies, and habitat assessment techniques.

The chapter also highlighted problems that exist in the interface between science and policy, drawing in particular from a 1995 analysis of this issue by the National Academy of Sciences in the United States. Cultural differences between scientists and policy makers were described, as were the problems posed by such differences in outlook, orientation, and patterns of behavior. Specific ways to improve science-policy interactions were presented.

....

Chapter 8

Formulation and Approval of an ICM Program

Introduction

In chapter 5, we discussed a number of factors that could affect the decision to undertake an ICM program in a given nation. This chapter assumes that such a decision has been made by an appropriate government body and outlines the issues to be faced and the steps to be taken in formulating an ICM program for the nation's coastal zone or a portion of it. (We prefer the term *program* for the product of this design effort. A plan, as we think of it, usually does not embody the institutional and legal aspects associated with implementation to the same degree as does a "program.") The chapter is divided into six sections, addressing the program formulation process; identification of initial problems, issues, and opportunities to be addressed; formulation of goals, objectives, and strategies; establishment of boundaries for the management area; assessment of existing institutional and legal capacity for ICM; design of the intersectoral-intergovernmental coordinating mechanism and the ICM office; and formal approval of the ICM program by participating governments.

The Program Formulation Process

Several options exist for the formulation of an ICM program. For example, it could be prepared by any of the following entities or a combination thereof:

- An international consultant or consulting team with experience in the coastal planning and management field (typically as part of a donor-assisted effort).

- A local in-country consultant or consulting group.

- A local university group.

• The agencies involved in the coastal and ocean management area.

• A group especially designated to undertake the ICM effort composed of representatives from the various government agencies (local, provincial, national), with one of the agencies serving as a technical secretariat.

Some variant of the first option has commonly been used in developing countries, largely because in-country capacity in the ICM field is still lacking in many
instances. For example, Fiji's national environmental strategy was prepared by a
consultant from IUCN, an international NGO (Fiji 1993). Although such international consultants can be very important and useful in the insights they bring to
the ICM process and the lessons they can draw from relevant experiences in other
countries, the program they produce may not have the same legitimacy or stand-
ing as one that has been produced "in-house." Moreover, training and capacity
building of local expertise will be limited if there is sole reliance on outside
experts. And, of course, outside experts will not have the detailed knowledge of
the country that local people have.

The appropriate mix of participants in the formulation of an ICM program will
vary according to the peculiarities of each case, particularly the extent to which
some ICM expertise is already present in the country. In our view, the last option
in the foregoing list—a group especially designated to undertake the ICM effort
composed of representatives from the various government agencies, with one of
the agencies serving as a technical secretariat—supplemented with local consultants or university personnel and international consultants, if needed, provides a
good basis. Such a combination has several advantages: the agencies involved will
develop a proprietary view of the ICM process and hence will be more likely to
adopt and implement the program; the international consultants will contribute
their conceptual frameworks and comparative experience; and the local consultants and/or university personnel will continue to build their capacity in the field
through hands-on experience (assuming their existing capacity is limited).

The program formulation team will need to be interdisciplinary in composition,
with competence in planning, resource management, resource economics, coastal
processes (geological), marine biology, physical oceanography, and legal, regulatory, and institutional aspects. The precise expertise needed in the group will, of
course, depend on the mix of problems the ICM program is to address. Ideally, the
team should be led by someone who has a "generalist" perspective, who can
mobilize the contributions of the various disciplines, and who represents a level
of government above the level of the individual ministries or departments—for
example, a national planning office, the prime minister's or president's office, or
a national office of management and budget—thus helping to endow the ICM
effort with a stamp of approval from a high political level and avert interministerial conflicts.

The program formulation team should be charged with developing an ICM program for the nation within a specified, relatively short length of time, such as one year. Of course, the time required will depend on the extent to which preliminary work has been done and the scientific information already available regarding coastal areas, resources, and communities. If little or nothing has been done by way of coastal planning and management, an additional year or more could be required. In terms of the functioning of the team, it is desirable to have the individuals "seconded" to the team on a full-time basis during the program development process. The team will coalesce faster and mutual learning and respect will develop more rapidly if participants work together on a full-time basis.

Identification of Initial Problems, Issues, and Opportunities: Setting Priorities

No aspect of the ICM program formulation process is more important than the setting of priorities. This is the means by which the program's initial targets are selected. What are the most urgent coastal management problems facing the nation in question? What problems pose the highest risks in the coastal zone? What coastal and ocean resources are most in need of improved management? Where do the greatest opportunities lie for new or expanded economic development? Do certain coastal areas demand earlier attention than others? All these questions need to be examined early in the program formulation process.

Development and Analysis of Coastal Profiles: Assessment of Issues

As discussed extensively in chapter 7, the formulation of an ICM program must be based on a solid scientific and technical assessment of the coastal and ocean problems facing a particular nation or region. In general, systematically derived information from the natural sciences, the social sciences, coastal engineering, and institutional-management analyses is essential. The mix of disciplines and diagnostic work needed will, of course, vary according to the mix of coastal issues being addressed. As noted in the World Bank's guidelines for ICM (World Bank 1993), information and data of the following types are needed in the ICM program formulation process:

Coastal Resource Base

- Existing coastal resources (beaches, wetlands, estuaries, mangrove forests, coral reefs, etc.)

- Present use of coastal resources (fishing, recreation, mining, etc.)

- Present status of coastal resources (including qualitative assessments of the health of the coastal ecosystem)

- Potential for present and future use

Social Organization in the Coastal Zone

- Existence and character of human settlements (villages, towns)

- Economic basis for human settlements

- Existence of indigenous peoples and their traditional coastal activities

- Social issues

Existing Environmental and Resource-Related Programs

- Environmental regulatory programs

- Fisheries management programs; other resource management programs

- Protected area programs

- Beach and erosion management programs

- Pollution control programs

- Other environmental management programs

Institutional, Legal, and Financial Capacity

- Relevant national-level institutions

- Relevant regional- or provincial-level institutions

- Relevant local institutions

- Survey of legal authorities relative to coastal and ocean activities

- Existing capacity-building efforts, including those funded by external sources

Relevant information gathering and lesson drawing from analogous cases in other countries are also appropriate at this point (see chapter 5).

Problems and opportunities to be addressed first by the ICM program are likely to have been in the forefront of public and government attention for some time, and hence a fair amount of information and data about these issues will probably be available in various locations. As discussed in chapter 7, one of the challenges in formulating an ICM program is locating such data as do exist and bringing the data together in a useful form. The environmental and resource agencies of the national, provincial, and local governments may well have some or most of the needed information. To the extent that local universities have conducted research or monitoring studies of the coastal environment and its resources, they will also have relevant data and information. Donor agencies, development banks, intergovernmental organizations, and other sponsors of resource studies, inventories, and management approaches may also be repositories for data and information collected as part of programs they have funded.

Considerable emphasis has been placed on this part of the ICM process by Chua Thia Eng and his colleagues in the ASEAN-US project in Southeast Asia, which involved ICM work in Malaysia, Thailand, Indonesia, Brunei Darussalam, the Philippines, and Singapore. In their early work with six nations in that region, they emphasized the importance of relying on information in hand and using rapid appraisal techniques to permit interpretation and extrapolation as necessary (Chua and Scura 1992). Occasionally, the lack of even the most basic data about a coastal issue or resource might require a new field effort to obtain the minimum necessary information, but this is likely to be the exception rather than the rule. Furthermore, an ICM program could include, as part of its initial management phase, the acquisition of data and information to fill particularly important gaps.

Programmatic Scope: One Issue or Multiple Issues?

A typical issue early in the formulation of an ICM program is the question of programmatic scope. How broad should the ICM program be? Should it address the full range of coastal issues or should it be limited to a smaller set of issues more immediately relevant to the particular coastal areas involved?

In some cases, one particular issue or set of issues may have driven the decision to embark on the ICM program. In the mid-1980s, for example, a pressing coastal erosion problem motivated Sri Lanka to begin an ICM program. Obviously, in such cases, the triggering problem will have a prominent place in the initial ICM program. But invariably, other coastal issues relate to the triggering issue and thus will need to be built into the ICM program as well. In the case of coastal erosion, aspects such as beach and dune management, sea-level rise, management of coastal structures, management of coastal sand mining, and planning and management of shoreline use all must be addressed in a properly integrated approach to the coastal erosion issue. Hence, although the initial motivation for ICM may be narrow and manifested in a single issue, the interconnectedness of coastal and ocean phenomena and the interplay between various uses and users will inevitably require that a larger set of coastal issues and potential adverse effects be factored into the decision-making process.

The availability of resources (funding, personnel) is always a key factor in determining the scope of an ICM program. If adequate resources in the form of staff and money are available, the program should be as comprehensive and inclusive as possible. If resources are limited, however, as is more typically the case, the program's initial focus might be on one or several core coastal and ocean issues only. Depending on the nature of the coastal zone in question, these might include the following:

- Protection and management of coastal resources such as beaches, dunes, and coral reefs.

- Protection of important coastal habitats such as wetlands, mangrove forests, sea grass beds, and mudflats.

- Protection of coastal water quality.

- Promotion of a coast-dependent economic use (e.g., tourism, aquaculture).

- Improvement of public access for recreational purposes.

- Reduction in losses of life and property due to coastal hazards (storms).

- Management of coastal beach erosion.

- Management of the use of coastal space, including the restoration of urban waterfronts.

Ultimately, ICM programs should address coastal fisheries issues as well, but this often takes some lead time and a considerable amount of preparatory work. The traditional view prevalent in many countries that fisheries management should be dealt with as a separate sector will take some time to overcome. In particular, fishery interests will have to come to see that their goals (sustainable fisheries) can be achieved only through a comprehensive, integrated management program—one that involves management of coastal water quality and coastal habitat as well as management of fishing effort, stock abundance, and the like.

Beyond its role in protecting coastal resources and regulating coastal uses, ICM should be seen as a tool for facilitating and promoting sustainable economic development in the coastal zone. Opportunities could include tourism, offshore sand and gravel extraction, aquaculture, and underutilized fisheries. As one of its

functions, ICM can assist in determining the environmental and economic feasibility of potential new uses and activities in the coastal zone and provide positive
incentives and encouragement for such developments, since ICM is aimed at max-
imizing the benefits that flow from coastal and ocean resources and ensuring that
those benefits continue on a sustainable basis.

Thus, the answer to the question of whether to start with one issue or multiple
issues must be "It depends"—on the problems at hand and on the triggering forces
motivating ICM. Although it may be appropriate and politically wise to begin
with one issue that has political salience, the essence of ICM is that other interconnected issues must eventually be added.

Geographical Scope: National Approach or Pilot Project?

Early in the formulation of an ICM program, the question of geographical scope
of the program must also be addressed. Should the initial ICM program be a pilot
effort applied to only a small segment of the country's coastal area where, for
example, the most severe coastal problems are in evidence or the most feasible
economic opportunities are present, or should ICM be developed for the entire
coastal zone? In some cases, it is clearly appropriate to begin in a single part of
the coastal zone, due to either the urgency of the coastal issues in that area or the
readiness of the local community to improve the situation or take advantage of a
new opportunity. Indeed, in a number of cases, the initiative and the early action
to move toward better integrated coastal management has come entirely from the
local community, with the national government entering into the process later.
Obviously, a pilot ICM program built on such local initiatives makes sense. It also
makes sense to begin on a small-scale, pilot basis if resources (staff and funding)
are in limited supply and if, indeed, certain coastal areas demand earlier attention
than others. In Colombia in the early 1980s, for example, the Atlantic coastal
region in the northern part of the country was coming under strong pressure for
both tourism and industrial development, whereas the Pacific coastal region
(largely jungle) in the western part of the country was experiencing little or no
such pressure. Hence, it made sense to begin the ICM effort on the Atlantic coast
(Knecht et al. 1984).

Starting an ICM program on an incremental basis, on the other hand, carries
some problems and risks. First, the pilot area may not encompass a sufficiently
large area to capture the major influences, both physical and socioeconomic, on a
particular coastal system. Second, if the pilot program takes an appreciable length
of time to be completed and implemented or its positive benefits are slow to be
seen, momentum and enthusiasm for ICM as a whole could be lost. Third, while
effort is being devoted to a particular pilot site, significant pressures may build up
to develop other parts of the coast in an ad hoc and rapid manner, unfettered by
national coastal policies and guidelines. Fourth, a common assumption underly-

ing the pilot program approach is that the program will "scale up" to include other coastal regions or the country as a whole. There is not, however, much positive evidence in the literature for the scaling up assumption. For example, in the ASEAN region where a number of ICM efforts were begun in the late 1980s through the ASEAN-US program, only some of the pilot programs actually scaled up to involve other regions or parts of the coastal zone (Chua and Paw 1996).

In our view, it is generally wise to proceed from the bottom up and from the top down at the same time. That is, it is important both to develop national policies and guidelines applicable to the nation's entire coast and to conduct more intensive, focused pilot efforts in a particular coastal site or sites. Efforts at the two levels should reinforce each other as well as prevent unfettered development of coastal areas not covered in the pilot projects. Involvement of the national government, at least in setting broad policies and guidelines for use of the coast, should prove useful in the administration of the pilot effort. National-level support and interest in coastal issues should be helpful in at least three ways: (1) by making national-level agencies operating at the pilot site more willing to participate and cooperate in the ICM effort, (2) by helping to ensure continued funding, and (3) by making it more likely that if the pilot effort is successful, it will be replicated in other coastal regions of the nation.

The Importance of Having Early Positive Results

In the initial determination of the scope and content of an ICM program, consideration should be given to the desirability of having early positive results, even from a fledgling ICM program. Because many of the program's goals and objectives may take several years to achieve, having some shorter-term goals can be a distinct advantage. The participants in the ICM program, the politicians who support the program, and the broader public being served all need to see that progress is being made—that some of the coastal problems are being successfully addressed and that the new management program is beginning to bear fruit. With this need in mind, it is generally a good strategy to include some issues on which early progress is likely to be made.

Formulation of Goals, Objectives, and Strategies

Goal setting is of major importance in ICM. Without clear, unambiguous goals and objectives, a management program is without direction. Furthermore, evaluation and midcourse improvements become impossible if the goals a management program is striving to achieve are ambiguous or nonexistent. Indeed, such goals should be not only clear and unambiguous but also quantitative and measurable to the maximum extent possible. In our judgment, setting such goals is one in a series of steps that can add needed rigor and discipline to the coastal management process. Other steps include the following:

- Setting goals that reflect what is needed, not simply what can be done at present.

- Seeking cause-and-effect relationships to link management actions and program goals.

- Creating the capability to systematically monitor and evaluate the extent to which the agreed goals are (or are not) being met.

Goal setting for an ICM program is a very important policy issue—one to be formally decided by the legislative body involved (see the discussion of formal government approval later in this chapter). However, the groundwork for that decision making is laid during program formulation. After the target coastal issues have been selected by the program formulation team and the necessary diagnostic analyses have been completed, the next step is to decide exactly what is to be achieved with respect to each of the targeted issues—to what degree the coastal water quality is to be improved and by to what extent a degraded wetland is to be restored and by when, what forms of tourism development should be promoted and with what kinds of infrastructure, and so forth. Provisional goals of this type should be drafted for later debate and approval by the formal decision-making bodies. Ideally, a policy analysis of each issue should be followed by a series of options for goals to be achieved. Generally, one option should be singled out as the "recommended option" and a discussion of the rationale for its selection should be presented.

Establishment of Boundaries for the Management Area

Designation of the management area, another key task of the program formulation process, involves setting the geographical boundaries within which ICM will take place. Exactly where the boundaries are set, of course, depends on what is being managed—that is, on the programmatic and geographical scope of the ICM program. If a major emphasis is to be placed on managing coastal water quality and if much of the marine pollution originates in upstream watersheds, the inland boundary of the management zone may need to be set somewhat farther inland than might otherwise be the case. Similarly, if the program is to address ocean resource issues related to fisheries management and/or offshore oil and gas development, its seaward boundary should extend far enough into the Exclusive Economic Zone (EEZ) to embrace the targeted activities.

In the early days of coastal zone management, management zones tended to focus mainly on the land-water interface—the shoreline—and the immediately adjacent land and water. Coastal zones for management purposes tended to be narrow, often with their landward width limited to a thousand meters or less and their seaward width coterminous with the nation's territorial waters. These early programs were usually limited in scope as well. In the United States, for example, the emphasis was on land use in the narrow zone abutting the shoreline and on protection of coastal resources such as beaches, dunes, and wetlands. This relatively narrow view of coastal management was partly due to the fact that U.S. laws assigned the responsibilities for protecting coastal water quality and for administering the federal coastal zone management program to different agencies—the Environmental Protection Agency and the National Oceanic and Atmospheric Administration (NOAA), respectively. Clearly, given the intimate connection between coastal water quality and coastal land use, they must be managed in an integrated fashion despite the fact that different laws and agencies are sometimes involved. Coastal nations that are newly organizing their resource management programs would do well to avoid this additional complexity and assign both responsibilities to the same government unit.

Additional factors that should be considered when setting the inland and seaward boundaries for ICM are discussed in the sections that follow.

Inland Boundary

In a conceptual sense, the inland portion of the coastal zone should be large enough to encompass all land the use of which could affect the resources and waters of the coastal zone. In an ideal world, the entire watershed or catchment basin would be included in the management zone, since much of what happens in the watershed eventually affects the coastal ocean through runoff into streams, rivers, and estuaries that eventually flow into the sea. However, it is often politically difficult to extend coastal management authority to great distances inland. A common way to circumvent this difficulty is to accept a narrower zone for coastal management purposes and to use other laws and programs to achieve the necessary control in the upland portions of the watershed. Many governments, too, have found it more practical to use an existing administrative or political subdivision boundary rather than attempt to define a new boundary. For example, some entities, such as the state of Massachusetts in the United States have designated the coastal road nearest the water's edge as the inland boundary of the coastal zone. Others have used fixed distances, such as a thousand yards from the shoreline (California) or the inland boundary of the coastal county (Washington State). Three considerations are important in setting an inland boundary: it must be politically feasible; it must be administratively workable; and it must embrace the bulk of the activities to be brought under the aegis of the ICM program.

Seaward Boundary

As discussed in chapter 2, some 27 percent of all respondents reported their nation had adopted a seaward boundary of 12 nautical miles (21 percent) or 3 nautical miles (6 percent), since this is the area in which the coastal state has essentially complete jurisdiction and control, save for innocent passage of ships. Most activities related to the use of ocean resources also take place in their zone. However, with improving technology, offshore oil and gas development is now taking place in deeper water and at greater distances from the shoreline, causing some coastal nations (8 percent of all responses in our sample) to extend their coastal planning and management concerns beyond the territorial sea and well into their 200-nautical-mile exclusive economic zones. Seventeen percent of the survey responses reported their nations had set the boundary at an arbitrary distance from the tidal mark; 23 percent reported that the boundary varied according to use; and 15 percent reported that the seaward boundary had not yet been determined.

Where nations set their seaward boundaries will depend on the emphasis they wish to give the marine area in their ICM programs. If fisheries operations or offshore oil and gas development are to be managed as an integral part of the ICM effort, clearly the seaward boundary should be set at a sufficient distance to include the marine areas in which these activities take place. The outer edge of the 200-nautical-mile EEZ may be appropriate in these cases. Nations have the option, of course, of setting the seaward boundary at a lesser distance from the shoreline—say, 12 nautical miles—with the intent of managing the more distant marine activities under existing sectoral management regimes brought within the ambit of the ICM program by means of a specially created coordinating mechanism. As mentioned earlier, the same approach could be used to extend the reach of the ICM program to activities in the upper portions of coastal watersheds.

A tension present in the setting of management boundaries for ICM should be taken into consideration. On the one hand, narrow boundaries, especially on the landward side, serve to heighten public awareness of the special character of the coastal zone and its need for special management. This is particularly true when management boundaries coincide with physical manifestations of the coastal area, such as sandy beaches, dunes, wetlands, harbors, estuaries, and the like. On the other hand, a broader landward management zone, although it will subject more activities to coastal management, will extend ICM into areas that are not visibly coastal in character, possibly making public understanding of the concept of coastal area management more difficult.

Finally, it should be stressed during the program formulation process that the boundaries of the coastal zone will define an area that is being given *special planning and management attention*—in terms of both regulation and possible development assistance and technical advice. Inclusion in the coastal zone should be seen as something positive. For example, farmers and others involved in agriculture in the coastal area would probably be eligible for various kinds of assistance to help them reduce their contribution to nonpoint-source pollution of coastal streams, rivers, and estuaries. Indeed, with respect to environmental management, present trends in developed countries, at least, are away from the traditional "command and control" regulatory approach and toward greater use of market-based approaches such as incentives, subsidies, and tax advantages.

Assessment of Existing Institutional and Legal Capacity for ICM

Virtually every coastal nation, from the smallest Pacific island governments to large developed nations, has some sort of coastal management activity already in place. Typically, these involve programs for management of fisheries activities, protection of sensitive habitats such as wetlands, mangrove forests, and coral reefs, and, perhaps, management of a system of national parks. In addition, more and more often, there is a department of the environment or an environmental unit responsible for dealing with air and water pollution and solid waste. These programs are typically organized on a sectoral basis, with separate departments or ministries for fisheries, natural resources, the environment, and so forth. Local staff are often supplemented by visiting consultancies of various durations, and the nature of the programs undertaken is often influenced by the wishes of donor institutions or nations.

Thus, an early step in formulating an ICM program is assessing the existing capacity of the nation or coastal community to undertake the program. Obviously, it is imperative to identify (and, indeed, reinforce and build on) program elements that are performing well and to pinpoint weaknesses and gaps. Six aspects of the existing management programs need special scrutiny:

1. Adequacy of the laws, decrees, and regulations under which the present management programs operate.

2. Adequacy of administration and execution of the program.

3. Adequacy of access to needed expertise (legal, scientific and technical, public administration, economic).

4. Adequacy of available resources (funding, trained staff, facilities).

5. Effectiveness of the programs (enforcement, compliance, etc.).

6. Public participation in the programs (existence of public hearings and an appeals mechanism, transparency of the process).

The aim of this assessment is to determine the extent to which existing management programs and the institutions operating them can play a role in ICM. In devising an ICM program, it is desirable to build on successful existing programs to the maximum extent possible and to minimize the creation of new institutions. Hence, reliable information regarding the effectiveness of existing programs is crucial at this stage. It is important that the assessment be both objective and sufficiently detailed to clearly identify the reasons for any observed shortcomings. For example, are the laws improperly written? Is enforcement too lax? Is the staff shorthanded or inadequately trained? Are necessary facilities lacking? Is the necessary scientific and technical advice lacking?

A separate assessment needs to be made of existing mechanisms to coordinate or harmonize the coastal and ocean activities of several departments or ministries. A national planning or development office, a national management and budget office, or the president or prime minister's office sometimes performs such functions. If such a mechanism exists, its effectiveness in identifying and resolving coastal and ocean disputes, overseeing and coordinating government-wide planning, and integrating policy should be assessed. As discussed in chapters 2 and 6, a key institutional element of a successful ICM program is an effective interagency coordinating mechanism. If one already exists or can be fashioned from another program that is working, a new mechanism need not be established. Note, however, that considerable care must be taken in judging the effectiveness of an existing coordinating mechanism. Generally, the convening organization (the parent body) will want the mechanism to be seen as effective, and some participating departments or ministries may also believe that this is in their interest. Other departments, however, may have a substantially different view of the success of the mechanism. Therefore, a range of different perspectives may need to be sampled in order to get an accurate view.

Design of the Intersectoral-Intergovernmental Coordinating Mechanism and of the ICM Office

The coordinating mechanism is the key institutional element of a successful ICM program. A critically important function of an ICM program is effective interagency coordination, the aim of which is to harmonize the policies and activities of the separate, specialized line agencies. As discussed in chapters 2 and 6, given the nature of the jurisdictions involved, this coordination is generally required in two dimensions—horizontally, among various sectoral agencies (fisheries, land use, coastal erosion management, oil and gas, etc.), and vertically, among various levels of government (national, provincial, regional, local). The functions of the coordinating mechanism are summarized in table 8.1. Although other organizational options exist, there are advantages in having the office that houses the coordination mechanism also serve as the lead agency in administering the ICM pro-

Table 8.1. Functions of the ICM Coordinating Mechanism

• Promote and strengthen interagency and intersectoral collaboration

• Reduce interagency rivalry and conflict

• Minimize duplication of functions of line agencies

• Provide a forum for conflict resolution among government sectors regarding coastal and ocean uses and, in the process, promote policy intergration

• Monitor and evaluate the progress of ICM projects and the overall program

gram. As such, it would direct the implementation and operation of the ICM program as well as oversee the harmonization and coordinating functions outlined here.

As discussed in chapters 2 and 6, several institutional options exist for the ICM coordinating mechanism. Among them are (1) use of an existing national planning, budget, or coordination office; (2) formal establishment of an interministerial or interagency coastal coordinating council or committee; and (3) designation (by the chief executive) of an existing ministry or department as the lead agency for ICM with responsibility to operate the coordinating mechanism and generally oversee the ICM program. Further options are discussed in Boelaert-Suominen and Cullinan 1994. In our judgment, the formal establishment of an interministerial or interagency coastal coordinating council is preferable, especially if it is chaired by someone representing a level of government above that of the ministries or line sectoral departments. The use of an existing national planning, budget, or coordination office, though having some efficiency, might not be as effective as a dedicated interministerial council, since an existing office would presumably have a number of other responsibilities and duties. Designating an existing department or ministry as lead agency could also lead to conflict between it and the sister agencies it is to coordinate.

As discussed in chapter 6, it can be useful to have incentives available to encourage the cooperation of participating entities in the coordinating process. Such incentives can include financial assistance, the addition of staff, legislative mandates for joint administration, and so forth.

Consideration of New Management Measures

In assessing the existing coastal management program as a part of formulating the ICM program, it may be determined that gaps exist, present management measures have not been effective in meeting their goals, or new needs have emerged. New management approaches may therefore need to be considered. Although the new measures clearly have to be consistent with the laws and traditions of the nation involved, they may not have to be invented totally anew. With nearly thirty years of experience already amassed in coastal zone management worldwide, it is likely that most coastal resource management problems have already been faced in one coastal country or another. Certainly, a wide range of management measures are already in place with respect to coral reefs, mangrove forests, sea grass beds, coastal wetlands, beaches and dunes, and estuaries. Similarly, a host of regulatory programs exist to deal with the placement of hard structures to control coastal erosion and to require new construction in the coastal zone to be set back certain distances from the water's edge. Even coastal zone problems only now coming over the horizon, such as accelerated sea-level rise associated with global warming, are already being dealt with in some coastal zone management programs. Our point here is that a rapidly growing body of expertise and information is being amassed by the scores of nations now involved in coastal management. Thus, sometimes it is only a question of knowing whom to contact in order to learn of some relevant experience and practice that may considerably help a particular nation in its ICM program design. Electronic mail discussion groups such as Coastnet, organized by the Coastal Resources Center at the University of Rhode Island, may be quite useful in this respect.

The Resulting ICM Plan

The ICM program formulation phase typically ends with the finalization of an ICM plan. The plan should contain a clear and understandable description of the ICM process being recommended and how it is proposed to function. In general, the plan will include the kind of elements shown in table 8.2.

Ideally, the main elements of the ICM plan resulting from this collaborative process will be well known to most of the potentially affected interests by the time the plan is completed. Representatives of important coastal and ocean users' groups and other stakeholders most likely will have participated in the discussions that led to the decisions embodied in the ICM plan. Preferably the planning process will have been sufficiently open and transparent to allow all points of view to be heard. In any event, with completion of the ICM plan, the stage is set for the formal approval process and implementation of the program.

Formal Approval of the ICM Plan by Participating Governments

ICM involves a multipronged, strategic approach to resolve coastal problems and address coastal opportunities. It can be expected that a range of management actions will be proposed in the ICM plan. Some will involve strengthening of existing institutions and empowerment of local authorities, and others will involve

Table 8.2. Elements of an ICM Plan

• A clear description of the coastal area to be managed

• A clear description of the problems to be addressed and the goals and objectives to be sought

• A clear description of the policies and principles that will guide the program

• A statement of the initial management actions to be taken

• A description of the proposed institutional arrangements, including assignment of responsibility for various parts of the programs, e.g., the interagency coordinating mechanism and supervision and support for the ICM program as a whole

• Funding and staffing requirements

• A listing of formal actions needed for official adoption of the plan and a suggested timetable for completing those actions

ways to increase the effectiveness of current resource management programs and the creation of new management efforts to fill important gaps. The ICM plan will recommend goals, policies, and strategies for management of the country's coastal and ocean resources and space. It probably also will recommend new institutional arrangements among the various government agencies involved and suggest ways to create an effective interagency coordinating mechanism. Securing approval of an ICM plan that contains strong policies, meaningful goals, and effective coordination and harmonization mechanisms is not always an easy task.

We believe that prospects for timely approval by participating governments will be increased if the following conditions are met:

• The proposed ICM program is succinctly described in clear and understandable terms. (What is it? Why do we need it? What will it do?)

• The benefits (economic, environmental, social) that will flow from the ICM program are described in tangible and meaningful terms.

• The proposed program is clearly and visibly endorsed by the users of the coasts, the public, and interested NGOs.

• Key legislative (parliament) and government leaders have received periodic reports on the progress of the ICM program formulation effort from its inception.

• The costs (political, financial, administrative) of implementing and operating the ICM program are clearly spelled out and ways to cover such costs are suggested.

In laying out the timetable for development of an ICM program, care must be taken to anticipate and include the vagaries of legislative (parliamentary) calendars. Some such bodies meet only every two years or consider legislation of one type or another only at fixed times during particular sessions. Obviously, it cannot be assumed that a national or provincial legislature is continuously open to receive and act on new ICM legislation—far from it. In most cases, hearings are required and favorable committee action is needed before full legislative consideration of a new measure can take place, and, of course, in many countries this must occur in two houses of parliament, at least doubling the amount of time required. An initial high level of enthusiasm for ICM can be substantially reduced if legislation necessary for the program's implementation languishes in the legislative process for two or three years. This suggests that the need for new legislation or for legislative changes should be kept to a minimum.

Indeed, it is useful to include interim alternative arrangements that can be put in place while any necessary legislation is under consideration. For example, pending the legislative creation of an interagency coastal coordinating council, the nation's chief executive (prime minister or president) could, by executive order, establish such a council and put it into operation, and as long as legislation is submitted to create such a council by law, it is unlikely that anyone would challenge the legitimacy of the interim arrangement. Depending on the precise nature of the ICM program being recommended, most aspects could probably be implemented administratively and confirmed later by appropriate legislation. Thus it would be desirable to build the possibility of interim implementation into the ICM program from the outset.

Summary

This chapter outlined the decisions coastal policy makers face as they formulate an ICM program and get it approved. A common first step is the assembling of an interdisciplinary team to design the program and, generally, to produce an ICM plan. Although the appropriate mix of participants in ICM program formulation will vary according to the specifics of each case, a good model, in our view, is a team composed of representatives from the key coastal and ocean agencies, supported by in-country local consultants, and perhaps aided by international consultants who can contribute knowledge about comparable cases. ICM formulation efforts that depend solely on external consultants may not be successful in the long run.

The next step is identification of coastal problems through diagnostic analyses and the setting of priorities. One important decision at this stage involves the programmatic scope of the ICM program—how broad it should be and whether it should address the full range of coastal issues involved or be limited to a smaller set of issues. Another key issue is the program's geographical scope—whether it should encompass the country's entire coastal zone or focus on one or more localities. Here, we counsel that it is generally wise to proceed from the bottom up and from the top down at the same time, since inevitably both the national and local levels of government will need to be involved. Also highlighted is the importance of showing early positive results in order to maintain momentum for the ICM effort.

The next important steps are the setting of goals for the management program and boundaries, landward and seaward, for the management area. Factors to consider in boundary setting were delineated, with emphasis on whether to have the ICM program encompass a wide geographical area or a narrow one. Further steps in formulating an ICM program are assessing the existing legal and institutional capacity and designing an ICM coordination mechanism and administrative office. These key decisions are likely to have a significant influence on the ultimate success of the ICM effort. The chapter ends with some suggestions for enhancing the prospects for timely approval of the ICM effort by participating governments.

Chapter 9

Implementation, Operation, and Evaluation of ICM Programs

Introduction

This chapter discusses three very important stages in an ICM program: implementation of the program; operation of the program; and evaluation of the program's performance. Once the ICM plan has been formulated and formally approved by the participating governments, implementation of the program can begin. When implementation is complete and the new laws, coordination arrangements, and procedures are in place, the program becomes operational; that is, it begins to perform the functions for which it was designed. The operation phase of the program continues indefinitely or until a new program replaces it. Periodically, during the operational life of an ICM program, the program's performance should be evaluated and improved as necessary.

The Implementation Process

Once an ICM program has been formulated for a country's coastal area (or a portion thereof) and the necessary government approvals have been secured, the next step is implementation of the approved program, that is, actually putting it in place. We use the term *implementation* to refer to the actions that must be taken to put an ICM program into operation. These typically include such onetime or start-up activities as enactment or amendment of legislation, preparation of new or revised regulations and procedures, formal establishment of any new institution or interagency coordinating mechanism, securing and training of any additional personnel that may be required, and the like. Note that although some authors include the operation phase of an ICM program in implementation, we choose to limit implementation activities to those onetime tasks necessary to prepare a new program for its long-term operation phase.

Even though problems and difficulties should have been identified in the program formulation phase and solutions worked out before the program is formally approved, implementation is often a difficult and time-consuming process. Some of the reasons are as follows:

- Securing the necessary legislative and legal changes can be controversial and time-consuming.

- Obtaining the needed resources (additional funding and staff) can be difficult.

- Policy gaps must be identified and satisfactorily filled.

- If a great deal of time has elapsed since the program was formulated, changes may have occurred that invalidate part of the proposal.

- Recommended institutional changes, including creation of the interagency coordinating mechanism, could be slow in coming because of bureaucratic inertia and resistance to change.

The overall timetable for the ICM effort should be prepared with these kinds of problems and potential delays in mind, although if at all possible, the start of the program's operational phase of the program should not be unduly delayed by factors such as these. Ways of getting around these problems, at least on an interim basis, are discussed later in this chapter.

This section is divided into four parts, giving practical guidance on securing the necessary legal, legislative, and regulatory changes; putting the institutional arrangements in place; obtaining the necessary resources; and involving the affected interests in the process.

Securing the Necessary Legal, Legislative, and Regulatory Changes

As mentioned in chapter 8, it is desirable to keep legal and legislative changes to a minimum in the ICM plan. If there is any way to accomplish a given goal other than enacting a new law or changing an existing one, it should be pursued. Legislative change is time-consuming, and outcomes are sometimes difficult to predict. Anyone who has followed the legislative process knows that the legislation that finally emerges (if, indeed, it does) can be far different from what was initially proposed. In the worst cases, those originally suggesting legislation must reverse their positions and oppose it because of consequential changes made during the legislative process.

Sometimes, however, new legislation is needed. To be effective an ICM program may need new legal authority to regulate or manage coastal activities that heretofore had not been regulated. New authority may be needed, for example, to establish and enforce set-back lines requiring newly constructed buildings to be located a safe distance behind the primary dune lines. In such cases, the legislation may move along more quickly if care is taken to show legislators exactly why the new laws are needed, how they fill specific gaps in the national framework for resource management, and why they are a critical element in the new ICM effort. Moreover, it is best if the legislative needs of the evolving ICM program are shared with key members of the legislature or their staffs as soon as they are recognized, since the element of surprise is usually not appreciated by members of the legislative branch of government.

Putting the Institutional Arrangements in Place

One of the more sensitive issues in ICM involves institutions (national, state, and local government agencies) and the relationships among them. It is well known that agencies jealously guard their missions and the responsibilities and resources that accompany them. Indeed, survival of the agency depends on its keeping the mission and resources intact (or better yet, expanding them). Anything that threatens the mission or the resource base tends to be resisted with great vigor and tenacity. Such fights over "turf" (mission) can waste resources and can extend over long periods of time, sometimes persisting under the surface well after they seem to have been settled. These bureaucratic realities often make meaningful collaboration and cooperation among government agencies difficult, if not impossible. Yet interagency cooperation and policy integration are at the core of a successful ICM program. The need for effective collaboration is more acute in ICM because of the existing sectoral fragmentation wherein each ocean activity tends to be controlled by a separate government agency and each level of government tends to have its own set of coastal and ocean agencies. Successful integrated coastal management requires that all these agencies and the activities they oversee be brought under the influence of a single coordinating or harmonizing mechanism or process. Designing this mechanism or process is a critical task of the program formulation phase.

In the typical pre-ICM situation, each sectoral agency has a specific (though usually somewhat narrow) set of responsibilities and a relatively free hand to go about meeting them. In the absence of a formal coordinating mechanism, the extent of interactions among agencies and the activities they oversee depends primarily on personal relationships and chance. Ad hoc arrangements of this type may be adequate to a point—until there is sufficient use of the coastal zone and adjacent ocean by a variety of different interests to require a better organized approach to coastal management.

The Coordinating Mechanism

With the creation of a coordinating or harmonizing mechanism (hereinafter called the interagency coordinating mechanism), personnel in each of the hitherto "independent" agencies may perceive their agency as losing some power or responsibility to the coordinating mechanism, possibly a coastal coordinating council. Although technically this may be true in a narrow sense (the coastal coordinating council may have to concur with the agency's decisions in large and consequential cases involving coastal erosion use), another way to describe the new arrangement is to say that the agency, as a member of the coastal coordinating council, now has a voice in *all* important coastal and ocean issues of the nation, not just those in its narrow issue area. Individual agencies should see the establishment of a coastal coordinating council and their membership in it as a way to achieve the following:

- Increase their effectiveness in achieving their own missions and goals.

- Helping government as a whole meet broader, overarching national or regional goals.

- Increase the efficiency and effectiveness of government as a whole.

If a coastal coordinating council has been selected as the interagency coordinating mechanism, care must be taken to create it in a way that maximizes its chances of success. The following attributes, in our judgment, can contribute to a coastal coordinating council's ultimate effectiveness:

1. The council is composed of directors of the coastal and ocean agencies or their formally designated policy-level alternates.

2. The council is chaired by an appointee of the chief executive (president or prime minister) that is of a higher rank than the members themselves.

3. The council has access to sufficient resources to hire the staff assistance and technical support needed to carry out its work. (Creation of an operating arm of the council or of a technical secretariat is important in order to implement the council's decisions and carry out the day-to-day work of coastal management.)

4. The council was formally established by means of the usual legal or legislative procedures used in the country (e.g., legislation, presidential order, or decree).

In our experience, failures of such coastal coordinating councils can be traced to the following problems:

1. Key coastal and ocean decisions begin to be made outside the council's deliberations.

2. Council members lose interest in attending meetings and send low-level representatives in their place.

3. The chief executive loses confidence in the council.

4. The council is unable to obtain adequate staffing and financial resources.

Hence, for a coastal coordinating council to be effective, it must be firmly and visibly supported by the highest levels of government (president or prime minister and line ministers or department heads), it must have a clear and legally supported mandate, and it must be given the resources to do its job.

The ICM Office or Administrative Entity

The other important institutional issue associated with establishment of an ICM program relates to supervision of and support for the program. Supervision and support tasks could include the following:

- Establishment and oversight of new zonation schemes or other new management programs that do not fit within the mission of existing line agencies.

- Oversight of the environmental impact assessment process.

- Coordination of planning activities of the various line agencies.

- Supervision of performance monitoring and ICM program assessment.

- Training and human resources development.

- Transnational and transboundary issues: maintenance of relationships with adjacent countries and with related regional and international programs.

There are two options with respect to the institutional "home" for supervisory and support activities. The first and most obvious option is to combine these activities and the responsibilities of the interagency coordinating mechanism in a single organizational unit. The second option is to assign the supervision and support functions to a separate unit—perhaps an existing agency or department already involved in these types of activities. If a reasonably well staffed and competent agency is already in existence (e.g., a department of natural resources or a department of environment) and already performing some of these tasks with respect to coasts and oceans, consideration should be given to assigning at least the support tasks to that agency. Otherwise, combining all these ICM-related tasks and assigning them to a single new unit, such as an interagency coastal coordinating council, would appear to make the most sense.

Obtaining the Necessary Resources

In most countries, funding is already dedicated to some sort of coastal resource management program. ICM generally calls for changes in the way these funds are spent rather than the appropriation of large new sums of money. Some additional funding will, of course, be required for new activities such as conduct of coastal diagnostic work and operation of the interagency coordinating mechanism, as well as enhancement of performance monitoring and assessment activities. With this in mind, during the formulation of an ICM program, attention should be given to ways of making it financially self-sustaining; that is, the program should be designed to bring in sufficient funds to pay its operating costs.

One obvious way to make the program self-sustaining is to institute user fees for the use of various coastal and ocean resources or space. Alternatively, a portion of the fees collected in the leasing of government-owned shore land or submerged land could be allocated to ICM operating costs. Similarly, a portion of the national government's revenues from offshore oil and gas development or from access fees paid by foreign fishing interests could be used for this purpose. If the funding mechanism is related to coastal and ocean use, as such use increases and the need for better integrated coastal management grows, more funds will become available.

There will, of course, be additional expenses associated with the ICM program. Although most of the interagency group assembled for the ICM effort can be seconded from parent agencies or departments, a nucleus of new staff may be needed. Moreover, expenses will probably be incurred in obtaining and processing data and information needed in the program formulation process. Agreement by the relevant agencies and departments to provide needed data and information free of charge and in an expeditious manner will, of course, help in this regard. Costs will also be incurred in printing and disseminating the draft and final versions of the ICM plan and in conducting public hearings on them. Donor agencies or development banks may offer funding to assist developing countries in the program formulation phase, especially if the countries agree to design and implement ICM programs that will become financially self-sustaining over time.

Significant amounts of new funding could be required if the ICM program includes field projects designed to restore degraded wetlands or replant destroyed mangrove forests or to construct port and harbor facilities, beach recreation facilities, and the like. Costs associated with restoration or capital improvement projects such as these will generally have to be covered by special means such as capital raising through bond issues or donor contributions.

Involving the Affected Interests in the Process

A comprehensive ICM program will affect a wide range of nongovernmental coastal and ocean interests or stakeholders, including commercial and recreational fishers, those concerned with ocean recreation (scuba divers, boaters, etc.), offshore oil and gas developers, marine sand and gravel users, aquacultural interests, farmers using land in the coastal zone, environmental and conservation groups, coastal land developers, coastal tourism operators, military personnel, and so forth. Everyone who uses the coastal zone or the coastal ocean will potentially be affected by the ICM program. Unveiling a completed ICM plan to these stakeholders at the end of the program formulation process and asking for their comments, suggestions, and approval at that time is not the preferred strategy. Obviously, it is much better to involve representatives of these interests from the very outset of the ICM development process, whether as members of the ICM program formulation team or as members of a committee advising the team. One satisfactory approach would be to establish a users' advisory committee and invite its chairperson to become a full-time member of the program formulation team, thus providing for a more or less continuous interface between coastal and ocean users and the program development process.

As mentioned earlier, there are many good reasons for ensuring that those likely to be most affected by an ICM program are involved in its development. First, the users of the coastal zone and its resources are in the best position to be aware of both the coastal zone's problems and the opportunities that exist for its sustainable development. Their insight and experience should help inform the ICM program formulation process. Furthermore, those most intimately involved with the coastal zone are apt to know which approaches for strengthening management will work and which will not. And, of course, the coastal and ocean users will want to have a voice in recommending the policies and goals for the ICM program. Users' groups will be very interested in the concept of a coastal coordinating council, since they will want to ensure that their interests—fisheries, tourism, oil and gas development, and the like—are not compromised. Engaging users' groups in the design of the coordinating mechanism can help produce a politically viable proposal. Including their input early in program development can, of course, complicate and delay the process, but this is a risk that generally must be taken. Failure to fully involve such users' groups early in the process carries a much greater risk—the risk that when brought into the process later, they will oppose the entire procedure and the ICM plan that has been produced.

Management Tools and Techniques in the Operation Phase

We turn now to the "bottom line" in ICM—the operation of the program and the achievement of results. If the preparatory work has been done well and if a good ICM program has been formulated, approved, and implemented, putting the program into operation should begin to produce the desired outcomes. But operating a program that has been designed to be dynamic and adaptable is not a routine matter. Procedures must be in place to collect, assemble, and interpret incoming information—information that might come as part of a proposal for a new use of the coastal and ocean area or that might result from the performance monitoring activity. Based on new information, management measures may be altered or discarded and replaced with new approaches; new development assistance projects or restoration projects may be undertaken; or any of countless other changes to the program may be made.

Given the likelihood that much coastal management will continue to take place under the aegis of sectoral management agencies (fisheries, coastal land use, water quality, etc.), the emphasis of the ICM program will be on such areas as coordination, harmonization, resolution of conflict, integration of coastal and ocean policy, filling of gaps in management, and monitoring and assessment of performance. ICM programs can also play a key role in decision-making processes involving large, consequential coastal or ocean development projects. In such cases, the program's contribution should be to ensure that good science and appropriate analytical techniques are used in the decision making and that potential impacts on other coastal ocean users, coastal communities, and the environment are taken into account.

A considerable number of tools and techniques are, in principle, available to the ICM program in its operation phase. Decisions as to which management measures to employ will depend on the extent to which they are authorized by legislation, their technical suitability, their acceptance by the general public, and the like. Several of the more frequently used management techniques are described in the sections that follow.

Zonation

In zonation, the coastal area is divided into geographical zones for management purposes. On the land side, zones can be determined on the basis of distance inland from the water's edge, with the greatest restrictions placed on use in the zone closest to the water's edge and successively less restriction imposed in zones located farther inland. Turkey's coastal management law of 1992, described in box 9.1, is an example of this type of zoning scheme. Zoning can also be based on elevation above sea level, with the greatest restrictions placed on the lowest zones; see box 9.2 for an example of this approach in the Virgin Islands of the United States. Alternatively, zones can be designated according to degree of risk of storm surges and other storm effects, again largely determined by elevation and topography. On the water side, zonation is commonly used to separate uses and prevent conflicts among them as well as to preserve sensitive areas. Marine zoning practices are most advanced in the context of marine protected areas, as demonstrated by the case of the Great Barrier Reef Marine Park in Australia.

However, in some countries, particularly developed countries, the concept of zoning has taken on a negative connotation, viewed by some as an improper attempt by government to restrict the use of privately owned land. Zonation

Box 9.1. Zoning of Coastal Land in Turkey

Turkey's Shore Law (4.4.1990, Amendment 1.7.1992) defines three types of zones in order to protect the sea, lake, river shores, and shore strips and to utilize the shore for the benefit of society.

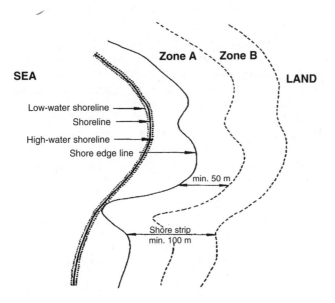

Shore

The shore is defined as the area between the shoreline and the shore edge line. It is illegal to excavate in this zone, and to mine sand, gravel, and the like at scales that might alter the shore.

The following are subject to land use permits:

1. Infrastructural and other facilities such as piers, ports, harbours, bridges, lighthouses, fisheries, installations, or breakwaters that affect either shore protection or utilization of the shore for the public interest;

2. Facilities for shipyards, ship-dismantling plants, and aquaculture, which cannot be located inland because of the nature of the activity.

Zone A

Zone A is defined as the first 50 meters (164 feet) width of the shore strip. Apart from structures that can be built on coastal shore as described above, no building of any kind is allowed in this zone; it is used only for pedestrian access for sightseeing and recreational purposes.

Zone B

Zone B encompassees the remainder of the shore strip; it is at least 50 meters wide. It can contain roads and recreational and tourism facilities open to public use; public waste treatment plants can be built there, subject to land use permits.

Source: Ozhan 1996a.

Box 9.2. Zoning Plans in the Virgin Islands of the United States

The Virgin Island Coastal Zone Management program for the Virgin Islands of the United States employs a two-tier approach to boundary delineation. The first tier generally below 10 feet in elevation, "comprises those areas with the strongest relationship to coastal waters," such as mangrove forests, coral reefs, offshore islands, and cays; the second tier comprises the upland areas of the three main islands.

The areas are managed by means of a land and water use plan, at the heart of which is a comprehensive zoning mechanism. The land and water use categories are called districts. Those districts with specific application to coastal nearshore and offshore areas are detailed as follows.

District	Environment	Use
Preservation and scientific study	Mangrove forests, reefs	Limited recreation
Conservation, preservation	Urban waterfronts, beaches, reefs, and traditional uses	Agriculture and mariculture, traditional docking and fishing, water-dependent recreation, salt ponds, etc.
Water-dependent and related commercial and marine facilities	Developed shores, sand bottoms	Low-intensity port and marine commercial uses that require a coastal site (e.g, marinas, cruise ships, mining)
Water-dependent and related industrial marine facilities	Developed shores	High intensity port, commercial, and industrial uses that require a coastal site (e.g., container docks, power plants)

Source: U.S. Department of Commerce 1979.

schemes generally require national legislation to be put in effect. In some island contexts, as in some small-island countries of the South Pacific, zoning is viewed as unsuitable and too "Western" a tool. As K. Chalapan (1996) argues, for example, the concept of dividing an area among various uses is alien to traditional cultures in which entire villages collectively own areas of the land and adjoining sea. We would argue, however, that the de facto zones demarcating the lands and fishing grounds of various villages are, in effect, a form of zoning.

In accordance with the Great Barrier Reef Marine Park Act (1975), zoning plans have been prepared for the various sections of the Great Barrier Reef (GBR). As shown in box 9.3, Lizard Island, in the Cairns section of the GBR, contains the following types of zones (GBRMPA 1992):

1. General use zones: Provide conservation with reasonable use.

2. Habitat protection zones: Provide conservation and management of habitats, free from trawling and shipping; fishing and collecting are permitted.

3. Conservation park zones: Provide protection; fishing is permitted only by indigenous fishermen.

4. Buffer zones: Only limited fishing allowed—trolling and baitnetting for pilagic species, and traditional fishing, hunting, and gathering.

5. National park zones: Provide strict protection; fishing is permitted only by indigenous fishermen. Recreational activities are limited.

6. No structures subzone: Area left in natural state, largely unaltered by human works, and free from structures and permanently moored facilities.

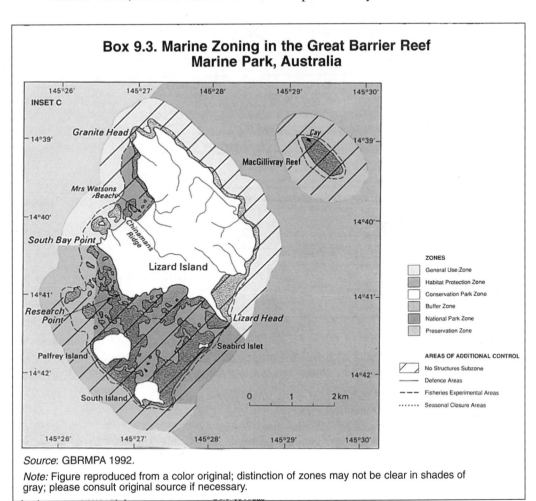

Box 9.3. Marine Zoning in the Great Barrier Reef Marine Park, Australia

Source: GBRMPA 1992.

Note: Figure reproduced from a color original; distinction of zones may not be clear in shades of gray; please consult original source if necessary.

7. Fisheries experimental areas: Provide areas for scientific research into the efforts of fishing on the living resources of the Great Barrier Reef Marine Park. Most fishing, collecting, and traditional fishing and hunting are not permitted.

Set-Back Lines and Exclusionary Zones

Set-back lines are lines in the shore zone seaward of which construction of buildings or other facilities is generally prohibited. See box 9.4 for the example of set-back rule for the coastline of North Carolina in the United States. Set-back lines are generally established by legislation to keep structures out of the active zone of the beach, where coastal erosion from storms and other causes can do

Box 9.4. Set-Back Rule for the Coastline of North Carolina, United States

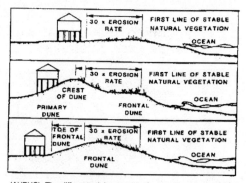

(A)(B)(C). The different minimum oceanfront set backs required when building small structures.

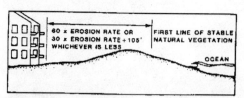

Minimum set-back required when building large structures on the oceanfront.

Adopted in 1979, North Carolina's set-back rule states that all new development is to be the farthest landward of the following alternatives:

• A distance equal to thirty times the long-term erosion rate
• The crest of the "primary" dune
• The landward toe of the "frontal" dune
• Sixty feet landward of the vegetation line

Source: North Carolina 1985.

great damage and where pieces of disintegrating structures carried by wind and storm waves can damage other structures. Increasingly, "dynamic" set-back lines, based on the annual erosion rate at that particular shoreline location, are being put in place. Thus, new structures may be prohibited in front of a set-back line located, say, thirty times the annual erosion rate back from the high-tide mark. Exclusionary zones are areas of the coast where construction of structures and other uses are prohibited because of the hazardous or sensitive nature of the area.

Protected Areas

Protected areas are areas of land or water that have been specially designated to protect some aspect of their fauna or flora. Often, the habitats of rare or endangered species are protected in this way. Management plans are devised that protect the targeted species or sensitive area and allow for uses that do not interfere with these goals. Protected areas (and the exclusionary zones mentioned in the previous paragraph) can be time dependent, with use restrictions applying only during particularly sensitive periods—for example, fish or coral reef spawning periods. As with set-back lines, protected areas are usually designated by special legislation. An example of the types of zoning found in a marine protected area— in this case, the Florida Keys National Marine Sanctuary in the United States—is given in box 9.5.

Special Area Planning

In certain circumstances, it is imperative to plan and manage an entire area as a unit (an ecosystem) regardless of government jurisdiction. This is often the case for bays, estuaries, and other coastal areas where the geography has created coastal ecosystems that are best managed as a whole. Occasionally, too, the present use of an area suggests special area planning as the best approach. This is often true for ports and harbors, where the interplay between marine uses and activities and land uses and activities requires a particularly well integrated approach. Plans for special area management typically are integrated into the coastal management program of the larger coastal area of which they are a part. Box 9.6 shows how this approach works in the context of the Hawaiian Islands of the United States.

Acquisition, Easements, and Development Rights

One way for government to control the use of a particularly important piece of coastal land is to acquire it through purchase. In some cases, as when the shoreline in question is uniquely valuable for, say, a public beach, outright purchase may be the best alternative. Box 9.7 describes such a situation in the state of California in the United States. Obviously, however, coastal land can be very expen-

---

Box 9.5. Zoning in Marine Protected Areas: Florida Keys National Marine Sanctuary, United States

Section 7 of the Florida Keys National Marine Sanctuary and Protection Act of 1990 requires the secretary of the U.S. Department of Commerce to prepare a comprehensive management plan to protect and manage significant marine areas in the coastal waters of the Florida Keys. The goal is to provide recreational benefits to the public over the long term and to preserve ecological, historical research, educational, and aesthetic qualities. The lead agency for management of this sanctuary is the National Oceanic and Atmospheric Administration (NOAA), which is preparing a general management plan.

The plan establishes five temporal and geographical zones to ensure the protection of sanctuary resources:

1. *Wildlife management zones* minimize disturbance to wildlife populations and their habitats by regulating access. A zone of this type could include no-access zones, no motor use zones, and idle speed zones.

2. *Replenishment reserves*, designed to encompass large, contiguous, diverse habitats and to protect and preserve all habitats and species within the zones.

3. *Sanctuary preservation zones* are necessary where a high degree of conflict exists between consumptive and nonconsumptive uses. Special attention is given to shallow, heavily used reefs.

4. *Existing management zones* delineate the jurisdictional authorities of other agencies and complement existing management programs.

5. *Special use zones* set aside areas for educational and scientific purposes and restorative, monitoring, or research activities.

Source: U.S. Department of Commerce 1993b.

---

sive, making acquisition a management technique that must be used with great care. In lieu of outright purchase, it is sometimes appropriate to purchase less than a full interest in the land—for example, to purchase easements that allow the government to prevent or allow certain activities on specific parts of the parcel of land. Easements are one way to provide public access to public beaches in heavily developed areas. Alternatively, development rights can be purchased wherein the landowner retains ownership of the property but is restricted as to the kind of new development that can be undertaken on the land. It is also possible to create a system that allows the landowner to transfer his development rights to a piece of property of less coastal sensitivity.

Box 9.6. Special Area Planning and Management in Hawaii, United States

Prior to developing Hawaii's CZM program, the state Shoreline Protection Act of 1975 designated special management areas (SMAs) extending not less than 100 yards inland from the high-tide line. The purpose of the SMAs is to maintain undeveloped portions of Hawaii's coast for recreational, scenic, educational, and scientific uses; to manage coastal development; and to encourage public participation in the planning process. The counties grant permits to coastal developers based on established guidelines for coastal use that ensure public access and preserve the coastal area's ecological character and scenic value. Although the size of the SMAs varies from county to county and each county has its own procedures for issuing development permits, the requirement and review process for SMA applications is consistent throughout the state.

Source: Hawaii 1990.

Box 9.7. Public Acquisition of Coastal Land in California, United States

Public acquisition is often used in coastal management as a means of protecting areas of critical environmental value or ensuring public access to areas with high recreational use value. In the United States, the California Coastal Plan, which established the California Coastal Zone Management Program in the early 1970s, sets out an aggressive acquisition policy for coastal lands.

Under California's plan, the decision to acquire lands is based on set priorities. The first priority is given to (1) lands that best serve the recreational needs of urban populations and (2) lands of significant ecological importance. Even higher priority is given to lands in either of these categories that have been proposed for development or use incompatible with the existing ecological or recreational value. The second priority is given to (1) open space along urban and suburban waterfronts where views and pedestrian access are limited, (2) areas where development impedes public access, (3) remaining areas of high recreational value, and (4) areas proposed for coastal reserves.

Techniques for acquiring land include the following:

- Public purchase
- Dedications from landowners as reasonable conditions of appropriate development
- Purchase and leaseback
- Scenic and open space easements
- Scenic restrictions
- Resource management contracts

Source: California Coastal Zone Conservation Commission 1975.

Mitigation and Restoration

In certain situations, a use of sensitive coastal land may be proposed that is clearly in the public interest but that will destroy valuable coastal habitat. Perhaps an urgently needed new bridge will, even with the most sensitive construction plans, destroy four hectares (ten acres) of coastal wetlands. When such a project must go ahead, mitigation should be required. In this case, the project's developer, the highway department, should be legally required to restore or create a wetland of at least equal value to the one being destroyed. Here, restoration would involve the renewal of a section of degraded coastal habitat to its earlier healthy and more productive state; creation could entail construction of a wetland, perhaps using clean material dredged from navigational channels and planted with suitable vegetation. (See box 9.8 for a description of coastal wetland restoration efforts in the state of Louisiana in the United States.) Both restoration and creation, however, are still in the experimental stage and require careful monitoring and oversight.

Coastal Permits

Coastal permits are an increasingly popular tool for coastal management. In this approach, a law is enacted that requires anyone who wishes to undertake an activity in the coastal zone (constructing a building, roadway, or seawall; building a

---

Box 9.8. Coastal Wetland Restoration in Louisiana, United States

In the United States, the state of Louisiana enacted a coastal wetland restoration program in 1989 to combat wetland loss estimated by the U.S. Army Corps of Engineers to occur at a rate of 100–130 square kilometers, or 40 to 50 square miles, annually. The state adopted a comprehensive plan utilizing funds from a Restoration Trust Fund (approximately $25 million per annum). Since 1989, twenty-two major restoration projects have been completed as well as a number of smaller projects such as wetland shoreline stabilization, water diversion and hydrologic management, and vegetative planting of wetland species. The lead agency for coastal zone management in the state, the Department of Natural Resources (DNR), Office of Coastal Restoration and Management, coordinates state efforts. Additional aid comes from federal sources.

   Although it is difficult to isolate the causes and effects of wetland loss in Louisiana's coastal zone, the DNR's restoration efforts reduced wetland loss to 65–77 square kilometers, or 25–30 square miles, in 1993.

Source: U.S. Department of Commerce 1993a.

pier; extracting sand or other minerals; etc.) to obtain a permit from the coastal management agency. The law set forth coastal policies that are to be followed in granting or denying such permits. Most proposals eventually receive permits, usually with conditions aimed at improving the proposal and making it less detrimental to the coastal zone and its resources. Occasionally, a permit is denied outright when the proposed project is directly at odds with established coastal policy. For example, the state of California in the United States has a coastal policy that new marinas are not to be built in the state's few remaining estuarine areas; rather, they are to be constructed from "solid" shorelines. Thus, an application for a permit to use an estuary for a marina could expect to be denied outright. Box 9.9 describes the permit system of California's San Francisco Bay Conservation and Development Commission.

Other management tools, are, of course, available. An operational ICM program can take advantage of the learning and experience regarding management

---

Box 9.9. Coastal Permits and the San Francisco Bay Conservation and Development Commission, United States

The first formal coastal management body in the United States, the San Francisco Bay Conservation and Development Commission (BCDC) was established in 1966 in response to increased filling of the bay for development and an associated loss of estuarine and wetland habitat. The BCDC set up an extensive permit system for the following activities:

1. Filling or dredging

2. Shoreline development

3. Substantial change in use (any large construction project or change in general category of use, as from agriculture to residential)

4. Developments within the Suisun Marsh, a critical wetland area

Two types of permits can be applied for: administrative permits for small projects and major permits for large projects. Administrative permits generally do not require a public hearing and can be approved by the BCDC office. Activities that require administrative permits include all routine maintenance dredging, placing of outfall pipes, routine construction activities on existing docks and wharves, and the like. Major permits, however, require a public hearing and often need approval by a number of government agencies, including cities, counties, and regional water quality boards. As a mitigation condition for granting permits for selective filling, large areas of previously filled land have been reopened to tidal flow, resulting in a net increase in water area of the bay.

Sources: BCDC 1985; Travis 1994.

tools and techniques that has been amassed over the past thirty years by staying in touch with other relevant ICM programs and by following the developing ICM literature.

Conflict Management in an ICM Program

Resolving conflicts among coastal and ocean users and agencies is typically a central function of ICM, but some ICM programs pay more attention to this function than do others. In our 1996 cross-national survey, we asked respondents to comment on the extent to which conflict resolution was an important aspect of their ICM programs. As noted in table 9.1, a large proportion of respondents (86 percent in developed countries and 95 percent in developing countries) reported conflict management as an important function. The conflicts respondents mentioned most frequently are noted in table 9.2.

For an ICM program to be effective in addressing conflicts, three elements are important: (1) efforts to understand the roots, causes, and consequences of coastal and marine conflicts through conflict "mapping" studies; (2) an established and transparent process for making decisions about the conflicts; and (3) the capability to adopt and implement measures to remedy injuries or damage to particular coastal and ocean users arising from coastal development or from the actions of other coastal and ocean users.

Understanding and Addressing Coastal and Marine Conflicts

Generally, there are two major types of coastal and marine conflicts: conflicts among users and conflicts among government agencies that administer marine or

Table 9.1. Importance of Conflict Management in ICM Programs in Selected Nations

Extent of Importance of Conflict Management	Developed (N=14)	Middle Developing (N=15)	Developing (N=20)
Very important or moderately important	86	87	95
Moderately unimportant or very unimportant	7	0	5
Uncertain	7	13	0

Source: Authors' 1996 cross-national survey.

Table 9.2. Types of ICM-Related Conflicts in Selected Nations

Developed	Middle Developing	Developing
Water quality	Water quality	Water quality
Erosion	Aquaculture	Port development
Tourism	Erosion	Fisheries
Fisheries	Coastal development	Aquaculture

Source: Authors' 1996 cross-national survey.

coastal programs. Conflicts among agencies and conflicts among users are often interrelated, and coalitions may pit particular users and agencies against other users and agencies. By *users*, we mean both direct users, those who actually participate in coastal and ocean activities (fishermen, subsistence users, offshore oil operators, and recreationists) or in support operations (fish processors), and indirect users, such as environmental groups concerned with coastal protection and members of the public who live in noncoastal areas.

Conflicts among users commonly occur for a variety of reasons: (1) competition for ocean space, such as occurs in aquaculture and fishing; (2) competition for the same resource, as when commercial and recreational fishermen pursue the same species; (3) competition for a linked resource, as when fishermen and marine mammals both pursue salmon; (4) the negative effects of one use on the ecosystem harboring another use, as in the effects of offshore oil development on fisheries concentration and reproduction; or (5) competition among users for similar onshore space or facilities in ports and harbors or elsewhere on the coast, as when fishermen and offshore oil developers compete for harbor space or recreational activities compete with aquaculture.

Government agencies also may conflict with one another for a variety of reasons. They may have different legal mandates and agency missions; they may have different agency styles and personnel with different training and outlooks; they may respond to different external constituencies who may demand opposing actions; or there may be a lack of information and communication among them.

As do other types of conflict, conflicts in the coastal setting often go through a series of stages. First, a precipitating event signals the emergence of a dispute. Specific issues may then be transformed into multiple issues or a generalized feud and the parties may become polarized. Finally, "spiraling" may occur as the conflict escalates and stereotyping and "mirror-imaging" takes place—"opponents come to perceive one another as the mirror opposite of their own exemplary characteristics" (Wehr 1979, 21).

To address conflicts effectively, coastal managers must understand the conflicts' root causes: are people disagreeing over values (the "what should be"), interests (the "who shall get what in the distribution of scarce resources or space"), or facts? Author Biliana Cicin-Sain (1992) hypothesizes that the more conflicts are rooted in differences over facts or interests (rather than values) the easier they are to solve, or the more tractable they are. Figure 9.1 arrays different types of conflicts on a tractability scale, categorizing conflicts according to their roots and the parties involved. On the left side of the tractability scale are the conflicts most difficult to solve—philosophical conflicts that pit, for example, environmental groups against oil developers, with the environmentalists holding that the oil developers simply should not be there and the oil developers opposing the environmentalists' right to have a say in marine operations. In such cases, the likelihood of negotiation and conflict resolution is low. At the opposite end of the tractability scale (the right-hand side) are conflicts that pit, for example, fishermen against oil developers, with both parties arguing over a factual issue such as the effects of oil exploration seismic operations on lobster pots. Conflicts such as these are much more easily resolved through the communication of appropriate facts.

Because conflicts in the coastal zone often involve the preservation of ocean and coastal resources and space and equitable distribution of access to them and benefits from their exploitation, coastal managers often must intervene to seek resolution. This is particularly important, we think, when a conflict poses a threat to public safety and order, when a conflict threatens the long-term viability of common property resources or involves irretrievable environmental damage, when participants in the dispute do not represent all interests involved, or when the conflict involves excessive duplication, waste, or inefficiency on the part of government.

The first challenge for the coastal manager in these cases is to understand why the conflicts are occurring and what kinds of consequences they may be incurring.

Figure 9.1. Tractability of Conflicts

	Types of Conflicts			
	Philosophical Conflicts	Potential Interaction Conflicts	Actual Interaction Conflicts	Imagined Interaction Conflicts
Roots of Conflict	Differences in values	Differences in facts, interests, possibly values	Differences in facts, interests	Differences in facts
Parties to Conflict	Indirect users/ direct users	Direct users/ direct users	Direct users/ direct users	Direct users/ direct users

←——————————————————————————————————→

TRACTABILITY

Most intractable   Most tractable

Source: Cicin-Sain 1992.

This task can be aided by the conduct of conflict-mapping analysis, whereby systematic discussions are undertaken with participants in the conflict, identifying their positions, the reasons for their positions, and areas of agreement and disagreement.

Conflict resolution methods vary greatly according to the socioeconomic and political contexts of particular situations. In traditional societies, turning to village leaders or elders may well be the most effective way to address disputes in the coastal zone. In developed Western societies, coastal conflicts are often resolved through the efforts of neutral mediators who assist coastal managers in bringing the parties together and negotiating a solution. An extensive literature addresses the use of mediation in resolving public disputes, and a number of lessons have been learned as to what approaches seem to work best (see, e.g., Bingham 1986). Table 9.3 outlines the tasks of the mediator in each phase of conflict resolution—from getting started, engaging in joint fact finding, and entering into negotiations to producing binding agreements and seeing these agreements implemented and monitored (Susskind and Cruikshank 1987).

In terms of lessons learned, for example, one study summarizing a decade of

Table 9.3. Tasks of the Mediator

Phases of Conflict Resolution	Tasks
Prenegotiation	
Getting started	Meeting with potential stakeholders to assess their interests and describe the consensus-building process; handling logistics and convening initial meetings; assist groups in initial calculation of BATNAs ("best alternatives to a negotiated agreement")
Establishing representation	Caucusing with stakeholders to help choose spokespeople or team leaders; working with initial stakeholders to identify missing groups or strategies for representing diffuse interests
Drafting protocols and setting the agenda	Preparing draft protocols based on past experience and the concerns of the parties; managing the process of agenda setting
Engaging in joint fact finding	Helping to draft fact-finding protocols; identifying technical consultants or advisors to the group; raising and administering the funds in a resource pool; serving as a repository for confidential or proprietary information

(continues)

Table 9.3 (*continued*)

Phases of Conflict Resolution	Tasks
Negotiation	
Inventing options	Managing the brainstorming process; suggesting potential options for the group to consider; coordinating subcommittees to draft options
Packaging	Caucusing privately with each group to identify and test possible traders; suggesting possible packages for the group to consider
Written agreement	Working with a subcommittee to produce a draft agreement; managing a single-text procedure; preparing a preliminary draft of a single text
Binding the parties	Serving as the holder of the bond; approaching outsiders on behalf of the group; helping to invent new ways to bind the parties to their commitments
Ratification	Helping the participants "sell" the agreement to their constituents; ensuring that all representatives have been in touch with their constituents
Implementation or post negotiation	
Linking informal agreements and formal decision making	Working with the parties to invent linkages; approaching elected or appointed officials on behalf of the group; identifying the the legal constraints on implementation
Monitoring	Serving as the monitor of implementation; convening a monitoring group
Renegotiation	Reassembling the participants if subsequent disagreements emerge; helping to remind the group of its earlier intentions

Source: Susskind and Cruikshank 1987.

experience with environmental disputes in the United States (not necessarily coastal disputes) found the following factors to be the most important in attaining successful dispute resolution (Bingham 1986): (1) parties must have an incentive to negotiate an agreement; (2) if disputes are "interest based," agreement is more likely; (3) surprisingly, the number of parties involved does not appear to affect the outcome (i.e., it is better to involve all potentially affected interests rather than limit the number of parties at the negotiations); and, most significant, (4) the most important factor in the likelihood of success seems to be whether those with the authority to implement the decision participate directly in the process.

In addressing conflicts, however, coastal managers must ensure that decisions are made through a public and transparent process, as stressed in the next section. Sometimes mediation in developed countries has been conducted through private negotiations, an approach that may not be altogether suitable when public coastal or ocean resources are at stake (Cicin-Sain and Tiddens 1989).

Coastal managers also must consider what measures may be needed and appropriate for addressing the negative effects of one coastal use over other coastal uses or resources. Some well-established measures include requirements that the negative effects be mitigated or offset or that compensation for the loss or damage be paid. Mitigation can become complex as in the case of developments that degrade coastal wetlands, such as port construction. Efforts may be made either to replace the lost functions of the wetland in the same watershed (the preferable option) or to create a new wetland in another area that fully compensates for the lost resource (Beck 1994; Etchart 1995).

Transparency and Public Participation in an ICM Program

Maintaining full transparency and a high level of public participation in the ICM process during the program formulation phase, the implementation phase, and the operation phase is of the utmost importance. In part, it is the openness and transparency of the process that ensures that all affected interests have an opportunity to be heard. This, in turn, has two advantages: first, the benefits and costs are likely to be distributed more equitably if all parties have a chance to participate, and second, decisions are likely to be better informed if all relevant points of view are heard and considered. The procedures used by an ICM program in its decision-making and policy-making activities are of considerable importance in building and maintaining public confidence in support of the program. A process that is seen as fully open, based on reliable information and good science, and accessible to all interested parties stands a much better chance of long-term success than does one that is difficult to access, does not encourage participation, and in which decision making goes on behind closed doors.

Consensus Building

Consensus building among government agencies, users' groups, and local communities is essential to ensure the social acceptability that will enhance the successful implementation and enforcement of ICM decisions. Two-way communication is essential to arrive at agreements about implementation strategies and compliance. The following strategies for ensuring public participation and building consensus are described in a 1997 set of ICM guidelines relating ICM to climate change (Cicin-Sain et al. 1997):

- An informed general public through media coverage and public fora, meetings and discussions of the proposed ICM project, and information education campaigns in the local language(s).

- Availability of accurate, timely, and documented official information on proposed ICM activities.

- An adequate community participation mechanism for providing input into ICM decisions and conducting ongoing monitoring.

- Proper feedback mechanisms to ensure that the outputs of public consultations and evaluations are incorporated into revised ICM plans.

- Transparent decision-making processes (permits, environmental impact assessments, etc.).

- Publicizing of plans made and actions taken in coastal areas, with emphasis on the local and regional benefits derived from national coastal management initiatives.

Types of Public Participation: "Advice Giving" and "Power Sharing"

Two major forms of public participation can be described. The first, which involves attendance at meetings and hearings and provision of written comments on ICM plans and the like, may best be described as "participation as advice giving" to government, with government retaining responsibility for ICM decisions and their implementation. This is by far the most prevalent approach to participation in ICM programs around the world. A general model for encouraging and sustaining this type of "advice giving" public participation is given in box 9.10.

Other forms of public participation are possible whereby the responsibility for some or all aspects of an ICM program are delegated to local communities or particular user groups. There is growing experimentation with this form of public participation. Devolution of power to local communities for the management of coastal resources is taking place in countries such as the Philippines, where a general program of government decentralization is in progress (Rivera and Newkirk 1997). In the case of fisheries in particular, in many countries there has been significant experimentation with the practice of comanagement, whereby the government and fishing representatives are empowered to manage fisheries jointly (Jentoft and McCay 1995).

Such "participation as power sharing" is highlighted in a special issue of *Ocean*

Box 9.10. Some Keys to Achieving Effective Public Participation in ICM Decisions

• *Early identification of the interests (groups and individuals)* that will be affected by pending management decisions is essential.

• Once identified, every effort should be made to *inform these groups and individuals* of the upcoming decision(s), the issue(s) involved, and how the decision(s) could affect them.

• *Input and comments should be solicited* and opportunities should be created for citizens to ask questions and to make statements at public meetings.

• *Public meetings*, ranging from formal public hearings to informal village gatherings, should be held in the coastal region where the proposed activity is to take place and at a time most convenient to those wishing to testify or otherwise make their opinions heard.

• *Full information on the proposal should be made available* in clear and understandable language sufficiently in advance of the public hearing or meeting to allow interested individuals to analyze the information and prepare their questions or comments. The meeting should be conducted in a fair and businesslike fashion, with all those interested given an opportunity to speak. Written statements should be allowed within a given comment period following the meeting.

• *Major ICM decisions should be carefully made with a full account of the information, data, and facts that affected the decision making.*

• The decision-making body (e.g., the coastal coordinating council) would be well advised to *issue its decision(s) in written form*, perhaps in the form of "findings." An interested reader would learn from these findings the basis of the body's decision and any assumptions that went into it. Ideally, such findings would first be issued in draft form, with the public invited to make comments and suggestions on the draft decision during a short comment period.

• Subsequently, the *final decision should be publicly issued*. Note that this somewhat elaborate procedure would be appropriate only for decisions of major or precedent-setting consequence. In fact, one of the early actions of an ICM agency should be to decide on working procedures such as these.

& Coastal Management focusing on community-based integrated coastal management, by Larry Hildebrand of Environment Canada (Hildebrand 1997). Hildebrand examines a number of different types of community-based management involving different degrees of power sharing among governments, users, and local communities. In Canada, for example, the national government has been experimenting with community-based ICM at a number of sites in the Atlantic provinces. Local communities with the technical assistance of the national government, are encouraged to engage in "visioning" exercises to consider and debate alternative future scenarios for their coastal communities and then are provided with technical assistance, through a government-supported local coordinator, to diagnose and act on agreed coastal priorities in their communities (Hildebrand, 1997). In another example, in the Dominican Republic ICM formulation and implementation are taking place in the Samana Bay area, led not by any government authority but by an external NGO (the Center for Marine Conservation) working in conjunction with local community leaders and NGOs (Jorge 1997).

Building and Maintaining Public Support: The Importance of Public Education and Awareness

An ICM program cannot survive over the long term without the support of the general public. Public support is needed, among other things, to undergird efforts to secure general compliance with the program. Indeed, it would be difficult to find enough money to pay for sufficient enforcement to make an unpopular resource management program succeed. Beyond this factor, the ICM program and the management programs of the line agencies that are part of it all need regular appropriations of funding year after year; laws and regulations need to be revised, amended, and, presumably, strengthened. All this requires the support of leaders in the executive and legislative branches of government, and without continued public support, government support will weaken and eventually disappear.

How can public support for an ICM program be built and maintained? The key, we believe, is in a strong, two-pronged public information and education program—the two prongs being the *what* and the *why* of ICM. The first element, the *what*, involves establishing early on and maintaining a good flow of information about the workings of the ICM program—what important problems ICM is intended to address, when the program was created and by whom, and who makes the decisions, on what basis, and using what information and data. An early operational task is the creation of mailing lists of leaders in all coastal communities affected by the ICM program. At the earliest opportunity after the ICM program becomes operational, an informational newsletter answering such questions and providing information about the ICM program's implementation schedule and other relevant events would be very useful.

The second element of the public information and education program involves the *why* of ICM. This should be seen as an exercise in "coastal awareness raising" and is primarily an educational endeavor. In most countries, the general population is only vaguely aware of the value of the coastal zone and its resources and of the adjacent coastal ocean. People's interaction with the coast is often limited to a summer vacation or holiday trip and the appreciation of "sun, sea, and sand." The sensitive nature of the coastal environment and the ecosystems it contains, the valuable ecological niches provided by the estuary and the great productivity found there, and the great ocean fisheries and their precarious condition are largely unknown to most citizens. It makes sense that as people become more aware of the value and uniqueness of the coast, they will support stronger management measures.

We believe, therefore, that a strong education program that includes an active outreach effort to all educational levels—from elementary school through high school and into the universities—is needed. Time and resources permitting, coastal and ocean curriculum materials should be developed and placed in cooperating schools and other institutions. Strong efforts should also be made to stay in contact with coastal users' groups, environmental and conservation organizations, and others concerned with the health and well-being of coastal area and its resources.

As an important part of coastal awareness raising, the ICM program could sponsor annual or biannual "coastal stewardship" awards wherein the best government and private coastal development and conservation projects are publicly recognized and rewarded. Such events would also highlight the fact that the ICM program is not devoted solely to regulations and prohibitions but also encourages and facilitates appropriate development. Enlisting the support of the local media (radio, television, and newspapers) should help guarantee that the awards, their recipients, and good coastal management gain the widest possible recognition. One can, of course, also visualize an event highlighting the worst coastal developments of the year. All things considered, however, this kind of activity is probably best handled by an environmental or conservation-oriented NGO—one that is independent of the governments involved in the ICM program.

Much experience has already been gained in the fields of public participation and public outreach. Those establishing such programs for the first time would be well advised to contact representatives of long-established coastal management programs in order to benefit from their learning in this regard. Some examples of public participation and outreach used in Thailand and Ecuador by members of the University of Rhode Island's USAID project, which has emphasized public participation in ICM, are shown in boxes 9.11 and 9.12.

Evaluation of ICM Programs

Evaluation is an activity undertaken to determine the extent to which a given program is meeting its goals. The evaluative activity may be a continuous, ongoing

---

Box 9.11. Success of Community Involvement in Thailand

Thailand has conducted successful community-involved coral reef management. For example, Phuket, located in the eastern Andaman Sea, is Thailand's largest island. Within 1.5 kilometers (0.93 miles) of shore in the island's western and southern bays, well-developed fringing coral reef habitats have been degraded or destroyed by improper recreational use, careless boat anchoring, illegal dynamite fishing, commercial collection of corals, siltation, and pollution.

Of Thailand's coral reef management strategies, the most important effort was to heighten public awareness and involve local communities in decision making regarding coral reef management. The following practical activities were carried out:

- Phuket Teacher College developed and tested a coral reef curriculum for local schools.
- Local residents and the provincial government organized and participated in community events such as a Coral Reef Day.
- Kodak (Thailand) Ltd. donated signs, posters, and brochures to increase public awareness of coral reefs among tourists.
- The Phuket Diving Association, the Kata-Karon Diving Group, Matlang Resort, the Phuket Rotary Club, and the Phuket Island Resort volunteered time and donated equipment to help install mooring buoys in Phuket area coral reefs.

This community-involved coral reef protection program has been ongoing since 1986 as part of the Phuket Island Action Plan sponsored by the University of Rhode Island and USAID. Efforts at community-based coral reef management continue to grow in Thailand with implementation of the National Coral Reef Strategy.

Sources: Hale and Lemay 1994; Lemay, Ausavajitanon, and Hale 1991.

---

process in which measures of program performance are obtained and systematically compared with program goals and objectives, or it may be undertaken periodically during a program's lifetime. Evaluation of government programs is especially important because they generally do not have the relatively easy-to-assess bottom-line, profit-related goals of private sector enterprises. Furthermore, a properly constructed evaluation not only will provide estimates of program performance, results, and effects but also will suggest *why* the program is not achieving the desired results if such is the case and point the way toward modifications

---

Box 9.12. Public Education and Participation in Ecuador

One of the primary objectives of the USAID Coastal Resources Center project in Ecuador was to involve and educate coastal communities in coastal zone management. Following the establishment of five ZEMs (zonas especiales de manejo, or special area management zones) in 1990, the ZEM coordinators began developing annual work plans for local public education activities. With special emphasis on public participation, the plans sought to create a common base of information, attitudes, and values among communities, users groups, and the general public. Actions that came out of these education programs included the following:

- Informal education programs for organizing and training community and resource use groups.
- Formal programs for creating school materials on coastal resource themes such as mangrove forests and fisheries, and for organizing school field trips and community projects.
- Outreach programs including radio, television, and printed matter with nationwide distribution.

As a result of these efforts, the ZEMs have functioned successfully. Communities, user groups, local authorities, and educational groups have come together and made the ZEM committees a useful and democratic means for resolving conflicts among users. Additionally, more than thirty user groups have organized themselves to participate in activities associated with the ZEMs.

Source: Robadue 1995

---

or corrections that will result in improved performance. Thus, conscious, structured evaluative efforts should be an essential component of programs such as integrated coastal management programs.

This having been said, the reality is that very few government programs benefit from regular evaluations. There are many reasons for this. First, few incentives exist on anyone's part to undertake serious evaluations. Personnel of government agencies often fear that objective evaluations will reveal weaknesses or inadequacies in their programs. Legislators who created programs may have a sense of pride in them, and donors who financially support programs want them to succeed and may be reluctant to discover otherwise. Stakeholders, interested in seeing improved resource management, prefer some action, however feeble, to none at all. In summary, those closest to the program in question generally will not seek ongoing objective assessments of its performance.

A second set of problems associated with the design and operation of government programs also makes evaluation difficult. Programs often have goals and objectives that are very vague or general and thus are not easily measured. In such cases it is difficult, if not impossible, to determine the extent to which goals and objectives are being achieved. In these instances, such evaluations as are attempted typically fall back on the use of indicators that measure effort rather than results. For example, the number of permits granted or denied might be used as an indicator of the performance of a coastal wetlands protection program rather than the number of acres of wetlands coming under protection or restoration. Some managers leave evaluation of the program until the very end of the process, ensuring that the results come too late to affect the program itself and only marginally affect related projects (Olsen, Tobey, and Kerr 1997). A good many managers skip formal systematic evaluations entirely, preferring to rely on anecdotal data and information if and as required.

In the sections that follow, we discuss the evaluation process and how it might be applied to ICM programs, addressing four questions:

- What should be evaluated?

- Who should do the evaluation?

- When should evaluation begin?

- How should evaluation be done?

What Should Be Evaluated?

Fundamentally, evaluations should be aimed at assessing the extent to which the ICM program is addressing and solving the problems that caused the program to be created in the first place. If one of the main problems triggering the need for ICM were degraded coastal habitats, the evaluation process should monitor the status of the coastal habitats at issue and the extent and rate at which improvement is taking place. If the ICM program was instituted in part to reverse a decline in important fishery stocks, the evaluation should focus, in part at least, on the extent to which the stocks are recovering. Of course, it may be difficult to say for certain that actions taken as part of the ICM program were responsible for increased fishery stock abundance, especially if other programs, such as those designed to improve coastal water quality, also affect the target resource. This problem of attribution is eased somewhat if all programs affecting a coastal region and its resources are under the umbrella of the ICM effort.

Frequently, even if the goals of the management effort are quantitatively expressed, the data and information necessary to measure progress toward these goals are not systematically obtained, either because they are technically difficult or expensive to gather or because of a lack of priority given to this kind of monitoring activity. In the absence of on-the-ground indicators of performance, analysts are typically driven to use process-related indicators. As mentioned earlier, effort-related or process-related indicators may or may not reflect the real performance of an ICM management program. A higher level of regulatory activity, a greater number of permits issued, creation of a larger number of resource inventories or maps, employment of an increased number of enforcement officers—all measures of effort—may not, in fact, mean improved coastal water quality, increased fisheries abundance, reduced coastal erosion, healthier coastal wetlands, or higher quality of life of coastal residents. Only monitoring that directly measures such parameters can provide the data necessary to clearly demonstrate the effectiveness of the management program.

To the extent that ICM programs are explicitly aimed at broader social and economic goals, for example, "sustainable resource use and an improved quality of life," it is necessary to recognize that such goals are likely to be achieved only after certain preliminary goals are reached. The U.S. Environmental Protection Agency, for example, recognizes first-, second-, third-, and fourth-order goals (Olsen, Tobey, and Kerr 1997):

First-order goals: Formalized institutional arrangements; formalized constituencies.

Second-order goals: Mitigation of selected adverse behavior; implementation of selected development actions.

Third-order goals: Improvements in social and environmental indicators.

Fourth-order goals: Sustainable resource use and improvement in quality of life.

Clearly, attainment of fourth-order goals will depend on attainment of first-, second-, and third-order goals. Evaluative processes will need to be designed to recognize this hierarchy of goals.

Who Should Do the Evaluation?

It is useful to differentiate between two kinds of evaluation or assessment, both of which are necessary and useful in ICM. The first type is the monitoring and assessment of program performance necessary for program operation and modification. This activity continually compares the program's results (performance) with its stated goals and makes the midcourse corrections and adjustments necessary to improve and strengthen performance. This type of monitoring and assessment is an integral part of the ICM program and therefore is internal to the program and performed by program staff.

The second type of evaluation is undertaken periodically, perhaps every two to five years, with the objective of providing outside interests (the legislature, stakeholders, the public) with an assessment of how well the program is performing—that is, how well it is achieving its goals. Although these evaluations are often done in house, they are most likely to be perceived as objective studies if they are performed outside the program. Evaluations performed within the program could be seen as self-serving, especially if they show a higher level of performance or greater beneficial results than is generally thought to be the case.

Independent outside evaluations can usually be satisfactorily performed by any of several different groups, depending on the national context involved, but could involve the following:

- A national academy of sciences.

- A planning and evaluation office of the government (assuming it is independent of the ICM program).

- A university group with suitable expertise.

- A group of credible NGOs.

Periodic evaluations of this type play an important role in gaining and maintaining support for the ICM program. Assuming these evaluations are seen as objective and credible, they are a very good way to keep the public informed as to the usefulness and overall effectiveness of an ICM program. Furthermore, they are important in securing the continued funding necessary to operate an ICM effort. Such evaluations, if they show good performance, are especially important when competition among programs for government funding increases.

When Should Evaluation Begin?

Operational monitoring and assessment is an integral part of the ICM process and should begin when the program begins. Although improvements as a result of the ICM process will take varying amounts of time to achieve depending on the problems being addressed, it is of utmost importance that premanagement and early management baseline data and information be obtained as early as possible. Only in this way will data be obtained against which improvements can be measured.

In terms of periodic evaluations, we believe a time frame of about five years should be sufficient for the effects of an ICM effort to begin to be apparent. Substantive on-the-ground results, however, may take appreciably longer to be seen and assessed. Clearly, it should be possible to evaluate the effectiveness and performance of an ICM program at the end of a ten-year period; lack of measurable results at this point should call into question the program's effectiveness.

How Should the Evaluation Be Done?

In its simplest form, a periodic evaluation should compare measured outcomes with the stated goals of the ICM program. Outcome indicators should be obtained (by observation, monitoring, and measurement) for each of the program's goals. To the extent that this is not immediately possible, process or effort indicators should be used to fill gaps until on-the-ground outcome indicators are available. Again, social and economic indicators need to be used where goals relate to sustainable development and improved quality of life.

One difficulty common to all efforts at evaluating public programs involves the question of attribution. How can one be certain that an observed improvement is directly related to management measures of the program being evaluated? Other programs could have produced the observed effects, or, in selected cases, unrelated changes in exogenous factors such as the economy could be responsible. Demonstrating a cause-and-effect relationship in evaluative efforts is a most difficult task.

Performance Monitoring and Program Accountability

Again, the existence of measurable goals, in contrast to goals that are expressed in general, ambiguous, and nonquantitative terms, greatly simplifies the task of performance monitoring. At the time these goals are established, monitoring programs to measure progress toward their achievement should be designed and put in place. The monitoring efforts must be sufficiently comprehensive to determine not only whether or not the goals are being met but also, if they are not, what has gone wrong with the assumptions or methodology so that appropriate midcourse corrections can be made.

Transparency in an ICM program means that the costs associated with achieving particular goals should be relatively easily known. For example, if one of the program's goals is to restore 200 hectares (500 acres) of degraded coastal wetlands annually, the costs of the management effort to bring this about should be available to the interested public. In this way, the public can weigh the costs of management against the value of the restored wetlands. The same type of information should be made available for all areas in which the ICM program seeks to achieve specific goals.

Hence, in our view, a key part of the ICM process is accountability. If an ICM program proposes to improve the management of coastal and ocean resources and space by means of a more integrated and better-informed decision-making process, the public should hold the program accountable for results consistent with these goals. The ICM program, for its part, should make available to the interested public the information that will allow judgments to be made as to accountability and overall performance. Just as an individual management program should account for what it has achieved with the resources entrusted to it, the overall ICM program should be expected to do the same.

Evaluation of ICM Programs: An Undeveloped Art

As essential as evaluations are in measuring the success of ICM programs and modifying them in view of the results, evaluation is the least developed phase of the ICM process, in terms of both methods and practice. As discussed in chapter 11, there is very little systematically collected evaluative evidence on ICM's effectiveness in improving the management of coastal resources and space. This is the case even in countries that have the oldest ICM programs (spanning more than twenty years), such as the United States, even though the practice of evaluating government programs is relatively well established there. In developing countries, J. C. Sorensen (1997) has noted, the evaluation of government programs is not well established, a factor that will impede the application of this practice to ICM.

As discussed in chapter 11, this is one of the areas in which ICM practice most needs to be improved in the next decade. Given the many ICM efforts taking place around the world and the considerable funding being expended on this work, it is time to conduct independent comparative studies of ICM programs to measure their effects, both positive and negative, intended and unintended. Some good first steps have recently been taken toward developing a database on evaluation. Examples include a major conference on lessons learned in ICM organized by Chua Thia-Eng and his colleagues at the GEF/UNDP/IMO Regional Programme for the Management and Prevention of Marine Pollution in East Asian Seas, held in Xiamen, China, in May 1996; ongoing efforts by the World Bank, UNEP, and UNDP to collect evaluative data on a number of cases (Hatziolos and Trumbic 1997; personal communication, Philip Reynolds, UNDP, May 1996); and efforts to develop indicators for measuring ICM performance (Olsen, Tobey, and Kerr 1997).

Summary

This chapter discussed challenges faced by coastal managers in the implementation, operation, and evaluation phases of ICM. The implementation phase includes the actions that must be taken to put the ICM program into operation—start-up activities such as enactment or amendment of legislation, preparation or revision of regulations and procedures, formal establishment of new institutions or interagency mechanisms, securing of additional personnel, and so forth. Typical challenges decision makers face at this stage include securing the necessary legislative and legal changes, obtaining needed resources, and identifying and filling policy gaps. Moreover, if an undue length of time has elapsed since the ICM plan was formulated, changes may have occurred that invalidate part of the proposal, and institutional changes may be difficult to put in place because of bureaucratic inertia and resistance to change.

In the operation phase (once new or altered ICM processes have been put in place), the emphasis of the ICM program will be on coordination, harmonization, conflict resolution, integration of coastal and ocean policy, filling of management gaps, and monitoring and assessment of performance. A wide variety of management tools and techniques are available to the ICM program in the operation phase, including zonation; establishment of set-back lines and exclusionary zones; establishment of protected areas; special area planning; acquisition, easements, and development rights; mitigation and restoration; and issuance of coastal permits. Which of these tools are used will depend on the extent to which they are authorized under legislation, their technical suitability, and their public acceptance.

Evaluation is essential in measuring the success of an ICM program and in modifying it in view of the results. Unfortunately, evaluation is the least developed phase of the ICM process, in terms of both methods and practice. This is one area in which ICM practice most needs to be improved in the next decade.

PART IV

COUNTRY CASE COMPARISONS AND LESSONS LEARNED

Chapter 10

Case Comparisons of ICM Practices in Twenty-Two Selected Nations

Introduction

In appendix 1, "ICM Practices in Twenty-Two Selected Nations," we provide narrative case studies of twenty-two nations in the developed and developing worlds, a subset of the twenty-nine nations included in our 1996 cross-national survey. The developed countries are Canada, the United States, the United Kingdom, France, the Netherlands, Spain, the Republic of Korea, and Australia; middle developing countries are Brazil, Turkey, Thailand, Malaysia, and Fiji; and developing countries are Ecuador, the People's Republic of China, Indonesia, Pakistan, India, the Philippines, Sri Lanka, the Federated States of Micronesia, and the United Republic of Tanzania. We recommend that those interested in examining these countries' ICM experiences read appendix 1 carefully.

In this chapter, we compare and contrast these twenty-two cases, relying on both the information in appendix 1 and the responses to our cross-national survey. To this effect, we have prepared a set of tables that compare the country cases on a number of variables. Our choice of variables was guided by our own hypotheses about factors that might make a difference in the patterns of ICM observed in each country. We start out by summarizing the background variables and socioeconomic context of the country, specifically the level of development as measured by per capita GDP, which may suggest both the types of problems present in the country and the goals coastal managers may be trying to achieve; the percentage of population on the coast (a higher concentration on the coast could predict the salience of coastal management as a national issue); and the diversity of the population along ethnic and religious lines, which may be indicative of participation in a particular ICM program—for example, one could hypothesize that the more diverse is a society, the greater will be the need for participatory processes in ICM to accommodate different perspectives.

Next, we summarize characteristics of each nation's political system: type of government (democratic, authoritarian, unitary, federal, etc.); colonial heritage, since there may be an institutional legacy from previous colonial powers; concentration of power among national-level institutions—for example, whether the chief executive (president or prime minister) has centralized control or faces significant challenges from the legislature, interest groups, provincial leaders, and so forth—and the degree of autonomy of subnational levels of government, such as provinces and local communities, or how much legal and actual power they have vis-à-vis the national government. The latter factor is influential, we believe, in determining whether an ICM program proceeds mainly at the national, provincial or local level or in some approach combining two or more levels.

We then compare the nations according to the maritime jurisdictions claimed by each with regard to the Law of the Sea Convention, especially the territorial sea and the Exclusive Economic Zone. We depict the major coastal and ocean issues present, issues that are related to each nation's geographical location, type of coast, population concentration along the coast, level of development and major economic activities occurring along the coast.

We also compare the cases using a number of variables related to government responses to ocean and coastal issues: Which is the primary level of government concerned with ICM? What are the nature and timing of ICM efforts? What has been the overall approach to ICM (top-down, bottom-up, or both)? What has been the specific approach to ICM (regulatory, incentive, planning, or consensus building)? To what extent have ICM efforts been implemented? How important has external assistance been in the ICM effort? What has been the influence of UNCED on the ICM efforts? How effective have the ICM efforts been? And, finally, has there been some movement toward integrated management?

Lest we raise readers' expectations unduly, we warn that for many of the countries discussed, there is scanty evidence regarding some of these questions, particularly the extent of implementation and the extent of effectiveness.

Case Comparisons

Tables 10.1, 10.2, and 10.3 summarize the information discussed in the previous section. To complete these tables, we had to make judgment calls, particularly for those categories not based on objective data. To increase the accuracy of the tables, we followed a twofold procedure. First, at least two of the main researchers involved in this study coded each country case (found in appendix 1) independently and then compared notes, discussing and reconciling differences. Second, each of the tables (as well as the narrative on the country case) was returned to one of the respondents of the survey for comment and feedback, and corrections were made accordingly.

Table 10.1. Case Comparisons of Eight Developed Countries

	Canada	United Kingdom	United States
Background variables and socioeconomic context			
Total population	28 million	58 million	263 million
Percentage of population on the coast	25%	no data	About 60%
Per capita GNP	$19,510	$18,340	$25,880
Major ethnic groups and religions	40% British Isles, 27% French, 20% other European, 1.5% Indian and Eskimo; 46% Roman Catholic, 16% United Church, 10% Anglican, 28% other. Significant recent influx of Asian immigrants, especially to the western coast	82% English, 10% Scottish, 2% Irish, 2% Welsh, 3% West Indian, Indian, Pakistani, and other. Largely Protestant, also Catholic, Muslim, Hindu, and other	83% white, 12% black, 3% Asian, 0.8% Native American
Political system characteristics			
Type of government	Confederate parliamentary democracy	Constitutional monarchy	Federal republic
Colonial heritage	British	Not relevant	British
Concentration of power among national-level institutions	Some pluralism	Some pluralism	Pluralism
Autonomy of subnational levels of government	Provinces and communities have some autonomy, but the federal government has exclusive jurisdiction over the marine environment and marine resources below the low-water mark	Counties have substantial autonomy	States have some autonomy; local governments have relatively little autonomy
Maritime jurisdictions	12-NM[a] territorial sea, 24-NM contiguous zone, 200-NM EFZ,[b] 200-NM EEZ	12-NM territorial sea, 200-NM EFZ	12-NM territorial sea (1988), 200-NM EEZ (1983)
Major coastal and ocean issues	Loss of wetlands, decline in fishery stocks, resource conflicts, pollution,	Pollution, damage to productive ecosystems, resource depletion, erosion	Nonpoint-source marine pollution, decline in fishery stocks, offshore oil

(continues)

Table 10.1 (*continued*)

	Canada	United Kingdom	United States
	offshore oil development	of coastal areas, lack of integrated policy, navigation in English Channel, fishery conservation, pollution control, strategic sea use, boundaries, offshore oil development, lack of planning regime for marine areas	development, closed shellfish beds, coastal erosion, coastal hazards (storms)
Primary level of government concerned with ICM	National, but provincial and local levels also play a role	Mixed (national and county)	National, state, and in some cases, local
Nature and timing of ICM efforts	1985, Fraser River Estuary Management Program; 1987, Great Lakes Water Quality Program; 1991, Atlantic Coastal Action Plan; 1996, Canada Oceans Act (includes ICM)	1980s–1990s, various marine sectoral laws; 1990s, NGO reports urging establishment of ICM policy and national and local coastal fora; 1992, House of Commons inquiry	1970s, various marine and coastal sectoral laws; 1972, Federal Coastal Zone Management Act (reauthorized at about five-year intervals, most recently in 1996)
Overall approach to ICM	Top-down (Canada Oceans Act), bottom-up (local consensus building)	Bottom-up	Top-down, bottom-up
Type of approach to ICM	Consensus building	Consensus building	Regulatory, planning
Extent of implementation of ICM	Partial implementation	Little implementation	Good implementation
Importance of external assistance	Not important	Not important	Not important
Importance of UNCED influence	Very important	Moderately important	Not important
Effectiveness of ICM	Community-based programs reportedly highly effective, but outputs of programs are not clear	Too early to tell	Moderately effective
Movement toward policy integration	Some in locally based programs	Some movement in the national and local coastal fora	Good at state level; mixed at national level

Table 10.1 (*continued*)

	France	**Netherlands**	**Spain**
Background variables and socioeconomic context			
Total population	58 million	15 million	40 million
Percentage of population on the coast	No data	60%	35%
Per capita GNP	$23,240	$22,010	$13,440
Major ethnic groups and religions	Predominantly Celtic and Latin, Teutonic, North African, Indo-chinese, Basque minorities; 90% Roman Catholic	Predominantly Dutch (96%), with Turks, Moroccans, others (4%); 34% Roman Catholic, 25% Protestant, remainder unaffiliated	Predominantly Mediterranean and Nordic; 99% Roman Catholic
Political system characteristics			
Type of government	Republic	Constitutional monarchy	Parliamentary monarchy
Colonial heritage	Not relevant	Not relevant	Not relevant
Concentration of power among national-level institutions	Very concentrated	Some pluralism	Pluralistic
Autonomy of subnational levels of government	Little autonomy	Municipalities and provinces have some autonomy	Some autonomy
Maritime jurisdictions	12-NM territorial sea, 200-NM EEZ	12-NM territorial sea, 200-NM EEZ	12-NM territorial sea, 200-NM EEZ
Major coastal and ocean issues	Pollution, tourism, land-based sources of pollution, control of urban and indust-rial development, degradation of productive coastal ecosystems, oil transportation, fishery depletion, marine scientific research, conflicts over fisheries (France vs. Spain)	Coastal defense, water pollution (heavy metals, organic compounds, nutrient loading), wetland loss, com-petition for coastal space, sea-level change, vessel-source pollution, adverse environ-mental effects of sand mining, offshore oil and gas develop-ment, sea-use zoning	Coastal erosion, tourism development, general urbanization, sedimentation, degraded water quality, loss of productive ecosys-tems, intergovern-mental duplication, marine pollution, competition for fishery resources
Level of government concerned with ICM	Mainly national	National, but regional and local water	National, but regional and local

(*continues*)

Table 10.1 (*continued*)

	France	**Netherlands**	**Spain**
		boards are also important	governments also play a role
Nature and timing of ICM efforts	1970s, Commission on CZM: "State of the Art"; 1975, Conservatoire du Littoral; 1983, Schéma de Mise en Valeur de la Mer; 1986, Loi Littoral (Seashore Act); 1995, new Secrétariat à la Mer (linked to prime minister's office)	1984, North Sea Harmonization Policy; 1991, Dynamic Preservation Strategy; 1995, Coastal Defense Act	1988, Shores Act; 1992, Regional Guidelines for Coastal Zone
Overall approach to ICM	Top-down	Top-down, bottom-up	Top-down, bottom-up
Type of approach to ICM	Regulatory—land use planning; acquisition of coastal property	Regulatory, studies	Regulatory
Extent of implementation of ICM	Little implementation; partial implementation of Schéma de Mise en Valeur de la Mer on the Mediterranean coast	Reportedly good implementation	Little implementation
Importance of external assistance	Not important	Not relevant	Of little importance
Importance of UNCED influence	Of little importance	Very important externally	Of little importance
Effectiveness of ICM	Uncertain	Reportedly effective	No data
Movement toward policy integration	Possible with new Secrétariat à la Mer—concerned with fishery conflicts, marine transport	Significant— provincial consultative bodies (coordinating mechanisms)	Little

Table 10.1 (*continued*)

	Republic of Korea	**Australia**
Background variables and socioeconomic context		
Percentage of population on the coast	40%	75%
Per capita GNP	$8,260	$18,000
Major ethnic groups and religions	Homogeneous; 48.6% Christian, 47.4% Buddhist	93% Caucasian, 7% Asian and other; 75% Christian
Political system characteristics		
Type of government	Republic	Federal parliamentary democracy
Colonial heritage	Japanese	British
Concentration of power among national-level institutions	Very concentrated	Some pluralism
Autonomy of subnational levels of government	Provinces have little autonomy; local governments have some autonomy	Substantial autonomy
Maritime jurisdictions	12-NM territorial sea (1978) 3 NM in Korea Strait, 200-NM EEZ (1996)	12-NM territorial sea, 200-NM EEZ
Major coastal and ocean issues	Use conflicts and environmental deterioration due to high rate of coastal area reclamation; establishment of heavy industrial complexes along southern coast and increase in coastal population; loss of fishery nursery grounds, threats to mariculture from water pollution, red tides, oil spills	Habitat destruction, resource depletion, use conflicts, government fragmentation, offshore oil development, marine protected areas, coastal fisheries
Primary level of government concerned with ICM	National	Provincial (state); also national and local
Nature and timing of ICM efforts	Late 1970s, Environmental Preservation Act, Marine Pollution Prevention Act; 1987, Marine Development	*State*: 1994–1995, Western Australia State Coastal Review; 1995, Queensland Coastal Management Bill
	Basic Law, Marine Development Council; 1994, pilot study on ICM in Chinhae Bay; 1996, Ministry of Maritime Affairs and Fisheries; 1996–1997, National	(ongoing), Victoria Coastal and Bay Management Act; 1995–1996, New South Wales Revised Coastal Policy; 1996, Tasmanian Draft State Coastal Policy

(*continues*)

Table 10.1 (*continued*)

	Republic of Korea	**Australia**
	Coastal Survey; by 1997 Coastal Management Act; by 1998, National Coastal Management Plan	*Federal*: 1980, 1991, 1993, Commonwealth inquiries; 1991–2001 Ocean Rescue 2000; 1995, State of Marine Environment Report; 1995, the Commonwealth Coastal Policy; 1996–1997, proposed National Oceans Policy
Overall approach to ICM	Top-down; some local activity	Bottom-up, with top-down assistance
Type of approach to ICM	Regulatory, planning, participatory	Regulatory on ocean uses, studies, consensus building, intergovernmental incentives
Extent of implementation of ICM	Partial implementation (at site of pilot study)	Implementation just beginning
Importance of external assistance	Not important	Not relevant
Importance of UNCED influence	Very important	Somewhat
Effectiveness of ICM	Too early to tell, but looks very promising	Too early to tell
Movement toward policy integration	Some: Chinhae Bay pilot case, creation of Ministry of Maritime Affairs and Fisheries	Some: commonwealth inter-departmental coastal commit tee, intergovernmental coastal reference group (federal and state governments), national coastal advisory committee (including community and industry), state and federal memoranda of understanding (MOUs)

[a] NM = nautical miles.
[b] EFZ = exclusive fishing zone.

Table 10.2. Case Comparisons of Five Middle Developing Countries

	Brazil	**Turkey**	**Thailand**
Background variables and socioeconomic context			
Total population	160 million	63 million	60 million
Percentage of population on the coast	38%	47%	70%
Per capita GNP	$2,970	$2,500	$2,410
Major ethnic groups and religions	55% Caucasian, 38% mixed Caucasian and African; 70% Roman Catholic	80% Turkish, 20% Kurdish; 90% Muslim	75% Thai, 14% Chinese, 11% other; 95% Buddhist
Political system characteristics			
Type of government	Federal republic	Republican parliamentary democracy	Constitutional monarchy
Colonial heritage	Portuguese	Not relevant	Not relevant
Concentration of power among national-level institutions	Some pluralism	Very concentrated	Very concentrated
Autonomy of subnational levels of government	Provincial level has substantial autonomy; local level has some autonomy	Provincial and local levels have little autonomy	Provincial level has little autonomy; local level has some autonomy
Maritime jurisdictions	12-NM[a] territorial sea, 200-NM EEZ	12-NM territorial sea on the Black Sea coast, 6-NM territorial sea on the Aegean Sea coast; 200-NM EEZ	12-NM territorial sea, 200-NM EEZ
Major coastal and ocean issues	Destruction of mangroves, coastal erosion, tourism, urbanization, shipyard development, nonpoint source pollution, multiple-use competition, offshore oil production, offshore fisheries	Rapid coastal industrialization and urbanization, pollution, tourism, public health, multiple-use competition, decline in commercial fishery stocks, pollution, nature conservation, endangered species	Coral reef destruction, mangrove deforestation, decline in fishery stocks, pollution

(continues)

Table 10.2 (*continued*)

	Brazil	**Turkey**	**Thailand**
Primary level of government concerned with ICM	National	National	National
Nature and timing of ICM efforts	1974, ocean planning; 1980s, environmental laws; 1983, Coastal Program (PROGERCO)	1980s, environmental laws; 1983, first version of Shore Law (new version enacted in 1990 and amended in 1992); 1988, special management areas; 1988–1989, ICM pilot study of Izmir Bay; 1994–1996, other pilot study projects	1991, National coral reef management strategy; 1992, ICM management plan Ban Don Bay and Phangnga Bay
Overall approach to ICM	Bottom-up, especially in terms of local coastal ecosystem management initiatives	Top-down; bottom-up in pilot study	Top-down; bottom-up in pilot studies
Type of approach to ICM	Regulatory, institutional	Regulatory, pilot study, workshops and conferences	Regulatory (for coral reef protection), participatory
Extent of implementation of ICM	Little implementation	Partial implementation	Partial implementation (coral reefs)
Importance of external assistance	Of little importance	Moderately important	Very important
Importance of UNCED influence	Of little importance	Moderately important	Moderately important
Effectiveness of ICM	Little data	Reportedly moderately effective	Reportedly moderately effective
Movement toward policy integration	Some, at national level	Some	Some, in pilot study

	Malaysia	**Fiji**
Background variables and socioeconomic context		
Total population	19 million	770,000

Table 10.2 (*continued*)

	Malaysia	**Fiji**
Percentage of population on the coast	70%	100%
Per capita GNP	$3,480	$2,250
Major ethnic groups and religions	59% Malay, Chinese Indian; 53% Muslim, remainder Christian, Hindu	49% indigenous Fijian, 46% Indian; 52% Christian, 38% Hindu, 8% Muslim
Political system characteristics		
Type of government	Constitutional monarchy federation	Republic
Colonial heritage	British	British
Concentration of power among national-level institutions	Some pluralism	Very concentrated
Autonomy of subnational levels of government	States have substantial autonomy; communities have little autonomy	Local communities have substantial autonomy
Maritime jurisdictions	12-NM[a] territorial sea, 200-NM EEZ	Archipelagic base- lines; 12-NM territorial sea, 200-NM EEZ
Major coastal and ocean issues	Erosion, mangrove loss, coral reef destruction, pollution from tourism and industry	Pollution from coastal develop- ment, mangrove deforestation and coral reef loss, fishery loss, vulnerability to climate change
Primary level of government concerned with ICM	National, provincial	National, local
Nature and timing of ICM efforts	1984, national coastal erosion study; 1992, national and coastal resources management policy; 1992, South Johore Coastal Plan	1993, National Environmental Strategy; 1995, ICM pilot efforts
Overall approach to ICM	Top-down, bottom-up	Top-down, bottom-up
Type of approach to ICM	Regulatory—for erosion and mangrove forests	Regulatory, educational
Extent of implementation of ICM	Partial implementation (erosion, mangrove forests)	Little implementation
Importance of external assistance	Moderately important (ASEAN/US)	Very important
Importance of UNCED influence	Very important	Very important

(*continues*)

Table 10.2 (*continued*)

	Malaysia	Fiji
Effectiveness of ICM	Appears good for erosion issues—unknown otherwise	Too early to tell
Movement toward policy integration	Some, in pilot study	Some, with implementation of National Environmental Strategy

[a] NM = nautical miles

Table 10.3. Case Comparisons of Nine Developing Countries

	Ecuador	China	Indonesia
Background variables and socioeconomic context			
Total population	11 million	1.2 billion	203 million
Percentage of population on the coast	45%	40%	60%
Per capita GNP	$1,280	$530	$880
Major ethnic groups and religions	55% mestizo, 25% indigenous; 95% Catholic	92% Han Chinese; officially atheist	45% Javanese, 14% Sudanese, several others; 87% Muslim
Political system characteristics			
Type of government	Republic	Communist state	Unitary republic
Colonial heritage	Spanish	Not relevant	Netherlands
Concentration of power among national-level institutions	Some pluralism	Very concentrated	Very concentrated
Autonomy of subnational levels of government	Some autonomy	Some autonomy	Substantial autonomy
Maritime jurisdictions	200-NM[a] territorial sea	12-NM territorial sea, 200-NM EEZ, continental shelf in Yellow Sea and East China Sea	Archipelagic base-lines; 12-NM territorial sea, 200-NM EEZ
Major coastal and ocean issues	Mangrove deforestation, estuarine pollution, fishery depletion (shrimp postlarvae)	Economic development of coastal resources, pollution, use and agency conflicts, reclamation and mangrove loss, coral and sand mining, fishery depletion	Overfishing and destructive fishing, habitat destruction, marine pollution, conversion of mangrove forests and wetland areas, coral mining

Table 10.3 (*continued*)

	Ecuador	**China**	**Indonesia**
Primary level of government concerned with ICM	National, local	Provincial	National, provincial
Nature and timing of ICM efforts	1981, first workshops; 1986, USAID/CRC program; 1988–1989, coastal resource management program; 1993, five special area management programs	1980–1986, 1988–1993, comprehensive surveys of coastal environment and resources; late 1980s, provincial CZM regulations and provincial oceanic administrations created in eleven provinces; 1994, pilot ICM project in Xiamen	1990, Conservation of Living Natural Resources and Their Ecosystem Act (marine parks); 1992, pilot study, integrated management plan for Segara Anakan-Cilacap; 1993–1997, 1998–2002, marine resources evaluation and planning project
Overall approach to ICM	Top-down, bottom-up	Bottom-up	Top-down, bottom-up
Type of approach to ICM	Regulatory, planning, educational	Regulatory, planning, educational	Planning
Extent of implementation of ICM	Partial implementation	Partial implementation	Partial implementation
Importance of external assistance	Very important	Of little importance	Very important
Importance of UNCED influence	Of little importance	Somewhat important	Moderately important
Effectiveness of ICM	Unknown	Unknown	Unknown
Movement toward policy integration	Little, pilot projects still small scale	A little, in pilot study	Some

	Pakistan	**India**	**Philippines**
Background variables and socioeconomic context			
Total population	131 million	936 million	73 million
Percentage of population on the coast	10%	25%	80%
Per capita GNP	$430	$320	$950
Major ethnic groups and religions	Punjabi, Sindhi, Pashtun, Baloch, Muhajir (from India); 97% Muslim	72% Indo-Aryan, 25% Dravidian; 80% Hindu, 14% Muslim	91.5% Christian Malay, 4% Muslim Malay; 92% Christian, predominantly Roman Catholic

(*continues*)

Table 10.3 (*continued*)

	Pakistan	India	Philippines
Political system characteristics			
Type of government	Republic	Federal republic	Republic
Colonial heritage	British/India	British	Spanish/U.S.
Concentration of power among national-level institutions	Very concentrated	Pluralistic	Very concentrated
Autonomy of subnational levels of government	Little autonomy	Some autonomy	Substantial autonomy, especially at the regional and community levels
Maritime jurisdictions	12-NM territorial sea, 200-NM EEZ	12-NM territorial sea (1967) 200-NM EEZ (1976)	100-NM territorial sea (285 NM in South China Sea); 200-NM EEZ (1978)
Major coastal and ocean issues	Effects of reduced freshwater flow on mangrove forests, shrimp aquaculture, and productivity; seawater intrusion; industrial pollution	Environmental degradation due to coastal urbanization and accelerated industrial development; pollution; coastal hazards and disasters; uncontrollable exploitation of mangrove forests and and coral reefs	Fishery depletion through overfishing, use of dynamite and habitat destruction; coral depletion through mining; loss of mangrove forests and wetlands through expansion of human settlements
Primary level of government concerned with ICM	National, provincial, some local	National	National, local
Nature and timing of ICM efforts	1994, IOC/UNESCO workshop on ICM	1974, Water Prevention and Control of Pollution Act (central and state water boards); 1981, Department of Ocean Development; 1982, ocean policy; 1991, Coastal Zoning Law, national program on marine sensing information service; 1996, beginning of ICM activitied at state level	1975, Fisheries Act; 1976, Coastal Zone Management Committee of National Environmental Protection Council; mid-1980s, ASEAN/US Coastal Resources Management Project, Lingayen Gulf pilot study; 1990–1994, fisheries sector programs (includes coastal management aspects)
Overall approach to ICM	Top-down	Top-down	First top-down, then bottom-up

Table 10.3 (*continued*)

	Pakistan	**India**	**Philippines**
Type of approach to ICM	Too early to tell	Regulatory	Regulatory, planning, participatory
Extent of implementation of ICM	No implementation yet—too recent	Partial implementation	Partial implementation
Importance of external assistance	Very important	Of little importance	Very important
Importance of UNCED influence	Very important	Of little importance	Of little importance
Effectiveness of ICM	No data	No data	Reported highly effective
Movement toward policy integration	A little, as a result of the workshop	Little	Some

	Sri Lanka	**Federated States of Micronesia**	**United Republic of Tanzania**
Background variables and socioeconomic context			
Total population	18 million	122,950	28.7 million
Percentage of population on the coast	34%	100%	No data
Per capita GNP	$640	Est. $726–$2,895	$140
Major ethnic groups and religions	74% Sinhalese, 18% Tansil; 69% Buddhist, 8% Christian, 8% Muslim	Nine ethnic Micronesian and Polynesian groups; 50% Roman Catholic, 47% Protestant	99% native Africans, refugees from African countries at war; 45% Christian, 35% Muslim, 20% indigenous beliefs
Political system characteristics			
Type of government	Republic	Constitutional government in free association with the United States	Republic
Colonial heritage	British	U.S.	British and German
Concentration of power among national-level institutions	Very concentrated	Pluralism	Some pluralism
Autonomy of subnational levels of government	Provincial level has some autonomy; community level has little autonomy	Substantial autonomy	Little autonomy

(*continues*)

Table 10.3 (*continued*)

	Sri Lanka	Federated States of Micronesia	United Republic of Tanzania
Maritime jurisdictions	12-NM territorial sea, 200-NM EEZ (1976)	12-NM territorial sea, 200-NM EEZ	12-NM territorial sea, 200-NM EEZ
Major coastal and ocean issues	Coastal erosion; coral and sand mining; degradation of coastal habitats; tourism; pollution; loss and degradation of archaeological, cultural, and scenic sites; fisheries depletion	Coastal erosion, sea-level rise, pollution from coastal development, inadequate sewage facilities, limited freshwater resources, fishery depletion, conflict between artisanal and commercial fisheries	Local community uses: reef blasting, use of dynamite in fishing, depletion of nearshore fish stocks, mangrove deforestation, seaweed farming, tourism development, land-based activities such as coastal agriculture and ranching, river impoundments, coastal deforestation
Primary level of government concerned with ICM	National	State	Federal (policy, legislation), state (community-based management)
Nature and timing of ICM efforts	1978, Coast Conservation Division; 1981, Coast Conservation Act; 1983, coastal permit system; 1988, Coastal Zone Management Plan; 1990, Marine Pollution Prevention Authority; 1995, special area management plan; 1996, second Coastal Zone Management Plan	1992, Kosrae State Coastal Management Program (development permitting, land use planning, coordination of other ICM functions, e.g., public education); Yap and Pohnpei CZM planning	1993, workshop on ICM in East Africa, Tanga ICM pilot project
Overall approach to ICM	Bottom-up, Top-down	Bottom-up	Top-down, bottom-up
Type of approach to ICM	Regulatory, special area management strategy, permit system	Regulatory	Regulatory, consensus building, educational
Extent of implementation of ICM	Full implementation	Little implementation	Partial implementation

Table 10.3 (*continued*)

	Sri Lanka	**Federated States of Micronesia**	**United Republic of Tanzania**
Importance of external assistance	Very important	Very important	Very important
Importance of UNCED influence	Important	Moderately important	Moderately important
Effectiveness of ICM	Moderately effective	No data	No data
Movement toward policy integration	Early move toward integration—creation of Coast Conservation Department (CCD), Conservation Department in Ministry of Fisheries (does planning and coastal engineering, issues permits for coastal activities), Coast Advisory Council (advises CCD, issues variances)	Some (National Marine Resources Division has a broad coordinating role, interagency committee is set up in reference to National Environmental Strategy)	Some

[a] NM = nautical miles

Patterns of Commonalities and Differences

In this section, we make some generalizations on the basis of the case comparisons. Examination of tables 10.1, 10.2 and 10.3 and review of data from the cross-national survey presented in earlier chapters suggest a number of commonalities and differences among the nations.

General Observations

Although each nation has pursued a unique path to ICM, many similarities among nations may be identified.

1. *Problems faced by coastal and marine managers exhibit similar patterns in nations around the world.* On the whole, environmental degradation, marine pollution, fishery depletion, and loss of marine habitat are problems faced by both developed and developing nations. Many nations experience conflicts among uses and among agencies administering coastal and marine programs, as seen in both the case studies and the results of the cross-national survey. In nations with similar geographical features and levels of development (for example, tropical developing nations), a commonly seen "bundle" of problems involves environmental degradation and pollution due to urbanization,

fishery depletion through overfishing and improper fishing methods (such as the use of dynamite), coral reef depletion and degradation, and loss of mangrove forests and wetlands through competing activities such as aquaculture, tourism, and expansion of human settlements.

2. *The evolution of each ICM program differs due to each nation's unique combination of geography, development issues, and political system; nevertheless, some common patterns can be identified.* Common patterns include the creation of environmentally oriented institutions (departments of environment, water quality boards, etc.) in the 1970s and 1980s, partly in response to the United Nations Conference on the Human Environment; the creation of sectoral coastal and ocean programs in the 1980s or earlier to deal with problems such as shoreline erosion, coastal pollution, and the need for marine protected areas; and growing concern with integrated management in the 1990s, especially following the Earth Summit. As discussed in chapter 6, the influence of UNCED on the creation of more integrated ICM institutions is more evident in middle developing and developing countries than in developed countries.

3. *In most nations, more than one level of government is involved in ICM, with the national government often having primary responsibility*, as found in responses to the cross-national survey. In the case studies, it is rare to find a nation in which ICM has been carried out by only one level of government. Most nations tend to employ a combination of top-down and bottom-up approaches—in our view, the way ICM should be carried out. The national government often begins by taking primary responsibility for ICM (top-down first) and then, in various ways, encourages subnational entities to engage in ICM activities (bottom-up later).

In countries with a federal system of government—such as Australia, the United States, and Brazil—relations between the states and the national government over marine and coastal resources are often conflictual; typically, much negotiation and bargaining takes place among the different levels of government.

In a growing number of countries—for example, Canada, the United Kingdom, the Philippines, Ecuador, Australia, China, Brazil, and some states in the United States, an important role in ICM is played by local authorities and communities. This is especially the case where coastal communities depend on marine resources that are facing significant declines, as on the Atlantic coast of Canada and in the Philippines.

4. *It is difficult to assess the extent of implementation of ICM.* Although more than half of the respondents to our cross-national survey reported that their nations were in the ICM implementation phase, it is difficult to determine exactly what this means. Our country case narratives are not sufficiently

detailed to capture the true extent of implementation. Whereas there is evidence in most of the countries of much ICM planning activity (especially apparent in pilot studies in developing countries with external assistance), there is little tangible evidence of on-the-ground changes in human behavior and in the status of coastal ecosystems. The evaluative data needed to make these assessments are notably lacking. It is possible, too, that in many cases on-the-ground changes are not yet taking place.

5. *A similar bundle of ICM tools and approaches is seen in many nations.* The specific ICM programs in various nations often combine regulatory approaches (e.g., prohibiting development in hazardous areas to control coastal erosion or regulating particular activities in marine protected areas); coastal and marine policy development (development of general goals, guidelines, and standards, typically at the national level, to guide the ICM management process); planning processes (examining conditions in a particular coastal area or for the nation's entire coast, envisioning future desirable states, and developing specific plans containing prescriptions for action); the conduct of studies to diagnose natural system and socioeconomic system issues and government capacity for ICM; consensus-building and participatory approaches to address particular coastal and marine use conflicts and other problems. The uses we have seen of another management tool—provision of incentives— involve financial and "consistency review" incentives provided by one level of government (typically the national level) to another (states or localities) to encourage planning and management, generally following guidance established by the higher level of government, as in the United States and Australia.

6. *The importance of external assistance for ICM varies among countries.* External funding assistance appears to have played an important role in the ASEAN countries we have examined—Thailand, Malaysia, Indonesia, and the Philippines; in the Indian Ocean countries Sri Lanka and Pakistan; in the Pacific island nations Fiji and the Federated States of Micronesia; in African states such as Tanzania; and in some Latin American nations, such as Ecuador. External assistance not only involves financial support but also often influences the ICM approaches and methods ultimately adopted by the recipient countries. The ASEAN nations mentioned, for example, appear to have emphasized a pilot program approach, mirroring the general approach followed in the ASEAN-US Coastal Resources Management Project and more recently in the GEF/UNDP/IMO Regional Programme for the Prevention of Marine Pollution in East Asian Seas.

7. *A movement toward greater integration (intersectoral, intergovernmental, spatial, and interdisciplinary) is evident in recent years in many nations.* The past several years have seen an increase in the number of mechanisms for coordination and consensus building—for example, new national-level insti-

tutions (the Republic of Korea's Ministry of Maritime Affairs and Fisheries); interagency coordinating committees (Fiji, the Federated States of Micronesia, Australia, and the Philippines); state-federal coordinating mechanisms (Australia and the Netherlands), local and national fora for discussion of coastal issues (United Kingdom); and local consensus-building processes (Canada). This trend very much mirrors the prescriptions found in Chapter 17 of Agenda 21.

Similarities and Differences among Developed, Middle Developing, and Developing Countries

Our comparisons of countries, based on both the country case studies and the results of the cross-national survey, suggest that there are more similarities than differences among developed, middle developing, and developing countries. Recalling our discussions of data from cross-national survey in past chapters (especially chapters 2, 6, and 9), in most cases all three sets of respondents gave similar answers regarding catalysts responsible for ICM initiation, types of ICM activities conducted, types of ICM boundaries established, primary level of government responsible for ICM, ICM steps completed, importance of conflict management in ICM, types of ICM coordinating mechanisms established, and roles of external participants (NGOs and industry) in the ICM process.

Similarly, there were no major differences among developed, middle developing, and developing countries on the longevity of ICM programs. Long-standing programs were found in both the developed world (United States, 1972) and the developing world (Philippines, 1976; Sri Lanka, 1978), and very recently established programs were also found in these different contexts (Canada, 1996; Korea, 1994 and 1996; Fiji, 1995).

Differences among the three groupings, however, were found on some variables:

- Regarding reasons for initiating ICM programs, middle developing and developing countries tended to cite environmentally related issues, whereas developed nations tended to cite issues related to economic opportunity. As discussed earlier, we think that this seemingly counterintuitive finding suggests that developing and middle developing nations, many of which have experienced rapid development of their ocean and coastal areas in recent years, have now become aware of the environmental costs of too rapid or inappropriate development.

- ICM activities were largely similar in all development contexts, with small exceptions—promotion of marine recreation was greater in developed countries, and promotion of ecotourism was greater in developing countries, especially those located in tropical areas.

• Respondents from middle developing and developing countries were more likely to cite the influence of UNCED prescriptions on the design of their ICM coordinating mechanisms.

Respondents from developed countries were more likely to report serious problems with policy integration (along intersectoral, intergovernmental, spatial, and science and policy lines) than were those from middle developing and developing countries. This could be due, we think, to a number of reasons: (1) in developed countries government bureaucracies are highly diversified and bureaucratic divisions are typically long-standing, deeply ingrained, and based on divergent legal mandates; (2) in a number of developing and middle developing countries, such as Thailand, Brazil, and Mexico, power at the national level is more centralized and hence the chief executive is able to exert greater harmonizing influence; and (3) in smaller developing nations such as Fiji and the Federated States of Micronesia, there are fewer government players and they are well known to one another. With regard to the science-policy interface, natural and social scientists in developing countries tend to be more aware of and attuned to economic development and social issues than are their colleagues in developed countries. Moreover, they typically depend on funding from government sources and thus are more likely to produce research results that are management relevant.

We now compare the ICM efforts of specific nations within the three groupings.

Developed Nations

Among the developed nations addressed in the country case studies—Canada, the United States, the United Kingdom, France, the Netherlands, Spain, Korea, and Australia—the United States and the Netherlands have the most experience in both coastal and ocean management. The **United States** was the first to create a national state-oriented program of coastal management through national legislation, in 1972. Implementation of this program has evolved over the years, with great variations evident among the thirty-one different coastal states and island territories according to various physical, socioeconomic, and political conditions present in each. The national government was the key catalyst in providing incentives (money and "consistency" review) to the states for both planning and implementation of coastal programs. In the program's first decade, state coastal programs tended to emphasize land-oriented and shoreline issues; in the mid-1980s and into the 1990s, a number of them began serious efforts to include ocean waters in their ICM programs. Most states limited their purview to state-controlled waters (to three nautical miles offshore), though some, such as Oregon, attempted to exert their influence in federal waters as well (beyond three nautical miles). Other national legislation enacted in the 1970s included a very complex body of ocean law embracing almost every conceivable ocean use from fishing operations and oil and gas development to marine mammal protection, establishment of marine protected areas, and protection of water quality. Indeed, the United States has perhaps the most elaborate legal framework in the world for management of ocean and coastal resources. However, even though significant achievements have been made in coastal and ocean management, serious problems of policy integration exist (Cicin-Sain and Knecht 1992). This is particularly the case in terms of intersectoral integration at the national level, where frequent conflicts among agencies are exacerbated by the absence of interagency coordinating mechanisms, and in the area of science-policy integration, wherein very advanced science capability is typically divorced from the management arena (Knecht and Archer 1993; NAS 1995). Spatial integration was considerably enhanced by 1990 amendments to the 1972 Coastal Zone Management Act emphasizing nonpoint sources of pollution and bringing together the major agencies involved in water quality protection and coastal management. Intergovernmental integration has been successfully achieved through a system of federal incentives (monetary incentives and "consistency" reviews).

The **Netherlands** has extensive experience and expertise in ICM, particularly regarding coastal defense (through hard structures), an area in which it is the world leader. The Netherlands also became a world leader in harmonization of national coastal and ocean policies with its establishment of interagency and intergovernmental coordinating mechanisms and naming of a lead agency. The Netherlands has had great success in building technical capacity for ICM and developing a close connection between science and management.

France and Spain generally follow regulatory approaches to ICM, and as discussed in the country case studies in appendix 1, problems exist in both countries in the intergovernmental dimension, with duplication and overlap among national, state, and local authorities. **France**, however, has been a leader in ocean sciences and technology and appears to be beginning to direct these efforts to the analysis of coastal zone issues, as seen in the work of the Institut Français de Recherches sur Exploitation de la Mer (IFREMER). **Spain** has completed excellent technical work in marine mapping, with very detailed assessment and mapping of baselines and internal waters (MOPU 1988a).

Canada began developing a national ICM program as part of its national oceans policy. It has had success in community-level ICM efforts and in consensus building among local stakeholders, particularly in the Atlantic provinces. Integrated coastal management was adopted as one of the underlying principles of Part II of the Canada Oceans Act of 1996. In this context, interested parties, authorities, and stakeholders work together toward agreement on common goals, plans, and policies affecting specific coastal and ocean issues. The **Republic of Korea**, after pursuing a very strong marine development program and an extensive program of coastal land reclamation, faced significant issues of marine degradation. Korea's national government has apparently embraced the Agenda 21 model of ocean and coastal management, perhaps to the greatest degree of all nations studied. The government is proceeding deliberately and systematically to merge ocean and coastal management into a new Ministry of Maritime Affairs and Fisheries while testing key ideas and concepts at the local level through a pilot study in Chinhae Bay.

Australia, long noted for its multiple-use management of the Great Barrier Reef, did not have an ICM policy or program at the national level until recently, notwithstanding its conduct of inquiry after inquiry documenting coastal problems. In recent years, this situation has changed significantly, with important ICM initiatives coming from both the commonwealth and state levels, including formulation of commonwealth oceans policy to enhance integration of activities. In terms of integrating the natural and social sciences, the Ocean Rescue 2000 program, a ten-year commonwealth program for conservation and understanding of the marine environment, provides a guide for integrating scientific data into coastal and ocean policy making.

Of all the developed nations considered, the **United Kingdom** perhaps lags the most in ICM, notwithstanding its long experience as a great maritime power and its variety of sectoral marine programs. Although there has been some government action and some strong spurring from the NGO community, no tangible ICM policies or programs are yet in evidence at the national level. At the local level, considerable experience exists in coastal planning, but there are major problems in connecting land-based and sea-based planning. Inquiries at the national level in 1992, combined with the formation of a national government interdepartmental group and the convening of coastal fora for discussion and consensus building at the local level, may well portend positive movement toward ICM in the near future.

Middle Developing Nations

With the exception of Fiji, characterized by a small population, the other middle developing nations examined—Brazil, Turkey, Thailand, and Malaysia—are all relatively large countries (especially Brazil) with growing economies. All have taken steps toward ICM. **Brazil** began coordination measures in CZM in the mid-1980s. **Turkey's** efforts in special area management, ICM pilot studies, and shore management began in the late 1980s and are ongoing. **Malaysia** began its ICM efforts with a national coastal erosion study in 1984 and continued them with the development of a national coastal resources management policy and an ICM pilot program, both in 1992.

Thailand undertook ICM work in 1991 with its development of the national coral reef management strategy. Among other ICM efforts in Thailand, ICM management plans prepared with external assistance for Ban Don Bay and Phangnga Bay in 1992 represent comprehensive attempts toward ICM implementation. In **Fiji**, national-level work on ICM, including a pilot project on Ovalau Island, was begun in 1995, following development of the nation's national environmental strategy in 1993.

All these countries combine bottom-up and top-down approaches to ICM. It appears that in every case, ICM development has been moderately or significantly influenced by the UNCED model.

Developing Nations

The developing countries examined in our case studies represent a diverse lot from all major geographical regions of the world—from the most populous, China, to one of the least populated, the Federated States of Micronesia, and from the more affluent, such as Ecuador, to one of the poorest, Tanzania.

Sri Lanka and Ecuador have some of the oldest programs in coastal management, and both have been influenced by external funding from the USAID program administered by the Coastal Resources Center at the University of Rhode Island. **Sri Lanka** represents a good example of how ICM may proceed in stages—beginning with one immediate, salient issue (coastal erosion in the case of Sri Lanka) and evolving over time into a more comprehensive program embracing many other issues in the coastal zone. **Ecuador**, too, started out in a limited way, with pilot projects, but emphasizing a national approach and the creation of a national coordinating committee to ensure national-level support of ICM.

China, with long experience in the ocean sciences through its State Oceanic Administration and oceanographic institutes, has tied marine development to its unprecedented drive for economic development in the past decade or so, occasioning many environmental problems in the process. China has conducted large-scale systematic surveys and assessments of the coastal zone and prepared a technical base for the establishment of ICM. Although ICM has not yet been undertaken at the national level, there are interesting examples of ICM experimentation in the provinces, such as the setting up of provincial ICM councils, and at the local level, as in the coastal city of Xiamen, where a major ICM demonstration project funded by the GEF/UNDP/IMO is taking place (Chua and Yu 1996).

Indonesia and the Philippines have both been influenced by external assistance from the ASEAN-US program, and both have experimented with ICM pilot efforts typical of the ASEAN-US approach. The **Philippines** has emphasized community-based management and tied fisheries operations to coastal resources management, a practice that is desirable but seldom seen in most countries. Although it is the largest archipelagic nation in the world, **Indonesia** only recently appears to have realized the potential of ICM. In 1995, it embarked upon a seven-year, two-track approach involving both ICM pilot projects at the local level and a strengthening of coordination and policy making at the national level (Dahuri and Dutton 1997). It has also enacted marine park legislation, and become engaged in major efforts at marine resources evaluation and planning and coral reef rehabilitation and management.

India has tended to pursue its own unique course of ICM development. It was one of the first countries to create a Department of Ocean Development (1981), presumably in response to the Law of the Sea negotiations, and to create a regulatory framework for coastal water quality (1974). In 1991, it enacted coastal zoning legislation and by 1997, using guidelines formulated by DoD, several provinces were beginning the process of developing coastal management programs (Sabramanian 1997).

Two of the developing countries discussed, Tanzania and Pakistan, are in the process of beginning ICM programs. **Tanzania** has recently taken first steps in ICM, with successful follow-up implementation. These steps have entailed establishment of an ICM pilot project in Tanga involving extensive community participation. **Pakistan** also has taken first steps in ICM through analysis of its coastal problems and development of a strategy for beginning ICM, in connection with an in-country workshop funded by the IOC in Karachi in November 1994. It appears, however, that little implementation of the steps outlined in the proposed strategy has yet taken place.

Summary

Drawing on the case studies of twenty-two countries in appendix 1 and on the results of our 1996 cross-national survey, this chapter compared the ICM practices of twenty-two nations, both developed and developing, on a range of variables. Among the variables examined were national background and characteristics, nature of coastal and ocean problems, nature of maritime jurisdictions, and nature of government responses in ICM, depicting the kinds of ICM approaches established and other factors including nature and timing of ICM activities, level of government involved, and extent of implementation and effectiveness. In general, there appear to be more commonalities than differences among the nations. The differences perceived appear to be related to factors other than level of development; for example, developed nations differ among themselves just as much or more than developed and developing nations differ as groups.

Chapter 11

Summary and Conclusions

Introduction

As the previous chapters have shown, integrated coastal management has been recognized by the international community as a framework of central importance and is being adopted as a practice in many countries around the world. As we define it, ICM is a continuous and dynamic process by which decisions are made for the sustainable use, development, and protection of coastal and marine areas and resources. ICM is a process that recognizes the distinctive character of the coastal area—itself a valuable resource for current and future generations. ICM is multipurpose oriented; it analyzes implications of development, conflicting use, and interrelationships among physical processes and human activities, and it promotes linkages and harmonization between sectoral coastal and ocean activities. The goals of integrated coastal management are to obtain sustainable development of coastal and marine areas, to reduce vulnerability of coastal areas and their inhabitants to natural hazards such as storms and accelerating sea-level rise, and to maintain essential ecological processes, life support systems, and biological diversity in coastal and marine areas.

We have dedicated many pages in the preceding chapters to defining the central characteristics of ICM—its scope, functions, goals, principles, typical activities, institutional forms, analytical methods, and management tools—as well as the dynamics of the ICM process, including initiation, formulation, implementation, operation, and evaluation. Generally, we have attempted to provide both examples and advice regarding practices and methods that are likely to be effective. Notwithstanding the complexity of the earlier discussions in the book, we believe that it is possible to pinpoint four key features which are central to the ICM approach. These are:

1. The use of a set of principles based on the special character of the coasts and oceans to guide ICM decision making.

2. The need to ultimately work at ICM from both directions—bottom up

(involving the local community level) and top-down (involving the national government).

3. The need to have a coordinating mechanism or mechanisms to bring together coastal and ocean sectors, different levels of government, users, and the public in the ICM process.

4. The need to have good (relevant) science available on a timely basis to inform the ICM decision-making process.

The need for ICM is directly tied to the rapidly increasing pressures being imposed on the world's shorelines and coastal areas. Populations are burgeoning in these areas as more and more people are drawn to the coasts for both economic and recreational purposes. Furthermore, pressures for development of coastal and ocean resources (offshore oil and gas, ocean minerals, fisheries, aquaculture, etc.) often compete with the need to protect and conserve particularly sensitive coastal areas, such as those with valuable and diverse habitats. Similarly, growing human settlements in the coastal zone place additional stresses on natural systems as pollution and sewage loadings increase. Clearly, the best hope for dealing with these multiple and interconnected problems is an integrated approach that brings all coastal sectors, levels of government, and coastal users into a rational goal-setting and decision-making framework such as ICM.

Throughout this book, we have stressed the need to tailor the ICM approach to fit the unique circumstances of each country that undertakes such a program. Countries differ a great deal in their resource bases, levels of economic development, systems of government, cultures and traditions, needs and expectations, and approaches to government regulation and resource management. A successful ICM program will incorporate these distinctions into its structure and, to the maximum extent possible, will build on existing organizations and arrangements, adding new institutions or procedures only to the extent necessary to fill gaps or meet new requirements—for example, to coordinate and harmonize coastal activities.

In this concluding chapter, we first conduct a "reality check" on ICM, raising several questions about the evolving practice of ICM at both the global and national levels. Next, we provide a summary of the practical guide to ICM programs we presented in part III, with observations on the dynamics of the ICM process and decisions coastal managers will need to make as they initiate, formulate, implement, operate, and evaluate ICM programs. We then provide some examples of good practices in ICM programs, although we point out the difficulties of doing this given the scarcity of evaluative data on the accomplishments and problems of ICM programs. Finally, we reflect on future prospects and challenges for integrated management of oceans and coasts.

A Reality Check Regarding ICM

At this point, we think it is useful to pause for a reality check. As mentioned earlier, the ICM concept has been widely adopted as the management framework of choice, both internationally and nationally. Is this justified? To explore this question, we pose and respond to five questions:

1. *Is ICM a planning and management methodology likely to meet the needs of most coastal countries?* We believe that ICM is a methodology sufficiently flexible to meet the different needs of most, if not all, coastal countries. It is a logical framework that contains the elements most experts agree are centrally important to sustainable use of the coastal zone and its resources. It is not geared to any particular political system or type of coastal zone. Nearly three decades of experience suggest that ICM is a workable methodology and it probably is the most appropriate framework for achieving long-term sustainable use of a country's coastal zone.

2. *In response to the call for ICM at the global level, are countries initiating ICM programs?* Extrapolating the results of our cross-national survey, most coastal countries report that they are undertaking ICM programs of one type or another or are planning to do so in the near future. The extent to which countries pursue full-scale ICM programs will undoubtedly depend on such factors as the following:

a. The perceived seriousness of existing coastal problems or opportunities (erosion, resource depletion, ecosystem degradation, new economic activities, etc.).

b. An awareness within the government structure of ICM and its benefits and the presence of user groups, local communities interested in initiating ICM.

c. Availability of necessary resources (trained staff, coastal information and data, finances).

d. The existence of political will within the government to take the necessary steps to initiate and put in place an effective ICM program.

3. *Is the emergence of ICM as the planning and management framework of choice for many international programs a reasonable and logical development?* It is clear that the specific goals of international agreements such as the Framework Convention on Climate Change and the Convention on Biological Diversity cannot be met in isolation. Similarly, achieving the goals of the Global Programme of Action on Protection of the Marine Environment

from Land-Based Activities requires the comprehensive, intersectoral approach called for in the ICM framework. In our judgment, it is appropriate and necessary that these new global initiatives specify ICM as the preferred framework within which to pursue their goals and objectives. Indeed, in most situations there will be a coincidence between a coastal country's own goals and those to which it is committed under various international agreements.

4. *Is appropriate guidance on ICM methodology being provided by the international programs that have designated ICM as the framework within which to meet some or all of their goals?* Only now are the secretariats and other institutions (e.g., expert groups such as the Subsidiary Body on Scientific, Technical, and Technological Advice to the Convention on Biological Diversity) created by the new international agreements beginning to focus on the ICM issue. It seems likely that under at least three of these agreements (the Framework Convention on Climate Change, the Convention on Biological Diversity, and the Global Programme of Action), guidance on ICM will be needed. In February 1997, an International Workshop on Climate Change and ICM produced a set of guidelines for an ICM framework to address such climate change issues as sea-level rise, increased erosion, saltwater intrusion, and increased storm frequency. ICM guidelines may also be developed as part of the implementation of the Jakarta Mandate dealing with the protection of coastal and marine biodiversity under the Convention on Biological Diversity. As discussed later in this chapter, the challenge will be to ensure that the various sets of ICM guidelines produced under these international agreements are generally consistent with one another and that they all share a common view of the ICM framework.

5. *Does a consensus exist on what ICM is and how to carry it out?* We believe the evidence presented in this book demonstrates the existence of a consensus regarding the main features of the ICM methodology. As shown, there is general agreement among the ICM guidelines produced so far by IUCN, UNEP, the OECD, and the World Bank. Some differences do exist, especially in areas of emphasis. But on the whole, the guidelines deal with such important issues as intersectoral and intergovernmental coordination and inland and seaward management boundaries in very similar ways. We expect that as use and application of ICM increase, a tendency toward even greater commonality will be manifested.

A Practical Guide to ICM Programs: Summary

In chapters 5–9, we provided advice for coastal policy makers about the many issues they will confront at different stages of the ICM process. Here we provide a summary of options and possible choices faced during three major phases of the ICM process: initiating an ICM effort, formulating an ICM program and getting it adopted, and implementing and operating the program.

Initiating an ICM Effort

Major issues faced by coastal policy makers as they begin an ICM effort are summarized in this section, as are possible strategies for addressing such issues.

Tailoring ICM to a Particular Country's Context

- It is imperative to tailor ICM to fit the physical, socioeconomic, and political context of a particular country.

- A study of physical and socioeconomic variables establishes which problems and opportunities are present in a coastal or ocean area, why ICM is needed, and what its goals and objectives should be. An understanding of the country's political system is crucial to determining how best ICM can proceed and by whom it should be carried out, and answers such questions as, Who needs to be convinced to establish ICM?; Who will design, implement, monitor, and enforce ICM?; and Whose behavior needs to change to make ICM effective?

- In understanding a political system for ICM purposes, two questions are central: (1) What is the degree of concentration of power and authority among national-level institutions? (2) What is the division of authority between national and subnational (provincial, local) levels of government?

- It may be useful for nations to draw lessons from similar ICM efforts in other countries. Care should be taken, however, to draw lessons from the countries most similar in coastal context, socioeconomic makeup, and political system characteristics. Whereas nations once outright copied and emulated other countries' programs and institutions, today a more common approach is for decision makers to engage in "synthesis" and "inspiration" methods of lesson drawing, shaping unique ICM programs that correspond to their countries' particular circumstances.

Getting Started

- Political will at the national and local levels is essential to establishing ICM. Recognition of the need for an ICM program can develop in many different places—the national government, the local community, the NGO community, a donor agency—but wherever it first emerges, it must find its way to and ultimately be accepted by both the national and local or community levels of government since both generally play important roles in ICM implementation.

• To enhance its political acceptability, ICM should be described not as supplanting sectoral programs but supplementing and strengthening them.

• ICM need not be presented as an elaborate, complex methodology that will require simultaneous full-scale implementation in all coastal areas and for the full range of coastal issues. Rather it should be made clear that ICM can be fashioned in such a way as to be implemented incrementally, beginning in locations most in need of integrated management or with issues of highest priority.

• Visible support for ICM by the constituencies most affected by coastal management measures—users of coastal resources and the concerned public—is important for generating the necessary political will.

• A well-written and well-reasoned document on the need for ICM generally needs to be prepared, showing (1) clear evidence of problems with existing sectoral approaches, such as escalation of conflicts; (2) new management needs, such as protecting biodiversity or addressing the effects of climate change; and (3) new economic opportunities that would benefit from a more integrated approach.

• Such a document should describe the ICM concept and outline the benefits that would be derived from its adoption, as well as how ICM would function in the government setting in question and steps needed to develop an ICM plan and put it into practice, including a timetable, an estimate of costs, and professional staff requirements.

• It is usually best to reserve politically sensitive decisions, such as questions regarding establishment of a lead agency, for later, when the scope and objectives of the program are better known.

• ICM advocates can expect to encounter, and must overcome, a number of barriers to ICM initiation, such as bureaucratic inertia, ideological opposition, and opposition from economic interests tied to the status quo.

• Support of coastal users, NGOs, and the public is crucial to ICM initiation. In this regard, it is useful to create a coastal users' group (users is broadly defined) that (1) includes representatives from all groups likely to be affected by ICM, (2) fully informs members about the ICM process, (3) is structured so members perceive that their suggestions and input are fully considered (i.e., are not a "rubber stamp" for already agreed-to government actions) and can actually affect the way the ICM program is defined and ultimately implemented.

• It is important to gain broad political support for ICM at the national level of government. Some conditions that facilitate this are formulation of the ICM effort by a high-level interagency group that has been named and is led by the highest level of government (e.g., the prime minister's office); is bureaucratically higher than the sectoral agencies; operates with written terms of reference and a timetable; has good representation from the relevant ministries and from provinces, local governments and communities, and NGOs; is run in a transparent manner; and is required to present its findings in a public forum.

• Although the length of time required to initiate and implement ICM will vary greatly from country to country, in our view, the initial work in deciding how to begin ICM (not the creation of an ICM plan or program itself) can take place in about a six-month period. ICM plan formulation, including assessment studies, could take a year or two. Tangible results from ICM should be seen by the end of a five-year period. If the process extends over too long a period without tangible results, momentum for ICM may well be lost.

Structuring a Phased Approach.

• ICM need not be implemented all at once and for a country's entire coastal zone. In many cases a phased approach may be more appropriate—for example, if urgent problems exist more in one part of the nation's coastal zone and less so in others; if one coastal area is institutionally, technologically, or politically ready for ICM and others are not; if funding and staff limitations dictate an incremental approach; or if it is desirable to conduct an ICM or pilot approach in a specific coastal area before applying it generally.

• If a phased approach is chosen, it should be structured carefully so that later increments can benefit from the first effort. Even though the area covered by the initial effort may be small, care should be taken to ensure that the intergovernmental and intersectoral coordinating mechanisms are adequately comprehensive. The institutional, technical and political changes made as part of the pilot effort can be scaled up to ultimately handle the country's entire coastal zone, but tests to judge the success of the pilot effort must be built in as part of the process.

• If human and financial resources are limited, ICM programs can be simplified to include only the following components: (1) harmonization of sectoral policies and goals; (2) cross-sectoral enforcement mechanism; (3) a coordination office; and (4) permit approval and Environmental Impact Assessment (EIA) procedures (FAO, 1991).

Addressing Intergovernmental Issues:
National- and Local-Level Involvement

- Responsibility for the management of coastal and ocean resources rarely falls on one level of government. In most countries, national, provincial, and local governments have some form of jurisdiction or management control over coastal lands and waters. Therefore, it is essential to reach an understanding of the relative roles of national, provincial, and local authorities in ICM.

- National and local governments may sometimes be in conflict over ICM because they have different responsibilities, legal authorities, and priorities and respond to different constituencies. In our cross-national survey, 41 percent of respondents indicated that the nature of the intergovernmental relationship (positive or negative) varied according to the issue, whereas 20 percent of respondents reported a competitive or "hands off" relationship. Among developing countries, 30 percent reported that national-level institutions had little to do with or were generally competitive with state- and local-level institutions.

- Local government resistance to ICM can be expected if a national government's proposal is viewed as shifting power or authority away from the local level, reducing the amount of discretion available to the local government, or imposing additional costs or other burdens on the local government without providing commensurate benefits.

- National government resistance to an ICM proposal that originated at the local level may be expected if the local plan was developed in isolation from the national level or if national concerns were ignored.

- Each level of government, however, brings unique expertise and perspective to the ICM process. The local level can contribute the most detailed understanding of the local coastal zone and its problems, constraints and limitations that will affect the choice of solutions, data and information on the local coastal zone, and support of coastal user groups and the community. The national government, in turn, can contribute specialized data and expertise on various sectors of coastal activity (fisheries, wetlands, etc.), capacity to harmonize sectoral activities through a coordinating mechanism, funding assistance (in some cases), and ties to relevant global and regional coastal and ocean programs.

- Fruitful intergovernmental partnerships may be built through the following means: (1) identification and pursuit of common interests, such as reducing loss of life and property due to coastal hazards, rather than focusing on ques-

tions of ownership and control; (2) identification and use of unique expertise, talent, and data that exist at the two levels; (3) deferral of difficult
issues, such as those involving jurisdiction and division of management
responsibilities and revenues, until a history of working together has been
established; and (4) use of respected outside expertise for difficult issues.

• In one model for an intergovernmental partnership in ICM, the national government, with the advice of local governments and affected stakeholders,
formulates and legislates broad coastal policies and goals for the nation, and
the local government develops plans and actions for their coastal zones that
are consistent with and incorporate these national coastal policies. The local
government then operates a regulatory system consistent with its coastal
plan.

Addressing Intersectoral Issues

• The problem of achieving intersectoral coordination was the integration
challenge cited most often by respondents to our cross-national survey. In
three-fourths of the cases, ocean management and coastal management are
not administered by the same organizational unit.

• Although the problems associated with lack of intersectoral integration are
generally serious, they should not be exaggerated; not every interaction
among sectors is problematic and in need of management. Moreover, the
costs of achieving integration should be kept in mind, and it should be
recalled that integrated management generally does not replace sectoral
management but instead supplements it.

• Integration should be viewed as a continuum rather than an absolute. The
goal is to move away from a situation in which agencies don't talk to one
another and toward a situation in which a forum for coordinating and harmonizing policies exists and is used frequently and effectively.

• Incentives are needed to bring about continued collaboration among sectoral
agencies. Some factors that may enhance the likelihood of cooperation
among agencies include financial incentives, legal mandates, perception of
a shared problem, shared professional values, perception of political advantage, desire to reduce uncertainty, and availability of a forum for cooperation.

• Some options for achieving intersectoral coordination include creation of a
special interministerial coastal coordinating council or commission, assignment to an existing high-level planning, budget, or coordination office (provided it is at a level above that of the line ministries or agencies); and formal designation of an existing line ministry to act as lead ministry for the

ICM program. About one-half of the countries in our cross-national survey had an interagency or interministerial commission as the national-level coordinating mechanism, and in almost half of these cases, a lead agency for ICM had been named. Fifty-eight percent of respondents also reported that creation of an ICM coordinating mechanism was highly or somewhat related to Agenda 21 recommendations, with middle developing countries (78 percent) being most influenced by UNCED.

• Three factors tend to enhance the effectiveness of an integrated coastal management process: (1) the coastal management entity and process should be at a higher bureaucratic level than the sectoral agencies, to give it the necessary power to harmonize sectoral actions; (2) the effort should be adequately financed and staffed with its own people; and (3) the planning aspect of integrated coastal management should be integrated into national development planning (where such planning is taking place).

• Various government entities generally have legal authority over various aspects of management of coastal lands and waters. Legal analysis is needed to ascertain whether existing legal authorities in use by sectoral agencies are adequate to be used for broader coastal management purposes as well. To the extent that the existing legal framework is found to be deficient, new legal authority may have to be sought to fill specifically identified gaps.

Obtaining and Utilizing Scientific Data and Information

• One of the strengths of the ICM process is its reliance on accurate information and data. ICM must be based on sound scientific information; otherwise, ICM decisions will not stand up to political challenges raised by, for example, economic interests whose development activities may be affected by the ICM process.

• Different types of information are needed at different stages in the ICM process. For example, during the formal adoption, implementation, and operation stages of ICM, it is useful to have information on the costs and benefits of ICM, regulatory and management measures, conflict resolution techniques, and relevant analytical methods.

• The coastal zone and adjacent ocean constitute a complex and dynamic environment in which a number of physical, biological, geological, and chemical processes take place. Coexisting in the coastal zone are large numbers of people joined together in a wide variety of social institutions which themselves are made up of social processes and behavior. Much has yet to be learned regarding the behavior of both the physical and social systems making up the coastal zones. Therefore, one of the important challenges of ICM is the design of a decision-making system that can function in the presence of gaps in information and understanding.

• Some rules for making ICM decisions in the presence of scientific uncertainty include the following: (1) do not make decisions that have irreversible consequences; (2) do not make decisions that seriously threaten the resource base over the long term; and (3) do not make decisions that could reduce the options of future generations for utilizing coastal and ocean resources.

• The precautionary principle should be followed regarding proposed new activities. In the absence of convincing evidence to the contrary, a conservative regulatory and management approach should be taken. In such cases, the burden of proving that no adverse environmental effects will occur should fall on the developer. The government, acting on behalf of the public interest, should not have to demonstrate that harm will occur in order to regulate.

• Another strategy for coping with sparse data on the impacts of a proposed coastal or ocean development activity is to require that the needed data be collected as a condition of approval of the initial development activity. Where adverse environmental effects are a concern, it may be possible to design a monitoring program to detect any such effects before serious degradation takes place. The development activity could then be scaled back or terminated, depending on the circumstances.

• A continuing challenge in the management of natural resources generally centers on the science-policy interface. Improvements in resource management usually depend on improvements in understanding of the processes involved, yet obtaining these gains in scientific understanding has proven to be difficult and slow. Managers and policy makers seem to have difficulty motivating the scientific community to carry out the needed research because it is perceived as "too applied." Similarly, researchers complain that they do not get clear messages from policy makers as to what is needed and that managers and policy makers do not use much of the information scientists have already provided.

• Although environmental impact assessment (EIA) should undoubtedly be a part of an ICM process, it cannot by itself substitute for ICM. Environmental impact assessment tends to operate on a project-by-project basis and hence is not a good tool for comprehensive areawide planning. Moreover, preparation of an EIA by no means guarantee that the developer will take the least environmentally adverse alternative. It merely requires that he or she follow the process and prepare an accurate and complete assessment of potential impacts.

Formulating an ICM Program and Getting It Adopted

Major challenges faced by coastal decision makers as they formulate an ICM program and get it approved are summarized here. Possible strategies to address such challenges are also discussed.

Formulating an ICM Program

- An ICM plan is best formulated by an interagency team formed for this purpose. All key agencies involved in coastal or ocean activities should be represented on the ICM program formulation team. A range of disciplines should also be represented (or available), including planners, resource managers, coastal scientists (geologists, marine biologists, physical oceanographers), and social scientists (experts in legal and institutional issues, economists, cultural anthropologists).

- If possible, the team should work on a full-time basis over a specific period (about a year) to produce the ICM plan. Team members could be seconded from other agencies for this period.

- Some steps the team will have to take in the program formulation process include (1) identifying initial problems, issues, and opportunities to be addressed in the ICM program and setting priorities among these; and (2) setting the programmatic scope of the initial ICM effort: what bundle of issues should be included? ICM programs might include the following:

   Protection and management of coastal resources such as beaches, dunes, and coral reefs.

   Protection of important coastal habitats such as wetlands, mangrove forests, sea grass beds, and mud flats.

   Protection of coastal water quality.

   Promotion of a coast-dependent economic use (e.g., tourism, aquaculture).

   Improvement of public access for recreational purposes.

   Reduction of loss of life and property due to coastal hazards such as storms.

   Management of beach erosion.

   Management of the use of coastal space, including restoration of urban waterfronts.

• A major initial decision in the ICM process concerns the geographical scope of the ICM program: should the program cover the country's entire coastal zone or be limited to a pilot project in a smaller area?

• Formulation of goals, objectives, and strategies is clearly an important aspect of ICM. The goals and objectives of ICM must be set carefully to reflect the real problems and opportunities present in the coastal zone. Setting measurable goals will make evaluating the program's performance much easier.

• Another important early decision in the ICM process involves the setting of boundaries (landward and seaward) for the management area. The area must be sufficiently large to encompass the uses and resources that need management, on both the land side and the ocean side.

• Appropriate design of the intersectoral/intergovernmental coordinating mechanism is a key decision in the ICM process. Functions of the mechanism include strengthening interagency and intersectoral collaboration, reducing rivalry and conflict, providing a forum for conflict resolution among sectors and ocean uses, and monitoring and evaluating the progress of the ICM program.

• An ICM plan will most likely include the following:

A clear description of the coastal area to be managed.

A clear description of the problems to be addressed and the goals and objectives to be sought.

A clear description of the policies and principles that will guide the program.

A statement of the initial management actions to be taken.

A description of the proposed institutional arrangements, including assignment of responsibility for various parts of the program, such as the interagency coordinating mechanism and supervision and support of the overall ICM program.

Funding and staffing requirements for the program.

A listing of the formal actions needed for official adoption of the plan and a suggested timetable for those actions.

Getting Approval for the ICM Program

• Timely government approval of an ICM plan will be aided by the following:

The ICM program being proposed is succinctly described in clear and understandable terms (What is it? Why do we need it? What will it do?).

The benefits (economic, environmental, and social) that will flow from the ICM program are described in tangible and understandable terms.

The proposed program is clearly and visibly endorsed by users of the coast, by the public, and by interested NGOs.

Key legislative and government leaders have been kept informed of the progress of the ICM plan formulation effort from its inception and have received periodic progress reports.

The costs (political, financial, and administrative) of implementing and operating the ICM program are clearly spelled out and sustainable ways to cover such costs are suggested.

Implementing and Operating an ICM Program

Issues faced during the ICM implementation and operation stages are summarized in this section.

Resolving Issues in ICM Implementation

• *Implementation* refers to the actions that must be taken to put the ICM program into operation: start-up activities such as enactment or amendment of legislation, preparation of new or revised regulations and procedures, formal establishment of new institutions or interagency mechanisms, securing of additional personnel, and the like. Decision makers may face a number of problems at this stage. For example, securing necessary legislative and legal changes and obtaining needed resources may be difficult; policy gaps may have to be identified and filled satisfactorily; if an undue length of time has elapsed since the plan was formulated, changes may have occurred that invalidate part of the proposal; or institutional changes may be difficult to put in place because of bureaucratic inertia and resistance to change.

• With regard to the interagency coordinating mechanism (such as a coastal council), the following attributes are usually important: the council is composed of directors of the coastal and ocean agencies or their high-level alternates; the council is chaired by an appointee of the chief executive and is of higher rank than the members themselves; the council has sufficient resources to hire its own staff and technical support (creation of an operating arm of the council or a technical secretariat is important to implement the decisions of the council and carry out the day-to-day work of coastal management; the council is formally established by the usual legal or legislative procedures used in the nation (e.g., legislation, presidential order, decision by local community leaders, etc.).

• Symptoms of problems in the operation of an interagency council are the following: key coastal and ocean decisions begin to be made outside the council's deliberations; council members lose interest in attending meetings and send low-level representatives in their place; the chief executive loses confidence in the council; the council is unable to obtain adequate staffing and financial resources.

• The major functions of an ICM office (created as a technical secretariat for the interagency council or as a separate office in a lead ocean or coastal agency) are to establish and oversee new zonation schemes and other management programs that do not fit within the mission of existing line agencies; oversee environmental impact assessment processes; coordinate the planning activities of the various line agencies; supervise the performance monitoring and program assessment functions; engage in training and human resources development; develop an active public participation program; and address transnational and transboundary issues.

• From the outset, attention should be given to finding ways and means of making the ICM program financially self-sustaining through, for example, the assessment of fees for the use of ocean and coastal resources and space.

• It is imperative that affected interests be involved in the ICM process from the outset, to better inform the ICM process as well as garner support for it.

Resolving Issues in the ICM Operational Stage

• The emphasis of the ICM program in its operation phase (once new or altered ICM processes have been put in place) will be on such areas as coordination, harmonization, conflict resolution, coastal and ocean policy integration, management gaps, and monitoring and assessment of performance.

• A wide variety of management tools and techniques are available to the ICM program in the operation phase; which are used will depend on the extent to which they are authorized under legislation, their technical suitability, and the public acceptance of the approach. Examples include zonation; establishment of set-back lines and exclusionary zones; establishment of protected areas; special area planning; acquisition, easements, and development rights; mitigation and restoration; and issuance of coastal permits.

- An essential function of an ICM program during its operation phase is the resolution of conflicts among coastal and ocean users and agencies. For an ICM program to be effective in addressing conflicts, the following elements are important: efforts to understand the roots, causes, and consequences of coastal and marine conflicts through conflict-mapping studies; the existence of an established and transparent process for making decisions about the conflicts; and the capability to adopt and implement measures to remedy injuries or damage to particular coastal and ocean users arising from coastal development or from the actions of other coastal and ocean users.

- Transparency in decision making and public participation are essential for the success of an ICM program. A process that is seen as fully open, based on reliable information and good science, and accessible to all interested parties stands a much better chance of long-term success than does one that is difficult to gain access to, that does not encourage participation, and decision making goes on behind closed doors.

- Some strategies for ensuring public participation and building consensus include disseminating information on the ICM program to the public through media coverage and public fora, meetings, and discussions in the local language(s); providing visible mechanisms for public participation in the ICM process; having proper feedback mechanisms to ensure that the outputs of public consultations are incorporated into revised ICM plans; and having a transparent approval process for permits, environmental impact assessments, and the like.

- A strong public education program that includes an active outreach effort to all educational levels and to the general public is essential in obtaining and maintaining public understanding of the special character of ocean and coastal areas and the need for special ICM management.

- Evaluations and assessments are essential in measuring the success of an ICM program and modifying them in view of the results. Unfortunately, of all the various phases of the ICM process, evaluation is the least developed, in terms of both methods and practice. This is one of the areas in which ICM practice most needs to be improved in the next decade.

Examples of Good Practices in ICM Programs

It is very difficult to put forward a general model of successful ICM because of the absence of objective evaluative data on any of the ICM experiences we have described in this book. The data on which to judge success are simply not there.

This is so, in part, because of the inherent difficulties in evaluating ICM programs. Few incentives exist on anyone's part to perform evaluation. Agencies often fear evaluation since a negative result may mean loss of resources or power. Donors similarly hope that their money has been well spent and may not wish to receive information to the contrary. Many stakeholders prefer some action, however feeble, to no action. Legislators who enacted a program (if legislative action is involved) have pride of ownership and also hope to see success.

Those responsible for coastal management programs may avoid potentially damaging evaluations by a variety of methods, as noted by S. Olsen, J. Tobey, and M. Kerr (1997). Some of these include (1) adopting vague goals; (2) selecting objectives that defy measurement; (3) selecting indicators that measure effort rather than results; (4) adhering blindly to the project's original objectives and strategies while refusing to adapt to changing circumstances; and (5) skipping formal evaluation entirely or leaving it for the end of the project, when it will have no effect on the design and operation of the original program and will only marginally influence other programs.

A recently completed survey sponsored by the UNDP Strategic Initiative for Ocean and Coastal Management and conducted by the Coastal Resources Center at the University of Rhode Island underlines the fact that most evaluations are limited to a single midterm and a final evaluation conducted by a single external reviewer or team selected or funded by the donor agency. These evaluations are often considered proprietary information and they are rarely published. Conceptual learning (i.e., identifying "lessons learned") is not a primary focus of these evaluations (Olsen et al. 1997).

The Absence of Evaluative Data on ICM Programs

Reviewing some of the best-known and oldest examples of ICM, we find that very little, if any, evaluative data exist for these cases.

United States. In this, the oldest case of ICM, there have been few efforts to evaluate the overall achievements of the program. Each state has prepared periodic reviews of program implementation for the federal government according to Section 312 of the Coastal Zone Management Act. However, these reviews have not proven very useful in determining overall success, since they tend to be akin to accounting reports, showing how federal funds have been spent and how federal grant conditions have been met. We conducted a study of perceptions of performance of U.S. state coastal management programs that provides some insights on how some external interests and experts rate the performance of ICM programs in the various states (Knecht, Cicin-Sain, and Fisk 1996). A major evaluation of the achievements of the U.S. CZM program is currently under way that will, we hope, yield valuable data (Hershman et al. 1997).

Some general conclusions can be drawn about the ICM process in the United

States—how many states have developed programs, the kinds of ICM institutions they have created, the kinds of ICM activities they have undertaken, the kinds of ICM boundaries they have drawn, the kinds of ICM plans they have prepared, how many and what kinds of coastal permits they have processed, the kinds of protective measures they have established, and so forth. However, because baseline data on the condition of the coasts were not collected at the outset of the program, it is difficult to assess on-the-ground improvements such as acreage of wetlands protected and miles of coastal public access acquired or protected or to evaluate whether management of coastal development has resulted in the maintenance of coastal ecological and human amenity values. A noted practitioner of ICM in the state of California, the California Coastal Commission's executive director, Peter Douglas, is quoted as saying, "The success of ICM is found in what you cannot see on the coast" (Lester 1996)—for example, the structures that were *not* built on vulnerable or sensitive coastal land.

The Great Barrier Reef Marine Park, Australia. Although the Great Barrier Reef Marine Park is frequently put forward as a model case of ICM, to our knowledge no overall external evaluation exists. However, evaluations of specific aspects of the program, such as indigenous participation, have at times been critical of the effort (see, for example, Bergin 1993).

Sri Lanka, Thailand, and Ecuador. ICM efforts in Sri Lanka, Thailand, and Ecuador have been conducted under the auspices of the USAID program directed by the Coastal Resources Center of the University of Rhode Island. To the best of our knowledge, although there have been in-house evaluations of these efforts, their results are not available in the open literature.

ASEAN/US Program. The ASEAN/US involved six countries in the ASEAN region—Indonesia, Malaysia, the Philippines, Singapore, Thailand, and Brunei Darussalem. Again, although the program was active from 1985 to 1991, a final evaluation is not available (Sorensen 1997).

CZM Initiatives in the Mediterranean Region. The World Bank, in collaboration with the UNEP–Mediterranean Action Plan coordination center in Split, Croatia, is conducting an assessment of nine coastal zone management initiatives operating in the Mediterranean region during the past decade. Programs in Albania, Cyprus, France, Greece, Israel, Morocco, Tunisia, and Turkey are involved. Results of the assessment were to be presented at a meeting of the Conference of Parties to the Barcelona Convention in late 1997 (Hatziolos and Trumbic 1996).

ICM Efforts in Europe. No systematic evaluation of efforts under way in Europe has been reported. Reports on selected national activities have been prepared by NGOs; S. Gubbay's reporting on ICM development in the United Kingdom is an example (Gubbay 1990, 1995).

Keeping in mind the difficulties inherent in evaluating ICM programs discussed earlier and the general absence of evaluative data on ICM programs, in this section we highlight ICM practices in various countries that we and others deem to be largely successful and that may be applicable to other countries' efforts. Note that these are our personal judgments and those of other professionals in the field rather than the result of systematic and independent evaluation. In several instances, we refer to cases in which a particular ICM practice is only in an emerging state and hence we cannot be sure of its ultimate effectiveness. We include such cases to signal their potential importance and the desirability of watching their progress.

Examples of Successful Practices Related to ICM Processes

In box 11.1, we note some examples of good practices related to ICM processes.

Examples of Successful Practices Related to Substantive Aspects of ICM Programs

In box 11.2, we note some examples of good practices related to various substantive areas of ICM.

Box 11.1. Successful Practices Related to ICM processes

1. *Achieving intergovernmental coordination: "consistency" in the U.S. coastal zone management program.* The national coastal zone management legislation of the United States includes a very useful provision requiring that actions of the national government that affect a state's coastal zone be consistent with that state's approved coastal zone management program. This provision has worked effectively in ensuring good intergovernmental coordination of ICM.

2. *Utilizing long range ICM planning and marine zoning: the Great Barrier Reef Marine Park Authority, Australia.* Water zoning has been used extensively in the management plan for Australia's Great Barrier Reef Marine Park. Different zones are delineated for different uses and different levels of protection and maps depicting these zoning districts are a principal management tool for the Great Barrier Reef Marine Park Authority. The authority recently undertook a "visioning" effort involving wide public participation to prepare a long range plan for the park with a time frame of twenty-five years. This is an exemplary model of proactive ICM planning.

(continues)

Box 11.1. (*continued*)

3. *Bringing ocean and coastal management together: the Republic of Korea.* Korea recently created a new Ministry of Maritime Affairs and Fisheries, which brings together coastal zone management at the national level with ocean programs such as those applicable to fisheries, navigation, and ports.

4. *Building public involvement in an ICM program: Ecuador.* In each of five ZEMs (zones especiales de manejo, or special area management zones) established as part of Ecuador's ICM program, great emphasis was placed on involving groups and the coastal community at large in various aspects of the program. Education, training, and outreach programs brought people of all ages into ICM.

5. *Determining the value of coastal ocean utilization: China.* In its pilot ICM program, the local coastal management authority in Xiamen, China, is placing a value on the use of ocean space for mariculture and anchorage purposes and charging users accordingly.

6. *Building a voluntary community-based ICM program: Canada's Atlantic Coast.* In the absence of a well-defined national ICM framework, local communities in Canada, working with the national government, are developing their own community-based ICM programs. Clearly, at some point, the provincial governments will have to become a formal part of the process if provincial-level interests and concerns are to be fully incorporated into ICM.

7. *Incorporating traditional management practices into ICM: American Samoa.* Land tenure follows a traditional Polynesian pattern in American Samoa, with a great percentage of coastal land in the hands of the villages. Given this situation, villages and their leadership structures were brought into ICM programs as full partners with the national and state governments. Villages are now responsible for monitoring and enforcing ICM management on their land and waters.

8. *Achieving national-level coordination: The Netherlands.* The Dutch have a very well established framework for intersectoral coordination at the national level. In 1984, the government of the Netherlands established a "harmonization policy" requiring coordination of all Dutch ministries involved in North Sea matters. The Ministry of Transport and Public Works acts as lead coastal agency at the national level.

Box 11.2. Successful Practices Related to Substantive Aspects of ICM Programs

1. *Erosion control, Sri Lanka.* Sri Lanka's ICM efforts were initially stimulated by a severe erosion problem along portions of its coasts. The program established boundaries of a coastal zone and set up a permit system for development within the zone. Coral mining was prohibited except for research purposes; and coastal communities were actively recruited to become involved in the process of developing managements plans.

2. *Control of nonpoint (land-based) sources of marine pollution: Chesapeake Bay, United States.* A partnership among three U.S. states (Virginia, Maryland, and Pennsylvania) and the federal government was created in the late 1970s to restore America's largest and most productive estuary. The program is an outstanding example of intergovernmental cooperation and of the successful application of good science to an important coastal management problem.

3. *Special coastal area management approach for protecting coastal resources: Turkey.* As a way to begin ICM, Turkey created an agency for specially protected area (SPAs) in 1989. Thus far, nine SPAs have been designated in the country's coastal zone. Special pilot programs are ongoing at two sites: the Bodrum Peninsula and the seaport city of Mersin. One of the main issues addressed by the programs is rampant tourism development and its deleterious effects on coastal waters and fisheries.

4. *Management of coastal hazards: North Carolina, United States.* North Carolina's coastal area has been visited by a number of large hurricanes in recent years. In an effort to place new homes and other structures farther from the water's edge and out of hazardous inundation zones, the state's ICM program has adopted construction set-back lines that are tied to the long-term erosion rates observed in the coastal zone. Multiple-family dwellings and other commercial buildings require a larger set-back (sixty times the annual erosion rate) than do single-family dwellings (thirty times the annual erosion rate).

5. *Community-based coral reef protection: Phuket Island, Thailand.* Lively and colorful coral reefs are one of the main tourist attractions in Thailand, yet 60 percent of the reefs are reported to be damaged or degraded. Residents of Phuket Island, the area's main tourist destination, established a community-based effort to protect their reefs. The local teachers college developed school curriculum materials; the community established a Coral Reef Day; the local branch of the Eastman Kodak Company distributed signs, posters, and educational brochures; and diving businesses helped procure and install mooring buoys at popular sites. This is a good example of bottom-up ICM.

Prospects and Future Challenges

We close with some observations on prospects and future challenges for ICM at the national and international levels.

ICM at the National Level

As shown in the previous chapters, particularly in the review of country activities in chapter 10, and as detailed in Appendix 1, considerable activity in integrated coastal and ocean management exists around the world, in both developed and developing countries. ICM experiments are taking place in many countries, some national in scope and some consisting of pilot projects in small areas. In developing countries in particular, the practice of ICM does seem to be influenced by the donor and consultant activity present. As we have stressed throughout this book, ICM seems to be most successful when it is undertaken at both the national level and the local level—that is, with both a top-down and bottom-up approach. We know of no case in which an ICM program based solely on one level of government has been successful over the long term. A sustainable ICM program, in our view, requires the resources, the knowledge base, and the commitment of the national government as well as the grounding in reality provided by the local community.

It is heartening to note that many countries in our cross-national survey report being in the "implementation stage" of ICM, but this needs to be looked at carefully case by case—to see what is actually being done, what effects the ICM effort is having on objective, on-the-ground indicators. It seems likely to us that the term *implementation* is sometimes used to mean the beginning of a planning effort or the initiation of resource inventories or mapping programs. As discussed earlier, we use *implementation* to mean the undertaking of start-up activities required to make an ICM program fully operational.

Unfortunately, as discussed earlier in this chapter, even with operational programs, little evaluative information is available. Clearly, more and better evaluations should be built into the ICM process and should incorporate external and unbiased expertise. In particular, donors should insist on such evaluations. Universities and scientific journals can play an important role, too, in documenting the effects of ICM efforts in particular areas and in encouraging the conduct of comparative studies of ongoing ICM programs.

Evaluations are particularly important because in some countries it is taking many years—ten or more—to get ICM off the ground. Although there may well be different stages through which ICM needs to evolve, the cost-effectiveness of ICM funding must be questioned if, indeed, it is not producing timely and adequate results.

In many countries, we have found a disjunction between coastal management and ocean management—that is, they involve different ministries, laws, outlooks, and so forth. This is understandable because of the evolution of ocean and coastal institutions. However, given the interconnectedness of the ocean and the coast, separate institutions, laws, and administrative systems for the coast and ocean pose problems and will increasingly do so, particularly as coastal management efforts involve activities farther and farther offshore. If not fully integrated, ocean and coastal management activities must at least be effectively coordinated, with good fora for periodic information exchange and consultation among ocean and coastal agencies. Increasingly, the boundaries between international, national, and local issues will become blurred, and although coastal managers cannot be all things to all people, they must have some understanding of the broader issues relating to oceans, including those involving resources of the high seas and resources that straddle EEZs and the high seas.

Some kind of harmonizing or coordinating mechanism is essential for ICM, and we are witnessing a great deal of experimentation with these mechanisms in various countries. In many countries, we are also seeing a growth in capacity building in ICM—new training opportunities and programs. This is more so in some areas of the world than in others; for example, in Africa it is notably lacking. In our opinion, training and teaching need to be better interconnected and need to build on each other and be rationalized. The United Nations University could play a role here. In this regard, networks need to be nurtured, and some of these are now being formed.

A vast amount of experience in coastal management techniques and approaches has been amassed since the first CZM programs were initiated several decades ago. Unfortunately, good intentions to the contrary, much of this valuable learning remains inaccessible to countries just now becoming involved. Ways should be found to put this information into a usable form and make it easily available to those who can benefit from it. A clearinghouse, similar to the one being established as part of the Global Programme of Action on Protection of the Marine Environment from Land-Based Activities is very much warranted. In fact, a recommendation to create such a clearinghouse emanated from the International Training Workshop on the Integration of Marine Sciences into the Process of Integrated Coastal Management, held in May 1997 and sponsored by UNESCO's Intergovernmental Oceanographic Commission and China's State Oceanic Administration.

ICM at the International Level

Even though the popularity of ICM is heartening—it is the framework of choice named in Chapter 17 of Agenda 21, in the Convention on Biological Diversity, in the Framework Convention on Climate Change, and in the Global Programme of Action—efforts must be made to ensure that these various fora interpret the ICM concept similarly. Otherwise, a country developing an ICM program will have a great deal of difficulty in reconciling the different visions of ICM.

In April 1996, the United Nations Commission for Sustainable Development reviewed progress achieved on oceans and coasts since UNCED, and indeed, much progress had been made: a binding agreement dealing with management of straddling and highly migratory stocks had been adopted; the Programme of Action for the Sustainable Development of Small Island Developing States had been approved; the Global Programme of Action on Protection of the Marine Environment from Land-Based Activities had been adopted; the Law of the Sea Convention had entered into force; and implementation of the Convention on Biological Diversity and the Framework Convention on Climate Change is proceeding successfully. Lately, however, we have observed a disquieting trend—the beginning of a tendency to undo the careful job of aggregation done in Chapter 17 of Agenda 21—to disaggregate the integrated framework into discrete areas of biodiversity, land-based sources, and so forth. This has been evident both among governments and among some NGOs. These tendencies need to be held in check: the integrated vision of Chapter 17 and of the preamble of the Law of the Sea Convention must be retained—otherwise, we are back to the beginning.

Every opportunity should be taken to reinforce the importance of integration in resource management. Perhaps the International Year of the Ocean 1998 and the work of the Independent World Commission on the Oceans offer a good platform for this purpose. An annual or biannual global oceans forum, as was suggested during the UNCED process, would also provide a very appropriate venue.

Sustainable development has become generally accepted as the overarching goal of coastal and ocean resource management. What does this ambitious goal entail? In our view, the most significant unsustainable activities existing or in prospect are overutilization of fisheries resources; loss of biodiversity in coastal and ocean ecosystems due to improper development; marine pollution caused by land-based activities; and loss of coastal habitats and resources due to accelerating sea-level rise. With the possible exception of overfishing, ICM can, in principle, address these problems and begin to move us toward more sustainable practices. But political will, backed up by management-relevant, convincing science, will be needed if such ICM programs are to be effective and sustainable over the long term.

Restoring overutilized fish stocks to sustainable levels of higher abundance will undoubtedly require a multifaceted approach. First, fishing pressures must be reduced over a sufficiently long period to allow stocks to recover. In some cases, assistance to those suffering economic hardship as a result of these reductions may be needed. Next, fisheries management must become more integrated in scope. Protection of fisheries habitat and the marine environment generally must become central goals in ICM programs that encompass important fishing areas. Politically acceptable ways must be found to adequately link coastal management and fisheries management. Ultimately, of course, the management of fisheries, the management of offshore oil and gas development, and all other coastal and ocean resource activities should be incorporated within comprehensive ICM programs.

As we look ahead to the next century, we can predict with some confidence that the migration of populations toward coastal areas will continue, further increasing the pressure on coastal ecosystems and the resources they support. In our view, the integrity of the world's coastal zones is even more threatened by the surging tide of coastal dwellers than it is by rising sea levels. Only by having strong and well-integrated management programs in place can we hope to protect and sustain the valuable coastal areas of the world.

Concluding Observations

Coastal nations have an unprecedented opportunity to set a new course toward sustainable use of the world's coastal and ocean heritage. Adoption of integrated coastal management programs is now seen as the best way to address a wide range of coastal and ocean problems, ranging from protecting coastal and marine biodiversity to dealing with the coastal manifestations of climate change. This means that in many countries, the time is ripe for the formation of a broad coalition of coastal and ocean interests to support the development of effective ICM programs for their countries.

But to be fully successful, ICM must be seen as much broader than the typical environmental protection program. At its base, ICM should be seen as a rational process for ensuring a sustainable future for the people who live and work in the coastal zone as well as a way to help ensure that the country as a whole obtains full benefits from its coastal and ocean areas.

Designing and implementing an effective ICM program is not an easy task. Agencies must overcome competitive tendencies and be willing to coordinate and harmonize their policies and programs. Policy makers must have the political will to put effective measures in place and provide the necessary resources. And coastal stakeholders must be willing to invest their time and energy in the effort. This having been said, there really is no other choice: the gifts the world's coasts and oceans provide can be ensured only in this way.

ICM Practices in Twenty-Two Selected Nations

Introduction

In this appendix we examine, to the extent allowed by the available information, patterns of ICM practice in twenty-two selected countries in the developed and developing worlds in a series of narrative country case studies. The sample of countries included in our discussion represents a subset of the countries included in our cross-country survey, as noted in table A.1.

To complete the descriptions of ICM practices in the various countries, we have relied on both published accounts of coastal and ocean management efforts and observations we made while carrying out research work, consulting, or lecturing in many of these countries. Information from our cross-national survey also helped to provide a guide to the evolution of ICM in each country.

Our discussion of each country's experiences is organized in four major categories: background: general country context; description of coastal and ocean problems; evolution of government responses; and assessment of government responses. Each category is discussed in turn.

Background: General Country Context

Here, we describe the major physical, social, economic, and political features of each country to summarize the context in which its ICM effort is cast. Our choice of indicators was guided by our own hypotheses about factors that might make a difference in the patterns of ICM observed in a given country. For example, the *level of development* (indicated by rate of economic growth and per capita gross national product) may suggest both the types of problems present in the country and the goals management may be trying to achieve. The *diversity of the population along ethnic and religious lines* may be indicative of how much participation will tend to accompany the ICM program; for example, one could hypothesize

Table A.1 Countries Included in Case Comparisons

	North and South America	Europe	Asia	Oceania	Africa
Developed countries	Canada ($19,510)[a]	United Kingdom ($18,340)	Republic of Korea ($8,260)	Australia ($18,000)	
	United States ($25,880)	France ($23,240)			
		Netherlands ($22,010)			
		Spain ($13,440)			
Middle developing countries	Brazil ($2,970)	Turkey ($2,500)	Thailand ($2,410)	Fiji ($2,250)	
			Malaysia ($3,480)		
Developing countries	Ecuador ($1,280)		People's Republic of China ($530)	Federated States of Micronesia[b]	United Republic of Tanzania[c] ($140)
			Republic of Indonesia ($880)		
			Islamic Republic of Pakistan ($430)		
			Democratic Socialist Republic of Sri Lanka ($640)		
			Republic of India ($320)		
			Republic of Philippines ($950)		

[a] Per capita gross national product (GNP) figures are derived from the World Bank's *World Development Report* (1996b). GNP measures the total domestic and foreign value added claimed by residents. It comprises gross domestic product (GDP) plus net factor income from abroad. Per capita GNP indicators in parentheses are 1994 estimates.

[b] Estimated to be lower middle income ($726–$2,895).

[c] Tanzania figures cover mainland Tanzania.

that the more diverse a society is, the greater the need for participatory processes in ICM to accommodate different perspectives will be (or, alternatively, the greater the firm authoritarian rule to suppress opposing views will be). Cultural and religious factors, such as individualistic or communitarian orientations, are important in understanding a people's interpersonal relations and orientations toward nature, and may be important in supporting (or not) a propensity for cooperation in political matters. The *percentage of population on the coast* could predict the salience of coastal management as a national issue (this may not be the case, however, in countries in which the center of political power is located far from the coast, as in Mexico).

We also provide a synopsis of each country's political system—its *type of government* (democratic, authoritarian, unitary, federal, etc.) and, if applicable, its *colonial heritage,* since countries that have gained independence from a colonizing country often take on the government organization of that country. The *concentration of power among national-level institutions* is examined: does the chief executive (president or prime minister) have centralized control, or does he or she face significant challenges from the legislature, interest groups, provincial-level leaders, and so forth, as in the United States? This variable could be useful in, for example, predicting the extent to which conflict and competition among national agencies may be brought under control through executive action.

Also important is the *extent of autonomy of subnational levels of government* (provinces and communities). How much legal (and actual) power do they have vis-à-vis the national government? This factor is influential, we would argue, in determining whether an ICM program proceeds mainly at the national, provincial, or local level or in an approach combining two or more levels.

By way of background, we also discuss the *maritime jurisdictions* claimed by each nation in accordance with the United Nations Convention on the Law of the Sea, especially the territorial sea and Exclusive Economic Zone. Although a country's declarations of maritime jurisdiction may indicate a growth in interest in ocean and coastal management, they may not be related to any efforts at domestic management of oceans and coasts. Proclamations related to the Law of the Sea Convention typically result from action at the highest level of government and in the foreign ministries and are, in some cases, motivated more by relations with other countries (e.g., nationalizing the maritime zone to prevent others from fishing there or preempting the claims by other nations to adjacent resources) than by domestic management considerations.

Description of Coastal and Ocean Problems

Here, we describe specific problems found in each country related to their geographical location, type of coast, concentration of population along the coast, level of development, and major economic activities occurring along the coast.

Evolution of Government Responses

In this section we describe the evolution of government responses to coastal and ocean problems in the form of laws, policies, programs, plans, government incentives, and efforts to bring about public participation and involvement efforts. As discussed in chapters 1 and 2, the approach to coastal and ocean issues in most countries has traditionally been sectoral and piecemeal; only in recent years have we seen moves toward integration of sectoral activities and toward area-based management. Therefore, our discussion adopts a relaxed view of integrated coastal management, describing major government actions vis-à-vis coastal and ocean affairs regardless of whether they are integrated or not.

Assessment of Government Responses

To the extent allowed by the available data, we assess government responses to coastal and ocean issues, discussing the following variables: *level of government primarily concerned with ICM, overall approach to ICM* (top-down or bottom-up), *implementation, importance of external assistance, importance of UNCED influence, effectiveness of ICM efforts,* and evidence (or lack thereof) of *movement toward integrated management*. Here, we repeat the disclaimer given in chapter 10: for many of the countries discussed, there is scanty evidence regarding some of these questions, particularly the extent of implementation and the extent of effectiveness.

We thus turn to a description of ICM practices in twenty-two selected countries. We begin with eight developed countries, followed by five middle developing countries, and nine developing countries.

Part I: Developed Countries

....

Canada

The pattern of ICM practice in Canada is described below. The information shows that Canada, after a relatively slow start, has significant ICM activity both at the community level and, with the passage of the 1996 Canada Oceans Act, at the level of the central government.

Background: General Country Context

Canada's enormous land area (nearly 10 million square kilometers), or about 3.9 million square miles, is surrounded by the longest coastline in the world (243,789 kilometers, or 151,484 miles) and an extensive offshore area (more than 16.8 million square kilometers, or over 6.5 million square miles). Despite these enormous land and marine holdings, Canada has a small population (28 million), of which less than one-quarter lives within 60 kilometers (37 miles) of the shore. Nevertheless, Canada has significant coastal and maritime interests: it declared a 12-nautical-mile territorial sea in 1970 and a 200-nautical-mile Exclusive Fishing Zone in 1977. On December 19, 1996, Canada's Parliament passed the Canada Oceans Act, establishing a 200-nautical-mile Exclusive Economic Zone and a 24-nautical-mile Contiguous Zone in accordance with customary international law and the United Nations Convention on the Law of the Sea (Canada 1996; and CIA 1995).

Canada borders three major oceans, the Atlantic, the Pacific, and the Arctic, and borders the Great Lakes along its border with the United States. As a result, it possesses a diversity and abundance of coastal resources and ecosystems. Unlike most coastal areas, much of the northern and central parts of Canada have a very low population density because of their harsh climate and arid tundra. However, significant arable lands in the southern prairie provinces make Canada one of the world's major producers and exporters of wheat, barley, and massive amounts of many other grains as well as cattle, other livestock, and minerals.

Canada's population is diverse in terms of both ethnic origin and religious preference. Forty percent of Canadians are of British Isles origin, 27 percent are of French origin, 20 percent are from other European countries, and 1.5 percent are indigenous North Americans and Eskimos. Forty-six percent of Canadians are Roman Catholic, 16 percent belong to the United Church, 10 percent are Anglican, and 28 percent follow other religions (CIA 1995). A recent wave of immigration, especially from Asia, is putting great pressure on Canada's western coast, particularly the southern mainland and Vancouver Island in British Columbia. This pressure also applies to the already densely populated northern shore of Lake Ontario.

Since World War II, the Canadian economy has broadened from an agrarian-based economy to an industrial and urban economy. The manufacturing, mining, and service sectors have been responsible for this shift as Canada has taken advantage of its abundant natural resources and skilled labor force. In 1994, Canada had an estimated per capita GNP of $19,510 (World Bank 1996).

Canada is a confederation of ten self-governing provinces and two semiautonomous territories (soon to be three as a result of land claim settlements in the Arctic) in a parliamentary democracy based on the British model. The British monarch is the nominal head of state, represented by a governor-general named by the Canadian government (the prime minister) and approved as a matter of course by the sovereign. The prime minister appoints the other federal ministers from members of his or her party in Parliament. The federal Parliament consists of a Senate (appointed by the prime minister on a regional basis) and an elected House of Commons, representative of the people on a demographic basis.

The ten provinces govern themselves by means of a legislature and premier. All powers of local governments are provided by the provincial governments. Constitutionally, the provinces own the resources down to the low-water mark and have full jurisdiction over their use and management. The federal government has primary jurisdiction over submerged lands below the low-water mark and offshore, although consultation is ongoing between the federal and provincial governments for certain offshore sectors (Hildebrand 1989, 32).

Description of Coastal and Ocean Problems

Many on-the-ground problems associated with the management of Canada's coastal resources relate to the country's vast areas of unmanaged and sparsely populated coastline. As a result, it has taken more than twenty-five years for coastal issues to gain sufficient public exposure to warrant government action. In the past decade, however, problems associated with contamination of coastal waters, loss of valuable wetlands, and growing resource conflicts in coastal areas have heightened. With the virtual collapse of several crucial Atlantic fisheries (northern cod and other groundfish) added to these problems, there has been a keen recognition by the Canadian government for a greater need to coordinate and integrate coastal and ocean activities among federal, provincial, and local governments and among coastal and ocean users' groups (Hildebrand 1996a, 505–507).

However, the size and diversity of the Canadian coastline make such management difficult.

Canada's Arctic coast encompasses roughly two-thirds of the country's coastline. Although dominated by the region's cold climate, the Arctic coast and offshore waters have intense springtime phytoplankton blooms that support critical polar ecosystems inhabited by whales, seals, polar bears, and seabirds. For this reason, many of the management efforts focus on preserving and protecting the area. At the same time, the area is a huge storehouse of hydrocarbons and minerals, replete with development potential. The natural resources of the Arctic coast are an integral part of the culture of indigenous residents, providing them with food, clothing, and other materials. Although the Canadian government has attempted to resolve conflicts among these three uses through a sustainable-use management approach, future conflicts may present an obstacle to ICM (Wright 1994).

The Atlantic coast of Canada has a wide continental shelf historically rich in fisheries resources, both nearshore and offshore in the Labrador Sea and its offshore banks. However, a recent combination of overfishing by Canadian and foreign distant-fleet fishermen and loss of coastal habitat as a result of coastal development and agriculture has depleted many commercial stocks.

The discovery of oil and gas off the coasts of Newfoundland and Nova Scotia in the 1970s and 1980s produced extensive exploratory activities and some resource development. Two provincial governments and the federal government formally agreed to share decision making regarding these activities and some of their revenues, although the federal government holds ultimate control.

Canada's Pacific coast is a complex system of rugged mountains, inlets, and fjords with numerous wetlands and estuaries. Productive coastal waters support large fisheries producing salmon, herring, and invertebrates. Other activities in the area consist of aquaculture and offshore oil and gas development. Coastal development and growth in British Columbia are placing increasing demands on the western shoreline, most markedly in the depletion of forestry and fisheries resources. Conflicts over resources are exacerbated by the lack of a provincewide planning approach; natural resource management control is fragmented among local, provincial, federal, and indigenous bodies (Hildebrand 1989).

Evolution of Government Responses

Government responses to emerging ocean and coastal problems are described below.

National Level

The need for coastal and ocean management was first recognized in Canada in the early 1970s through a series of workshops and government studies. After an initial period of government interest (including an oceans policy report in 1974), momentum for coastal and ocean management declined until 1976, when a land resources steering committee was created in British Columbia. The committee produced a two-volume report in 1978 that inspired the federal government to conduct a national symposium on coastal management in Victoria, British Columbia. The outcome of the symposium was a set of shoreline management principles and increasing interest and involvement on the part of the general public. However, there was little government reaction following the symposium, and interest in the concept of CZM waned at the federal and provincial levels over the next decade, although several local action programs were created. The oceans policy question was revisited temporarily following a 1987 report by the Department of Fisheries and Oceans, but it also was largely ignored by government decision makers (Canada 1996a). The political climate for coordination and cooperation in resource management between the federal and provincial governments has changed drastically in the past few years. Whereas adversarial relations between national and subnational units were the norm in the past, the federal–provincial relationship has matured, due in part to crisis conditions in many coastal areas, diminished administrative resources at all levels, and a recognition of the need to harness the skills and energy of community-based management (Hildebrand 1996a, 506).

In response to the new challenges facing the Canadian coast and ocean, the Canada Oceans Act (COA) was passed in Parliament in 1996. As enacted, the COA consists of three parts: addressing Canada's maritime zones; development of an oceans management strategy (OMS); and duties and powers of the lead ministry and provisions for government and management restructuring.

The Department of Fisheries and Oceans is the lead coordinating agency under the Canada Oceans Act, although there may be significant reordering of federal agencies and departments in the future. Tasks under the COA include developing an integrated coastal area management (ICAM) framework based on stakeholder consultations, intergovernmental dialogue, and direct feedback from the public; defining a mechanism for development of national marine environmental quality guidelines; and defining a system of marine protected areas. The federal ministry will, at minimum, provide resource information, management solutions, scientific expertise, and a forum for information exchange among provinces (Canada 1996a).

Subnational Level

Local regionally based ecosystem management schemes have operated successfully in Canada for many years. A local example is the Fraser River Estuary Management Program (FREMP), launched in British Columbia in 1985. The program attempts to ensure effective management of the estuary, given that a host of management agencies at both the federal and provincial levels conduct management activities, utilizing regional commissions for local input. The program's goals have been to provide viable opportunities for economic development in the estuary, to maintain water quality standards, to increase the productivity and ecological character of the estuary, and to enhance recreational sites. Regionally based ecosystem management schemes are shared with the United States in Georgia Basin and the Puget Sound on the western coast and in the Gulf of Maine on the eastern coast. Both cross-border initiatives are provincial–state ventures wherein the Canadian and U.S. federal governments are "nonsignatory observers" (Hildebrand 1989).

In other initiatives, a Frontier Area Management plan has been designed to maintain the unspoiled character of the far northern areas in the face of oil and gas exploration; the province of Quebec has joined with the federal government for management of the St. Lawrence River; and the Great Lakes Water Quality Program, a joint U.S.–Canada program, has been working since 1978 to "restore and maintain the chemical, physical, and biological integrity of the waters of the Great Lakes Basin ecosystem" (Hildebrand 1989, 51). Individual provinces (Nova Scotia and Newfoundland) have also begun to develop their own policies, and entire regions are examining the feasibility of regional frameworks for coastal management, such as an Atlantic Accord on Coastal Management. Although the province of Newfoundland was examining the feasibility of ICM in the 1980s, the subject has moved to the back burner. The provinces of Nova Scotia, New Brunswick, and British Columbia are now showing the greatest leadership and preparedness for ICM (Canada 1996a).

The Atlantic Coastal Action Program (ACAP), launched by Canada's federal government in 1991, works to assist coastal communities in developing "comprehensive management plans for the restoration and maintenance of coastal ecosystems" (Ellsworth 1994, 687). Other activities under the ACAP include public education and local action projects. The federal government acts only as a facilitator and enabler in the program, leaving the primary management responsibility at the community level. To participate in the program, community stakeholders (government organizations, NGOs, industry representatives, academics, and citizens) are required to form a nonprofit committee. The committees, in effect, become roundtables on the local environment and economy, and competing stakeholders must cooperate to make consensus-based decisions (Donalson 1994, 696–698).

Assessment of Government Responses

The inability of Canada's federal government to develop a national framework for ICM before 1996 fueled the rise of integrated community-based initiatives. Although these CZM "programs" are small in scope, focusing on specific rivers, estuaries, or embayments, they have evolved from remediation and cleanup efforts to more broadly based programs that address environmental and economic issues with a focus on sustainable development (Hildebrand 1996a, 507). Their success has also garnered much-needed political support for greater development of coastal policy at the province and federal levels in Canada.

The link of a strategy for integrated coastal area management with Canada's Oceans Policy will be one of the first formal linkages of the two activities in policy or law for a nation as a whole. Thus, the potential for spatial integration in the Canadian context is quite promising. At present, coordination of activities in Canada's coastal and ocean areas is somewhat fragmented and varies from region to region.

Intersectoral management at the federal level (among ministries) in Canada is perceived as highly successful, and with the implementation of a national Oceans Policy, it should continue to improve. Among the provinces currently pursuing coastal management schemes, integration of sectors at the provincial level is perceived as having moderate success. Intersectoral integration at the local level, particularly in the case of the Atlantic Coastal Action Program, has been successful in those coastal ecosystems in which regional management programs have been established.

Through programs like the St. Lawrence River program, the Frontier Area Management Program, and the ACAP, integration among different levels of government in Canada has been moderately successful. The strong intergovernmental component of the Oceans Policy should work to improve government linkages even further. However, Canada's size and the remoteness of many of its coastlines make integration of government activities difficult.

With twenty-five years' worth of feasibility studies regarding coastal management in Canada, considerable information and data are available for policy makers to utilize. The studies and reports cover a broad range of disciplines in the natural and social sciences and should be of great use in developing the Oceans Policy.

In terms of capacity building in ICM, Canada has established specialized in-country training of staff through new graduate programs in ICM at the university level. University-level training and capacity-building activities in coastal and ocean management have been provided for visiting students and professionals from the United States and overseas. Specifically, Dalhousie University's Marine Affairs Program, Memorial University of Newfoundland's new Advanced Diploma Program in CZM, and the Halifax-based International Ocean Institute's summer training program in UNCED and the Law of the Sea attract middle- to senior-level bureaucrats and interested parties from the private sector who wish to increase their capacity in the new paradigms of ocean and coastal management. Canada has been a world leader in the development of marine and coastal capacity in developing countries through its aid agency, the Canadian International Development Agency (CIDA). By design, bottom-up management programs in Canada such as the ACAP work to build the capacity of local managers and participating groups to address key issues and problems associated with coastal and marine management.

A number of technical support systems have also been developed in recent years to aid in building capacity. Specifically, marine resource inventories, databases, coastal and marine GIS atlases and maps, and ecosystem health studies provide policy makers with useful data and information.

Sources Consulted

Canada. Department of Fisheries and Oceans. 1996a. Integrated coastal area management: A Canadian retrospective and update. Presentation to the United Nations Commission on Sustainable Development, April, United Nations, New York.

———. 1996b. Minister Mifflin welcomes passage of the Oceans Act. News Release NR-HQ-96-100E. Ottawa, Ontario: Fisheries and Oceans Canada.

CIA (Central Intelligence Agency). 1995. *The World Factbook.* Washington, D.C.: Government Printing Office.

Donalson, C. 1994. An unholy alliance: Working with coastal communities—A practitioner's perspective. In *Coastal Zone Canada '94,* ed. R. G. Wells and P. J. Ricketts, 696–705. Halifax, Nova Scotia: Coastal Zone Canada Association.

Ellsworth, J. P. 1994. Closing the gap between community expectations and service delivery: Canada's Atlantic Coastal Action Program. In *Coastal Zone Canada '94,* ed. R. G. Wells and P. J. Ricketts, 687–695. Halifax, Nova Scotia: Coastal Zone Canada Association.

Hildebrand, L. P. 1989. *Canada's Experience with Coastal Zone Management.* Halifax, Nova Scotia: Ocean Institute of Canada.

———. 1996a. Canada: National experience with coastal zone management. In *Coastal Zone Management Handbook*, ed. J. R. Clark, 506–508. New York: Lewis.

———. 1996b. Head Coastal Liaison, Environment Conservation Branch, Environment Canada, Atlantic Region, Dartmouth, Nova Scotia, Canada. Response to ICM cross-national survey. Newark: University of Delaware, Center for the Study of Marine Policy.

Wright, D. 1994. The approach of the Department of Fisheries and Oceans to land use planning in Canada's Arctic coastal zone. In *Coastal Zone Canada '94,* ed. R. G. Wells and P. J. Ricketts, 31–45. Halifax, Nova Scotia: Coastal Zone Canada Association.

United States

As the information below shows, the United States was the first nation to formally initiate a coastal zone management program at the national government level.

Background: General Country Context

The United States is a federal republic with a strong democratic tradition. It has a population of 263 million (1995 estimate) and a population growth rate of 1.02

percent. Life expectancy is 72.8 years for males and 79.7 years for females. The country is composed of fifty states, one federal district, and a number of dependent territories, the most populous of these being Puerto Rico, the Virgin Islands, Guam, and American Samoa. In addition to the states, there are approximately 80,000 local-level governments operating under enabling legislation from their respective states. The United States is the fifth largest country in the world in terms of area and has the most powerful and technologically advanced economy with a GNP per capita of $25,880 (World Bank 1996). In terms of ethnic divisions, whites constitute 83.4 percent of the population, blacks 12.4 percent, Asians 3.3 percent, and Native Americans 0.8 percent (CIA 1995).

The national government has three branches: an executive branch led by a president, who is both chief of state and head of government, and a cabinet appointed by the president; a bicameral legislative branch with a Senate (six-year terms) and a House of Representatives (two-year terms); and a judicial branch headed by the United States Supreme Court. The U.S. political system is highly pluralistic, and emphasizes separation of powers among different levels of government, and the complementary roles of federal, state, and local governments.

Description of Coastal and Ocean Problems

Although it is clearly a continental country, the United States has a long coastline (19,800 kilometers, or 12,300 miles). Many of its largest cities are located on the Atlantic and Pacific coasts or on the Great Lakes; about 60 percent of the population lives in these coastal regions. Thirty of the fifty states are considered coastal, as are all the dependent territories. Historically, because the United States was a colony of a European power, its ocean interests on the oceans centered first on commerce and maritime transport, defense, and food production through fisheries. In the twentieth century, these largely maritime interests were joined by interests related to its position as a coastal country and to exploitation of its offshore resources. Significant oil and gas resources were found offshore from California (beginning at the end of the nineteenth century), in Louisiana in the 1930s, and in Alaska in the 1960s. The country possesses very rich fishing grounds, particularly in Georges Bank offshore from New England; in the Pacific Northwest and offshore from Alaska; and in the Gulf of Mexico. Given its valuable offshore domain, the United States was the first to launch the worldwide movement toward "enclosure" of ocean space by national governments with the Truman Proclamation in 1945, in which the country claimed jurisdiction over its continental shelf and offshore fishery resources.

Shortly after World War II, a surge of development occurred in the United States to meet the pent-up demand that had developed during the war years. One aspect of this building boom involved vacation homes at beaches as well as coastal resorts, hotels and motels, marinas, and the like. At first, large areas of coastal wetlands were sacrificed to make room for people who wished to be near the water. By the early 1960s, however, it was clear that something had to be done to preserve coastal areas. Valuable coastal habitats were being destroyed, access to the public shoreline was being cut off by extensive private developments, and ports and harbors needed to modernize and expand, but little coastal land was available. It was difficult, too, to find suitable coastal sites for needed power plants and other energy-related facilities. In response, a number of coastal states, such as California, Delaware, Michigan, and Washington, began enacting special coastal management legislation to protect and manage their coastal areas. Through the Coastal States Organization, an organization established in 1970 by the National Governors Association to promote the interests of the coastal states, they called for a federal initiative in this area.

The late 1960s and early 1970s also witnessed major changes regarding other coastal and ocean issues. The worldwide decline in populations of marine mammals, particularly whales and dolphins caught in the course of purse seine fishing for tuna, led to the mobilization of new groups to enact national legislation protecting marine mammals. The U.S. position as a major world fishing power was significantly eroded in the 1960s by the appearance in offshore waters of foreign factory trawlers with an awesome new capacity for harvesting and processing onboard a vast quantity of fishery resources. This development led to a movement to "get rid of the foreigners"; by 1976, a national law had been enacted establishing a 200-nautical-mile U.S. fishery zone.

Offshore oil development, which had been proceeding at a fast pace in the 1950s and 1960s through federal leases in Louisiana, California, and Alaska, came to an abrupt halt in 1969 with an oil spill offshore from Santa Barbara, California, that affected 150 miles of a coastline noted for its beauty and bountiful resources. This, the first spill to be witnessed by a large number of Americans through the medium of television, is credited by many as playing a major catalytic role in the rise of the environmental movement. Added to the Santa Barbara oil spill, books and studies on the deterioration of coastal resources such as wetlands and on growing water pollution in lakes, rivers, and ocean waters helped spur a powerful movement that led to the enactment of an unprecedented number of environmental laws in the 1970s, many of them oriented toward the oceans and coasts.

Evolution of Government Responses

The government response to the environmental concerns of the 1970s included the enactment of a plethora of largely single-purpose legislation addressing different ocean and coastal resources (Cicin-Sain 1982). The decade saw the enactment of a dozen major laws related to the environment and the oceans. The most prominent of these included the 1969 National Environmental Policy Act (NEPA); the 1972 Federal Water Pollution Control Act, later known as the Clean Water Act

(CWA); the 1972 Coastal Zone Management Act (CZMA); the 1972 Marine Mammal Protection Act (MMPA); the 1972 Marine Protection, Research, and Sanctuaries Act (MPRSA); the 1973 Endangered Species Act (ESA); the 1976 Fishery Conservation and Management Act (FCMA), and the 1978 Outer Continental Shelf Lands Act Amendments (OCSLAA). New, specialized regimes were created to deal with issues in each of these areas of concerns. Only the National Environmental Policy Act, which requires environmental impact statements (EISs) for all federal actions affecting the environment and the Coastal Zone Management Act were multiple purpose in nature, and both have served the United States well. The EIS process has been generalized to an environmental impact assessment process that many countries now employ as a mainstay of their environmental protection programs. The Coastal Zone Management Program, the first such national program to be adopted, also has served in some respects as a model for other countries considering coastal management initiatives.

Major accomplishments during this period included the creation of a new civilian oceans agency in 1970, the National Oceanic and Atmospheric Administration, which brought together a number (but not all) of the federal ocean-related programs, and a significant expansion of the national ocean science effort. This was in response, in part, to the recommendations of an influential ocean commission established to examine U.S. interests in the ocean, the so-called Stratton Commission, and its 1969 report, *Our Nation and the Sea.*

In the U.S. Coastal Zone Management Program, states are called on to develop and implement coastal zone management programs for their coastal zones in cooperation with the federal government and local governments (Knecht 1979; Kitsos 1981). The program is voluntary, but virtually all U.S. coastal states are participating, perhaps, in part, because of the incentives made available by the federal government. These include federal financial assistance in both developing and administering the program and a commitment, through "federal consistency" provisions, that actions of the federal government will be consistent with approved state programs. The CZMA legislation contains a series of standards that state programs must meet before they are approved and funded for implementation. For example, coastal policies contained in state CZM programs are to be enforceable and are to apply to both state and local actions. Recently, CZM in the United States has focused on such issues as coastal hazards (storms and hurricanes), wetlands protection, urban waterfront restoration, and management of nonpoint-source pollution (urban and rural run-off into coastal streams, rivers, and estuaries). The United States invests federal funds of more than $55 million per year in the Coastal Zone Management Program, with the coastal states providing significant matching funds.

The 1972 Coastal Zone Management Act (CZMA) also created a protected coastal area program, the National Estuarine Research Reserve System (NERRS), setting aside particular estuarine areas in the coastal states for research and monitoring purposes and as an aid in improving coastal management. A companion program, the Marine Sanctuaries Program, established as part of the 1972 Marine Protection, Research, and Sanctuaries Act created a system of marine protected areas along U.S. coasts (Foster and Archer 1988). The Marine Sanctuaries Program now encompasses fourteen sites, and the NERRS program twenty-two estuarine reserves. Both programs are administered by the Office of Ocean and Coastal Resource Management of the National Oceanic and Atmospheric Administration (NOAA), also responsible for administration of the CZMA.

Another coastal protection program—the National Estuary Program (NEP), enacted in 1987 in accordance with amendments to the Clean Water Act—is administered by the Environmental Protection Agency. This program provides federal funds to states bordering important estuaries in need of management. It establishes a complex process for development of comprehensive and coordinated management plans for the estuaries involving extensive participation by relevant stakeholders. A problem with this program has been, in a number of cases, the absence of effective implementation measures to put the estuary plans in effect once they have been developed.

With regard to other areas of U.S. ocean and coastal policy, implementation of the 1972 Marine Mammal Protection Act, which called for a permanent moratorium on the taking of marine mammals, led to significant declines in the mortality of marine mammals on U.S. coasts. Indeed, in some cases, great population increases took place, as with seals on the West Coast, occasioning significant conflicts with commercial fishermen.

Regarding fisheries, the 1976 Fishery Conservation and Management Act (later renamed the Magnuson Fishery Conservation and Management Act) ushered in a new era of fishery management in the United States by establishing a 200-nautical-mile fishery conservation zone—allowing foreign take only after domestic harvesting capacity has been fully employed and by establishing an elaborate system for domestic fishery management in the form of eight new regional fisheries management councils with the authority to prepare and implement management plans for stocks under their jurisdiction. Although the councils make these plans, the federal government has final authority regarding their approval. This regional system of governance, though exhibiting some very good institutional traits, such as opportunities for gaining extensive public involvement and for obtaining industry and scientific advice, has been unable to prevent a continued decline in fisheries, especially in the New England area. Declining stocks have forced recent large reductions in fishing efforts with federal economic assistance being made available to some coastal communities. The Magnuson Act was amended in 1996 (The Sustainable Fisheries Act) in an effort to strengthen the fisheries management regime and prevent further declines in stock.

With regard to the offshore oil and gas development, following a hiatus in the program due to the Santa Barbara oil spill, the Arab oil embargoes of 1973–1974 brought renewed federal offshore oil activities to U.S. coasts in an effort to gain independence from unstable sources of foreign oil. This led to sharp conflicts among the federal government, which favored expanding the pace of oil drilling; the coastal states, which favored a more paced approach; and environmental groups, many of which were openly opposed to additional offshore oil and gas activity. In recent years, these conflicts have increased to the point that major portions of the program are presently at a stalemate (Cicin-Sain and Knecht 1987; Kitsos 1994). A moratorium enacted by Congress currently prevents the federal government from undertaking further leasing on about one-half of the U.S. continental shelf.

Assessment of Government Responses

The United States probably has more coastal and ocean legislation than any other nation. Many of these programs represent significant advances in the protection and management of ocean and coastal resources; for example, there is an extensive state-based system for coastal management that in many cases has prevented inappropriate development in the coastal zone; point sources of marine pollution are effectively under control; and marine mammal populations have rebounded from the point of serious decline. On the other hand, serious depletion of fisheries continues and the offshore oil and gas development program is stymied in much of the country. U.S. coastal and ocean management must, therefore, be judged a mixed success. The coastal zone management program, for example, has both strengths and weaknesses. Among its strengths are its intergovernmental coordinating mechanism and its legal requirement for consistency of federal actions with state coastal policies. The public participation element of most state CZM programs is also generally strong. Recent assessments of twenty-four of the state CZM programs suggest that they are doing a reasonably good job of protecting coastal resources and providing public access but are doing less well at managing development in the coastal zone (Knecht, Cicin-Sain, and Fisk 1996). CZM in the United States also is uneven in the extent to which ocean resources and ocean uses (the "wet side") are covered under the program—for example, the seaward boundary of the coastal zone for domestic management remains at three nautical miles even though the U.S. territorial sea was extended to twelve nautical miles in 1988.

Overall, U.S. ocean and coastal policy is weak in its intersectoral coordinating activities, most notably at the federal level, where no effective coordinating mechanism exists. Moreover, no formal program for planning and management of the country's Exclusive Economic Zone (created by presidential proclamation in 1983) exists, save for the national Marine Sanctuaries Program, under which, as noted, sensitive and important ocean areas are set aside for special management.

Thus, a major challenge facing the United States in the 1990s and beyond is the creation of appropriate mechanisms at the national level for achieving intersectoral coordination and for providing review, oversight, and recommendations for change regarding national ocean and coastal policy. Some hopeful signs of initiatives in this area are evident at the time of this writing (fall 1997). The Ocean Principals Group—composed of leaders of the major federal ocean agencies, including the Departments of Commerce (NOAA), Agriculture, Defense, Energy, Interior (U.S. Geological Survey, Minerals Management Service), State, and Transportation (Maritime Administration, U.S. Coast Guard); the president's Council on Environmental Quality and Office of Science and Technology Policy; the Environmental Protection Agency; the Federal Maritime Commission; the National Aeronautics and Space Administration; the National Science Foundation; the Federal Emergency Management Agency; the U.S. Agency for International Development; and the Smithsonian Institution—has been meeting on a regular basis to prepare a series of background papers on the status of U.S. ocean and coastal policy, in conjunction with the International Year of the Ocean 1998. The Ocean Principals Group has also called on the United States Congress to ratify the United Nations Convention on the Law of the Sea, which it has yet to do. On the legislative front, policy initiatives have been evident as well. In September 1997, the Oceans Act of 1997 was introduced in the Senate, calling for two major actions: creation of a new National Ocean Council, composed of federal agencies, and creation of a Commission on Ocean Policy to conduct a comprehensive review of national ocean and coastal policy, much as the Stratton Commission did almost thirty years ago. These developments portend much-needed reform in U.S. national ocean policy. To be successful, however, they will need the support and concurrence of other important actors in U.S. ocean and coastal policy, such as coastal states, ocean industries, and environmental groups,

Sources Consulted

CIA (Central Intelligence Agency). 1995. *The World Factbook*. Washington, D.C.: Government Printing Office.

Cicin-Sain, B. 1982. Managing the ocean commons: U.S. marine programs in the 70s and 80s. *Marine Technology Society Journal* 6–18.

Cicin-Sain, B., and R. W. Knecht. 1987. Federalism under stress: The case of offshore oil and California. In *Perspectives on Federalism*, ed. H. Scheiber, 149–176. Berkeley: University of California Institute for Governmental Studies.

Foster, N., and J. H. Archer. 1988. Introduction: The national Marine Sanctuary Program—Policy, education, and research. *Oceanus* 31 (1): 5.

Kitsos, T. R. 1981. Ocean policy and the uncertainty of implementation in the 80s: A legislative perspective. *Marine Technology Society Journal* 15 (3): 3–11.

———. 1994. Troubled waters: A half dozen reasons why the federal offshore oil and gas program is failing—A political analysis. In *Moving Ahead on Ocean Governance: Roundtable Summaries,* ed. B. Cicin-Sain and K. A. Leccese, 36–46. Newark: University of Delaware, Center for the Study of Marine Policy, Ocean Governance Group.

Knecht, R. W. 1979. Coastal zone management: The first five years and beyond. *Coastal Zone Management Journal* 6: 259.

Knecht, R. W., B. Cicin-Sain, and G. W. Fisk. 1996. Perceptions of the performance of state coastal zone management programs in the United States. *Coastal Management* 24: 141–163.

United Kingdom

Background: General Country Context

The United Kingdom of Great Britain and Northern Ireland is made up of Great Britain (England, Wales, and Scotland) and Northern Ireland. Great Britain is separated from the European coast by the English Channel to the south and the North Sea to the east; its northern and western shores are washed by the Atlantic Ocean. The Irish Sea separates Great Britain from Ireland and Northern Ireland to the west. The United Kingdom comprises 15,300 kilometers (9,510 miles) of coastline, characterized by rolling plains in the southern region and rugged hills and mountains scattered with lochs and islands in the Northern Highlands of Scotland. The country is densely populated for its land area (only 241,590 square kilometers, or 93,278 square miles), with more than 58 million inhabitants (CIA 1995).

The economy of the United Kingdom is among the four largest in Europe. Great Britain has a long history as a trading power and financial center, a situation that persists to the present day. The United Kingdom's economy is capitalistic, with a small-scale but intensive high-technology agricultural base complemented by large reserves of coal, natural gas, and oil. Per capita GNP was $18,340 in 1994 (World Bank 1996). Primary energy production still accounts for 12 percent of the GDP, which is uncharacteristic for a developed country. However, the service sector accounts for the lion's share of GDP through banking, insurance, and business services. The rate of growth in GDP has been increasing of late, but questions remain regarding the United Kingdom's high rate of unemployment and how its intends to participate in the financial and economic restructuring of Europe (CIA 1995).

The United Kingdom is a constitutional monarchy. Its Parliament consists of a House of Commons (elected by the population) and a smaller House of Lords composed of hereditary landowners and nobles. The House of Commons generally draws up legislation that, if passed, can be delayed but not blocked by the House of Lords. English common law is the foundation of the legal system, which is based on statutes and precedent rather than a formal constitution. The political party holding the majority in Parliament appoints the prime minister and a cabinet of ministers to head the government. Administrative divisions in the United Kingdom are as follows: forty-seven counties and seven metropolitan counties in England, twenty-six districts in Northern Ireland, nine regions in Scotland, and eight counties in Wales. Of these, forty-eight counties (subnational units) are situated on the coast. With respect to natural resource issues, the majority of legislation comes from the national level, whereas day-to-day planning and implementation are carried out somewhat autonomously at the county or regional level.

Description of Coastal and Ocean Problems

The large population of the United Kingdom and the salience of coastal and maritime issues, historically in Great Britain, are responsible for the high intensity and variety of uses and activities that take place in the British coastal area. As such, there is a high demand for space and resources that has led to congested water space, overcrowding on the English coast, and extensive conflicts among coastal users' groups (Gubbay 1990, 5). There is also significant concern over issues such as loss or damage of productive ecosystems, increasing land- and vessel-based pollution of coastal waters, serious resource depletion, and erosion of coastal areas (Gubbay 1990).

Pollution of inland waters has been significant in England and Wales, frequently restricting the productivity of shellfish and finfish fisheries. Catches from a large number of industrial waterways, such as the River Thames, require cleansing or purification before sale, adding significant cost to the industry. Conflicts among offshore fisheries in the United Kingdom are also significant, involving competition among European Union fishing nations and conflicts between the United Kingdom's fishing and offshore oil industries over fishing grounds, gear damage, and pollution from spills (Mackay 1981).

Although terrestrial protected area management has been successful in the United Kingdom, the effectiveness of marine protected areas is limited, due in part to the diversity of protected area systems and the fact that they are administered by different organizations (Cole-King 1995).

The Scottish coast, although generally more remote and pristine than the English and Welsh coasts, is experiencing many of the same coastal and ocean problems, particularly the environmental risks associated with development of the offshore oil industry and marine transportation. Scotland has also experienced a decline in catch from inshore fisheries and pollution of recreational beaches (Gubbay 1995).

Evolution of Government Responses

Coastal planning in the United Kingdom, specifically with regard to England and Wales, is based on planning principles set forth in the 1947 Town and Country Planning Act (Waite 1981, 65). The town and country planning system regulates activities and uses in the landward part of the coastal zone down to the low-water mark (Gubbay 1990). Two tiers of government are generally involved, with county councils deciding on policy for development and regional or district councils drawing up more detailed local plans. However, only a handful of county councils have specific coastal plans; the remainder address coastal issues through larger plans with implications for coastal areas. Clearly, though, the main stimulus for integrated coastal zone management (ICZM) initiatives in the United Kingdom is at the local level. A *Directory of Coastal Planning and Management Initiatives in England* lists fourteen national, regional, and county strategies and plans; ten coastal management plans; thirty-one estuary and harbor management plans; and fourteen statutory local plans incorporating positive policies for the coastline (King and Bridge 1994).

At the national level, the Countryside Commission took an early interest in coastal management following its formation in 1968. It published several reports in the 1970s, including Coastal Heritage, which carefully examined coastal development in England and Wales in order to discover more effective ways to manage sensitive coastal areas (Gubbay 1990, 1–2). The commission identified thirty-four sites for inclusion in the "coastal heritage," a nonstatutory, voluntary coastal protection system created to preserve the natural amenities of the coast and benefit local landowners and visitors. At present, more than forty sites exist, with the commission providing financial aid to the community of landowners maintaining the sites (Williams 1992).

In reaction to growing problems in Great Britain's coastal zone during the 1970s and 1980s, an inquiry was conducted in 1992 by the House of Commons Select Committee on the Environment, which, as a result, recommended an integrated coastal strategy. The United Kingdom British government supported the idea of ICZM and published *Planning Policy Guidance on Coastal Planning for England and Wales* (United Kingdom, DOE and Welsh Office 1992) and a number of other documents in response to the committee's inquiry (United Kingdom, DOE 1992; United Kingdom, DOE and Welsh Office 1993a, 1993b). These took the view that existing arrangements were generally satisfactory (United Kingdom 1995). Despite this conclusion, several new institutional bodies were created, including an Interdepartmental Group on Coastal Policy, to link government departments involved with coastal management, and a Coastal Forum, made up of a wide-ranging membership of national sectoral bodies, to meet regularly and discuss coastal issues (United Kingdom 1995). The Department of the Environment, which is taking the lead in coordinating coastal zone management initiatives in England, serves as a secretariat for the Coastal Forum and has commissioned the preparation of a guide to good practice in ICZM and a review of the bylaw-making power of local authorities and others in the coastal zone. The Department of Environment has also published a summary of coastal management policy (United Kingdom 1995).

In Scotland, however, ICZM has been slower to take off; the Scottish office of the Department of the Environment remains to be convinced that ICZM is a useful way forward. Scotland has recently commissioned a study to examine coastal issues. *Scotland's Coast*, a discussion paper published by the Scottish office, proposes that the formation of local coastal fora be encouraged in Scotland, that the option of a Scottish Coastal Forum be considered, and that national guidance and advice be published on key coastal issues, including comprehensive national planning policy guidance for the coast (United Kingdom, DOE and Scottish Office 1996). A similar study was undertaken in Wales in 1992, although the Welsh government has yet to respond with formal guidelines. ICZM was the subject of several policy papers prepared by the Council for Nature Conservation and the Countryside in Northern Ireland in 1994. They are currently under consideration by Northern Ireland's office of the Department of the Environment (WWF 1995).

Assessment of Government Responses

With local jurisdiction extending down to the low-water mark, integration of land use and sea use is a significant shortcoming of coastal management in the United Kingdom. A 1990 report prepared for the World Wide Fund for Nature and the Marine Conservation Society (Gubbay 1990) highlights the United Kingdom's fragmented institutional structure for managing terrestrial and marine resources in its coastal zone and the need for greater integration of these activities in order to resolve conflicts among users.

Intersectoral integration in the United Kingdom is limited, but with the development of local and some regional ICZM initiatives, it has become more of a possibility. This is especially so at the local level, where county councils work with district councils to draw up coastal development plans and share the task of implementation (Waite 1981). However, with only a small proportion of purely coastal county plans, there is a need for greater integration of coastal issues into county and regional planning. At the national level, the interdepartmental Group on Coastal Policy has the potential to provide integration, but its role and mandate are unclear and there is no public record of meetings bringing together national-level departments for regular meetings.

In summary, with the establishment of national guidelines for coastal management, intergovernmental integration in the United Kingdom should improve. Presently, however, there appears to be poor coordination between national and county-based coastal planning.

The primary focus of marine scientific research in the United Kingdom is on ocean resources, such as fisheries and offshore oil and gas deposits. The Department of the Environment, for example, generally funds initiatives related to North Sea management, pollution from marine emissions, climate change, and environmental protection technology. As such, interdisciplinary integration of coastal management activities has been limited (Lennard 1988).

With regard to capacity building in ICM, the Joint Nature Conservation Committee (JNCC) has been carrying out a marine nature conservation review since 1987 to extend knowledge of benthic marine systems in Great Britain and to identify sites and species of importance in nature conservation (Hiscock 1996). A series of coastal directories providing an overview of coastal and marine resources and human activities at the national and regional levels are also being published by the JNCC. Graduate-level university programs in coastal and marine management have also been instituted, but there is a genuine need for greater research and information delivery to policy makers. The poor integration of land and sea uses exacerbates the problem, separating those who study and manage coastal lands (at the local level) from those who study and manage the sea (generally at the national and international levels). Integrating those two areas would enhance British capacity in ICM.

Sources Consulted

CIA (Central Intelligence Agency). 1995. *The World Factbook.* Washington, D.C.: Government Printing Office.

Cole-King, A. 1995. Marine protected areas in Britain: A conceptual problem? *Ocean & Coastal Management* 27 (1–2): 109–120.

Gubbay, S. 1990. *A Future for the Coast: Proposals for a U.K. Coastal Zone Management Plan.* Report prepared for the World Wide Fund for Nature and the Marine Conservation Society. London: World Wide Fund for Nature.

———. 1995. Integrated coastal zone management—Opportunities for Scotland. Briefing for the World Wide Fund for Nature, Scotland.

———. 1996. Coastal Management Specialist and NGO Advisor, Ross-on-Wye, England. Response to ICM cross-national survey. Newark: University of Delaware, Center for the Study of Marine Policy.

Hiscock, K., ed. 1996. *Marine Nature Conservation Review: Rationale and Methods.* Summary report. Peterborough, England: Joint Nature Conservation Committee.

King, G., and L. Bridge. 1994. *Directory of Coastal Planning and Management Initiatives in England.* Compiled on behalf of the National Coasts and Estuaries Advisory Group.

Lennard, D. E. 1988. Marine science and technology in the United Kingdom. *MTS Journal* 24 (1): 72–75.

Mackay, G. A. 1981. Offshore oil and gas policy: United Kingdom. In *Comparative Marine Policy: Perspectives from Europe, Scandinavia, Canada, and the United States,* Center for Ocean Management Studies, University of Rhode Island, 103–116. New York: Praeger.

Smith, H. 1996. Senior Lecturer, Department of Maritime Studies and International Transport, University of Wales, Cardiff. Response to ICM cross-national survey. Newark: University of Delaware, Center for the Study of Marine Policy.

United Kingdom. DOE (Department of the Environment). 1992. *Coastal Zone Protection and Planning: The Government's response to the Second Report from the House of Commons Select Committee on the Environment.* London: Stationery Office.

———. 1995. *Policy Guidelines for the Coast.* London: Stationery Office.

United Kingdom. DOE (Department of the Environment) and Scottish Office. 1996.

Scotland's Coasts: A Discussion Paper. Dd8433628. March. London: Stationery Office.

United Kingdom. DOE (Department of the Environment) and Welsh Office. 1992. *Coastal Planning Policy Guidance (CPPG).* September 20. London: Stationery Office.

———. 1993a. *Development Below Low Water Mark: A Review of Regulations in England and Wales. October. London: Stationery Office.*

———. 1993b. *Managing the Coast: A Review of Coastal Management Plans in England and Wales and the Powers Supporting Them.* October. London: Stationery Office.

Waite, C. 1981. Coastal management in England and Wales. In *Comparative Marine Policy: Perspectives from Europe, Scandinavia, Canada, and the United States,* Center for Ocean Management Studies, University of Rhode Island, 57–64. New York: Praeger.

Williams, A. T. 1992. The quiet conservators: Heritage coasts of England and Wales. *Ocean & Coastal Management* 17 (2): 151–169.

WWF (World Wide Fund for Nature). 1995. Integrated coastal zone management: U.K. and European initiatives. *Marine Update* 19. Surrey, England: World Wide Fund for Nature.

France

Background: General Country Context

Located in western Europe, France is bounded to the north by the English Channel, to the west by the Atlantic Ocean, and to the south by the Mediterranean Sea, making up a 2,783-kilometer (1,729-mile) coast; inclusion of the island of Corsica adds another 644 kilometers (400 miles) to this figure. The coastal area exhibits a wide range of physiographic and geographical features, including coral reefs, sea grass beds, coastal wetlands, and beaches and dunes. France claims a 12-nautical-mile territorial sea as well as a 200-nautical-mile Exclusive Economic Zone. The French continental shelf extends to the 200-meter (approximately 656-foot) isobath, or to the depth of exploitation (CIA 1995, 144).

In terms of terrain, the northwestern part of the country is generally made up of low lying or gently rolling hills; the inland eastern section and the southern area contain mountainous areas. France's land area, encompassing 1.4 million square kilometers (547,030 square miles), is densely populated, with more than 58 million inhabitants. The population is predominantly Celtic and Latin, with Teutonic, Slavic, North African, Indochinese, and Basque minorities. More than 90 percent of the population is Roman Catholic. Population growth has stabilized in recent years, despite a high life expectancy.

France has one of the most highly developed economies in the world, with sub-

stantial agricultural resources and a modern diversified industrial sector. Per capita GNP was $23,240 in 1994 (World Bank 1996). Fertile land supporting the
export of wheat and dairy products has placed France as the leading agricultural
producer in Europe. Industry supports more than one-fourth of the GDP, with
exports of steel, machinery, chemicals, automobiles, and other commodities.
However, a high unemployment rate (more than 12 percent) continues to be a
problem.

Following an unstable political and military history during and shortly after
World War II, a new constitution was approved to establish the Fifth French
Republic in 1958. Government control is largely centralized although the country
has twenty-two administrative regions. Coordination between the national and
local levels of government is achieved through a prefect, appointed by the national government to be a liaison. Executive power is held by the president, who
appoints a Council of Ministers, including the prime minister. Legislative power
is held by a bicameral Parliament composed of a Senate and a National Assembly. France is governed by a civil law system, and its judicial branch is headed by
the Constitutional Court (Cour Constitutionelle).

Description of Coastal and Ocean Problems

France's geographical setting (with five different coastal areas) and its proximity
to other countries make it highly susceptible to transboundary problems. On the
Mediterranean coast, problems associated with degraded water quality and
tourism impacts have prompted attention in the Mediterranean regional fora such
as UNEP's Regional Seas Programme. Given France's high level of coastal development, there is concern about degradation of productive coastal ecosystems and
preservation of the coast for tourism and leisure activities.

Regarding conditions offshore, concern over petroleum shipping and transportation in the North Sea became acute in the 1970s and 1980s following several major oil spills. In the most dramatic of these, in 1978 the *Amoco Cádiz* spilled
207,000 metric tons (230,000 tons) of oil, producing "black tides" on much of
France's northern shoreline (Aquarone 1988). With respect to oil and gas development, France has undertaken prospecting activities in the North Sea and the
Mediterranean Sea as well as in the proximity of its territories in the Pacific
Islands.

The French fishing fleet is small compared with the other fleets of Europe, but
catches are on the order of 630,000 metric tons (700,000 tons) per year. Fishery
stocks in waters adjacent to France, particularly in the Gulf of Biscay and the
North Sea, have been severely overexploited by French and Spanish fleets in the
past decade. Although the 1977 resolution establishing a common 200-nautical-
mile EEZ for members of the European Union remains in place, conflicts between
France and Spain over catch quota and fleet size continue. As a result of these
lowered catch numbers and conflicts in Europe, some French fishermen have

begun to redeploy in the Pacific and Indian Oceans in the EEZs of the island territories (Aquarone 1988).

Evolution of Government Responses

Coastal management in France began in the 1970s when the national government created a special commission to identify the opportunities and problems associated with coastal development with the aim of achieving state-of-the art solutions (Miossec 1996). This resulted in a broad umbrella of regulations that control all types of uses on the coast.

In 1975, following recommendations of the national coastal commission, the "Conservatoire du Littoral" was formed. Its objective was to acquire property along the shores of beaches and lakes to protect such lands from urban encroachment, preserve the ecological character of these areas and improve public access to them, and aid in the formulation of marine resource plans. The Conservatoire du Littoral remains a major coastal management agency today, although the two lead French coastal management agencies are the Direction de l'Environnement, which is responsible for zoning, land use, and protection of the environment, and the Direction de l'Équipement, which administers the plans of the SMVM, discussed in the following paragraph.

Two national laws in France regulate coastal management. The first, the Schémas de Mise en Valeur de la Mer (SMVM), was enacted in 1983. The SMVM, concerned primarily with zoning the adjacent marine environment, introduces a system of plans for enhancing and exploiting the sea. The second, the Loi Littoral (Seashore Act), passed in 1986, functions exclusively as a law for land use planning (Boelaert-Suominen and Cullinan 1994, 52).

Under the SMVM, areas of the coastal sea and adjacent land are zoned according to use, with options provided for future development. These areas are defined as "a geographic and maritime unity" of a multiple-use nature (Boelaert-Suominen and Cullinan 1994, 16). The SMVM serves as a template from the national government to communes (local government) and thus is superior to all local terrestrial zoning plans. However, setting up an SMVM system involves a feasibility study and report that is formulated with public and private consultation as well as a public comment period.

The Loi Littoral is specific to terrestrial coastal management, modifying the general French zoning and land use laws to take into account the special nature of the coastal area. It regulates development and other activities in the shore and beach area through control of urban expansion and protection of sensitive areas. The Loi Littoral both establishes set-back zones and maintains the natural character of the coastal zone which, as a part of domaine public, must be preserved (Boelaert-Suominen and Cullinan 1994, 68).

Whereas responsibility for coastal management is divided among several national institutions, management of marine and ocean areas was, for a time, carried out under an umbrella-like arrangement by the Ministry of the Sea (MOS). The MOS, created in 1981, coordinated all national maritime undertakings (shipping; fisheries operations; oil, gas, and mineral development; EEZs; scientific research; etc.), working with the relevant line ministries in the national government.

The MOS was dissolved shortly after its inception in 1981, following a change in government and agency reshuffling. However, a new institution was created by the French government in September 1995 called the Secrétariát à la Mer, which is directly linked to the Prime Ministry. Its primary purpose is to coordinate actions at the national level, specifically with regard to fisheries conflicts and French maritime transport policy.

France has always had a keen interest in territorial marine claims, declaring an EEZ in 1976 and working to shape a common fisheries policy in the European Economic Community in the 1980s. It continues to maintain a tight grip on its territorial island holdings in order to assert sovereignty over EEZs in the South Pacific and Indian Oceans and the Caribbean Sea (Aquarone 1988, 276).

Assessment of Government Responses

Several sources have noted that the French legal framework for managing its coastal zone is a complex combination of broad rules that make assessment of the whole system difficult (Miossec 1996). Be that as it may, the breadth of France's coastal management framework does give it the potential to address the country's entire spectrum of coastal and ocean uses. The Loi Littoral has been a useful means of directing coastal development away from areas prone to coastal erosion. The SMVM, though, has been less successful, with only one area actually zoned on the Mediterranean coast (Miossec 1996).

A significant problem with the French framework of coastal management is that the two major coastal laws, Loi Littoral and SMVM, are set up as a land use law and a marine planning law, respectively. The legislation does not describe the relationship between the two systems, and as such, there is very little coordination between them. If spatial integration is to evolve in France, there needs to be greater coordination of these two management philosophies.

On the basis of results from our cross-national survey, we estimate that intersectoral integration at the national level has been moderately unsuccessful. At present, the Ministry of Environment and the Ministries of Tourism, Transport, Agriculture, and Fishing are involved in coastal management. As such, bringing these ministries together has been difficult. However, the new Secrétariat à la Mer should improve coordination, at least with respect to fisheries and marine shipping management.

With a primarily nationally driven system, intergovernmental integration of coastal management is also deemed moderately unsuccessful. A greater government reliance on the SMVM system would work to improve linkages among the different levels of government managing coastal and ocean resources, but the limited success of that method undermines its potential. The Loi Littoral has also had limited success and localities still make the bulk of land use plans (Meltzer 1996, 5).

With respect to interdisciplinary integration, France has been a leader in marine sciences and technology for several decades. The bulk of this expertise has been utilized by the French government in shipping and fisheries management. The Centre National pour l'Exploitation des Océans (CNEXO), subsumed under the Institut Français de Recherches sur l'Exploitation de la Mer (IFREMER) in 1984, has long been a world leader in marine technology related to aquaculture and fisheries research. Similar expertise in coastal processes has been applied to coastal policy making and management but only in a moderately successful manner. Efforts are currently under way in IFREMER, however, to develop interdisciplinary methodologies for analysis of prototypical coastal settings, with pilot demonstration projects planned in both mainland France and its overseas territories (Denis et al. 1996).

Sources Consulted

Aquarone, M. C. 1988. French marine policy in the 1970's and 1980's. *Ocean Development and International Law* 19: 267–285.

Boelaert-Suominen, S., and C. Cullinan. 1994. *Legal and Institutional Aspects of Integrated Coastal Area Management in National Legislation.* Rome: Food and Agriculture Organization of the United Nations, Development Law Service, Legal Office.

CIA (Central Intelligence Agency). 1995. *The World Factbook.* Washington, D.C.: Government Printing Office.

Meltzer, E., et al. 1996. *International Review of Integrated Coastal Zone Management,* vols. I–III. Consultancy Report prepared by Meltzer Research and Consulting for the Canadian Department of Fisheries and Oceans, to be published by Government of Canada in 1998.

Miossec, A. 1996. Professor, Institut de Géographie et d'Aménagement Régional de l'Université de Nantes (IGARUN), Nantes, France. Response to ICM cross-national survey. Newark: University of Delaware Center for the Study of Marine Policy.

The Netherlands

Background: General Country Context

Despite having only 451 kilometers (280 miles) of coastline, the Dutch have turned a low-lying, marshy country into a densely populated nation in which 60 percent of the population inhabits coastal areas. Of the Netherlands' 15.5 million people, 96 percent are Dutch, with the remaining 4 percent Turks, Moroccans, and others. Major religious divisions are Roman Catholic (34 percent), Protestant (25 percent), and unaffiliated (CIA 1995). The Dutch inhabitation of the coastal zone has been facilitated by one of the most extensive coastal defense systems in the world whereby dams and dikes protect nearly one-half of the coastal area (IPCC 1994, 17). Bordered by the North Sea, the Netherlands claims a 200-nautical-mile Exclusive Fishing Zone and a 12-nautical-mile territorial sea.

The economy of the Netherlands is based on private enterprise, with trade and financial services providing 50 percent of the country's $275.8 billion GDP (CIA 1995). Industrial activity provides about 25 percent of the GDP, led by food-processing, oil-refining, and metalworking industries. Agriculture accounts for roughly 5 percent of the GDP, dominated by animal production (CIA 1995, 301). The 1994 estimate of per capita GNP was $22,010, indicative of a developed country (World Bank 1996).

The government of the Netherlands is a constitutional monarchy, with a national government led by a prime minister. The prime minister is assisted by a cabinet (or Council of Ministers). The legislative branch of government is bicameral, with a First Chamber elected by provincial councils and a Second Chamber elected by popular vote. The legal system is based on civil law; the judicial system is led by the Supreme Court. There are twelve provinces in the Netherlands and more than 700 municipalities. With respect to coastal management, activities are initiated at the municipal level under supervision from the provincial government. The national government acts to further supervise activities in a consistent manner and to resolve conflicts (Wiggerts 1981, 78).

Description of Coastal and Ocean Problems

The low-lying nature of the Dutch coastal zone makes it particularly vulnerable to coastal flooding and erosion associated with storms and sea-level rise. When the dramatic storm surge of 1953 inundated a large area of the southwestern Dutch coast, it destroyed coastal settlements and agricultural lands and killed nearly 2,000 people. The flood prompted a major government response to reinforce the country's protection from the waters of the North Sea. The Delta Plan was developed as a means of combating storm surges through the damming of waters from a number of estuaries (Koekebakker and Peet 1987, 125). It remains one of the most extensive coastal defense systems in the world today.

In the 1970s water pollution issues in the Netherlands, such as high concentrations of heavy metals and organic compounds and nutrient loading, have focused attention on the need for broader coastal management initiatives. Development of the coastal area for urban and industrial uses, recreation, and transportation continues to progress despite dwindling space in the coastal zone and continued erosion (Koekebakker and Peet 1987). Regional issues such as protection of the Waddenzee, one of the few remaining wetlands in northwestern Europe, is also a major coastal management concern (IPCC 1994).

Management of the North Sea remains an important issue. In particular, protection of North Sea water from vessel-source pollution, prevention of coastal erosion from proposed extraction of sand near the coast, and management of commercial fishing in coastal waters are of primary importance (Van Horn, Peet, and Wieriks 1985, 56).

Dutch interest in the North Sea is predicated on the fact that more than 60 percent of all North Sea navigation occurs in Dutch waters or en route to Dutch ports. The offshore oil and gas industry in the vicinity of the Netherlands has continued to expand despite several catastrophic oil spills in the 1970s and 1980s. Sites also exist for ocean dumping and ocean incineration (Peet 1987)

Information obtained from our cross-national survey indicated that loss of life and property associated with coastal hazards continues to be an issue of primary importance, followed by new or perceived economic opportunities and the need to coordinate coastal management activities among the national, provincial, and local governments (de Vrees 1996).

Evolution of Government Responses

Coastal management in the Netherlands is decentralized, with the bulk of management activities performed by local and regional bodies called water boards, or waterschappen. Composed of landowners and tenants whose votes are in proportion to landholdings, the water boards have legal authority to manage the coastline (i.e., set up coastal defenses) within the area under their jurisdiction. The provincial governments supervise their activities, and the entire system is overseen at the national level by the Ministry of Transport, Public Works, and Water Management, which acts as lead coastal agency (Koekebakker and Peet 1987, 127). Originally created solely for the purpose of coastal defense, the water boards (now numbering more than 800) have evolved a more broad-based agenda, carrying out planning and management duties for water-related activities in their areas.

In addition to its role as overseer to the water boards, the national government plays a lead role in coastal management with respect to issues of national significance. In 1991, it put forth the Dynamic Preservation Strategy, which set basic coastline positions to be maintained by law (de Vrees 1996). In 1995, the Coastal Defense Act was enacted by Parliament, which regulates the responsibilities of all parties concerned with coastal management. In order to coordinate activities better regionally and among different levels of government, Provincial Consultative Bodies were established, with representation from the national and provincial governments and the water boards. In some instances, municipalities involved with coastal defense activities and nature conservation agencies (NGOs) are also represented. The national government also took the lead role in marine policy with adoption of the North Sea Harmonization Policy in 1984. The main purpose of this policy is to coordinate the activities of all Dutch ministries involved in North Sea matters (Van Horn, Peet, and Wieriks 1985).

The North Sea Harmonization Policy contains three elements: a policy framework that coordinates government actions; an action program; and an institutional framework for preparing and implementing policy proposals resulting from the action program. The overall objectives of the policy include safe and efficient use of the North Sea, conservation of the ecology of the sea, and creation of a harmonious balance among the sea's various users (Peet 1987, 437). Some actions that have been carried out include the drawing up of a North Sea navigation policy and compilation of a wide-ranging plan for water quality protection. As with coastal management and defense, the Ministry of Transport, Public Works, and Water Management is the lead implementing agency with respect to North Sea policy (de Vrees 1996).

Assessment of Government Responses

P. Koekebakker states that the emphasis in coastal planning and management in the Netherlands has changed in recent years from a rather technocratic approach regarding coastal defense and land reclamation to a more integrated approach in which conservation also plays a major role (Koekebakker and Peet 1987). Implementation of the Coastal Defense Act should ensure increased consistency among local (water boards), provincial, and national coastal management actions. This intergovernmental coherence and consistency appears to be the most successful component of the Dutch system. However, these traditional decision-making bodies may have difficulty in addressing new areas of concern, such as conservation and recreation management.

In terms of intersectoral integration, planning and strategies at the national level have provided coherence among all levels of government, with particular success in the context of special area management plans. In the Waddenzee region, for example, all three levels of government cooperated in forming a policy that would protect the ecological character of the northern wetland area from various forms of industrial pollution (Koekebakker and Peet 1987, 131).

With respect to spatial integration, there is close integration of activities related to coastal defense between the coastline and adjacent coastal waters. The North Sea Harmonization Policy continues to evolve toward multiple-use management, but many of the problems in the North Sea, such as navigation and pollution control will depend not on national efforts but on international cooperation within the European Union (Van Horn, Peet, and Wieriks 1985).

The integration of coastal engineering and geomorphologic expertise into Dutch policy making is probably one of the best examples of science–policy integration. Even at the lowest administrative levels—that is, the water boards— each board has a technical staff to advise in decision making (Wiggerts 1981, 78).

Following the World Coast Conference sponsored by the Netherlands in 1993, the national government established the Coastal Zone Management Centre (CZM Centre) to promote capacity building in ICM. The CZM Centre, part of the

National Institute for Coastal and Marine Management (RIKZ), is charged with dissemination and exchange of information through the NetCoast global ICM information network; organization of training programs, conferences and special events; and provision of support for the development and improvement of ICM concepts, methodologies, and tools (NetCoast 1996).

Akin to its strong reliance on engineering solutions to coastal management problems, the Netherlands has developed a number of computerized modeling and decision support systems, including one for analysis of the North Sea—the Database Management and Modelling System (DMMS), designed to support policy makers and managers and build local capacity (Klomp 1993). Other capacity-building initiatives cited in the cross-national survey include specialized in-country training of staff, overseas training, hiring of new staff with appropriate qualifications, and establishment of new graduate programs at the university level. In addition, the Netherlands has long been a major participant in programs such as the Intergovernmental Panel on Climate Change and continues to build capacity through the IPCC's numerous global projects and research efforts.

Sources Consulted

CIA (Central Intelligence Agency). 1995. *The World Factbook.* Washington, D.C.: Government Printing Office.

de Vrees, L. 1996. Senior Project Manager, Coastal Zone Management Centre, The Hague, Netherlands. Response to ICM cross-national survey. Newark: University of Delaware, Center for the Study of Marine Policy.

IPCC (Intergovernmental Panel on Climate Change). 1994. *Preparing to Meet the Coastal Challenges of the 21st Century: Report of the World Coast Conference 1993.* November 1–5, Noordwijk, Netherlands. The Hague: Ministry of Transport, Public Works, and Water Management, National Institute for Coastal and Marine Management, Coastal Zone Management Centre.

Klomp, R. 1993. Lessons learned from the experience with policy decisions and supporting tools in coastal water management. Paper presented at MEDCOAST 1993, the First International Conference on the Mediterranean Coastal Environment, November 2–5, Ankara, Turkey.

Koekebakker, P., and G. Peet. 1987. Coastal zone planning and management in the Netherlands. *Coastal Management* 15: 121–134.

NetCoast. 1996. *A Guide to Integrated Coastal Zone Management.* World Wide Web site operated by the Ministry of Transport, Public Works, and Water Management, National Institute for Coastal and Marine Management, Coastal Zone Management Centre, The Hague, Netherlands. Internet: http://www.minenw.nl/projects/netcoast

Peet, G. 1987. Sea use management for the North Sea. In *The UN Convention on the Law of the Sea: Impact and Implementation,* ed. E. D. Brown and R. R. Churchill, 430–440. Honolulu: University of Hawaii, Law of the Sea Institute.

Van Horn, H., G. Peet, and K. Wieriks. 1985 Harmonizing North Sea policy in the Netherlands. *Marine Policy* 9 (1): 53–61.

Wiggerts, H. 1981. Coastal management in the Netherlands. In *Comparative Marine Policy: Perspectives from Europe, Scandinavia, Canada, and the United States,* Center for Ocean Management Studies, University of Rhode Island, 75–81. New York: Praeger.

Spain

Background: General Country Context

The kingdom of Spain lies in southwestern Europe and includes the Balearic Islands in the Mediterranean Sea, the Canary Islands in the Atlantic Ocean, and a few small sovereignties on and off the coast of Morocco. The country's 4,964-kilometer (3,085-mile) mainland coastline is bordered to the north by the Bay of Biscay, at the southwest tip by the Atlantic Ocean and the Strait of Gibraltar, and to the south and west by the Mediterranean Sea. Spain claims a 12-nautical-mile territorial sea and a 200-nautical-mile Exclusive Economic Zone (CIA 1995).

The Spanish climate is warm and dry, and the terrain is largely rough and hilly. Spain has a land area of nearly 500,000 square kilometers (190,000 square miles), of which 31 percent is utilized for agricultural purposes. Spain has approximately 40 million people (1995 estimate), 35 percent of whom live in coastal areas (MOPU 1988b, 1). The population is predominantly of Mediterranean and Nordic descent, and 99 percent are Roman Catholic (CIA 1995).

Spain has a per capita GNP of $13,440, indicative of a developed country (World Bank 1996). The Spanish economy is currently in recession, hindered by a 25 percent unemployment rate. Continued political turmoil has complicated the establishment of a stable budget and interest rates, labor law reform, and Spain's participation in the economic integration of the European Union (CIA 1995, 393). Nonetheless, there has been moderate growth in industrial output, tourism and other sectors in recent years. Agriculture accounts for 5 percent of GDP, with large-scale production of grain, vegetables, olives, wine grapes, and livestock. Major industries include manufacturing of textiles and apparel, food and beverages, metal and chemicals; automaking and shipbuilding; and tourism.

Spain is made up of seventeen autonomous communities. Like France, it possesses a civil law system headed by a Supreme Court (Tribunal Supremo). The king is chief of state under the 1978 constitution. He appoints the president (chief of government) and, on the latter's recommendation, the Council of Ministers. Legislative power is vested in the Cortes Generales, a bicameral legislative body made up of the Senate and Congress of Deputies. Each regional community has its own parliament and self-governing responsibilities (CIA 1995, 390–393).

Description of Coastal and Ocean Problems

Spain's coastal area, like those of other countries, is under pressure from increasing population density. Data on the growing Spanish tourism industry indicate that more than 82 percent of tourists visit the coastal area (MOPU 1988b, 1). As a result, roughly half of the coast has been developed or prepared for development. This has negatively affected the natural character of the Spanish coast through sedimentation, degraded water quality, and loss of productive coastal ecosystems, effects that are exacerbated by the increasing privatization of coastal lands (MOPU 1988b, 1–3).

The northern coast of Spain, with a large proportion of estuaries, coastal wetlands, and adjacent areas, has been most affected by increasing occupancy in the coastal zone; about 1.8 million people inhabit the area. The vast majority of the region's wetlands and estuaries have been drained and filled to support coastal development and low-productivity agriculture in the past fifty years. Only recently have the general public, administrative officials, and decision makers called attention to the need for revitalization and restoration of the natural character of the wetland areas (Rivas et al. 1994).

Coastal erosion is another major problem on the Spanish coast. In many ways, this problem is caused by shortsighted and unintegrated coastal defense systems, leading to the construction of artificial barriers that intercept the littoral flow of sand, structures that retard the natural oscillations of the beach, destruction of dune fields, and natural loss of beach materials to erosion (MOPU 1988a, 15–19).

Following a boom in the regional and distant-water fishing industry in the 1960s, the Spanish fishing fleet has been steadily reduced over the past three decades with the establishment of national Exclusive Economic Zones and the general reduction in commercial stocks around the world. Whereas in the past, 70 percent of Spanish landings came from within the 200-nautical-mile boundaries of other countries, the bulk of the catch today comes from high seas fisheries. Still, serious conflicts have arisen as a result of Spanish fishermen continuing to fish in areas under the national sovereignty of other states and accusations by France and the United Kingdom that Spanish fishermen are breaking quotas set by the European Union (Meltzoff and Lipuma 1986).

Because of its geographical setting, Spain is also highly exposed to marine oil pollution associated with oil tanker traffic. The Strait of Gibraltar and the Galicia coastal region are both major routes for oil tankers en route to Mediterranean ports.

With respect to management of the coastal zone, there exists an overlap of jurisdiction among national, regional, and local governments. Although the bulk of resource management jurisdiction is vested in the regional governments, the national government also has a role through the national Shores Act. Local governments participate in CZM through development of land use plans in beach and foreshore areas. The result is a complicated framework of coastal management arrangements and jurisdictions that appears duplicative and inefficient.

Evolution of Government Responses

As stated earlier, Spain's 1978 constitution sets out a framework for powers over coastal management. The national government exercises its jurisdiction through the 1988 Shores Act to "define, protect and regulate the use and government police power on the coastal public property and, in particular, the shores" (Suarez de Vivero 1992, 308). This broad statement specifically relates to management of public areas of the coastal zone, such as setting of coastal boundaries; concessions and authorizations of public lands; approval for use and protection of public lands (including fishing, mining, forestry, aquaculture, etc.); and regulations for the use of beaches, such as coastal defense and regeneration (Suarez de Vivero 1992, 310). The lead implementing coastal agency is the Ministry of Public Works, Transport, and Environment (MOPU).

As a direct result of the 1978 constitution, regional governments have exclusive jurisdiction over territorial and urban planning including, in many cases, coastal management and planning in the beach and foreshore area. However, none of the regions has developed the "necessary administrative means to carry out these jurisdictions to the fullest extent" (Suarez de Vivero 1992, 307–308). Regional governments also have jurisdiction over small commercial and pleasure harbors as well as fishing in inland waters, and they must manage marine resources in cooperation with the national government. Examples include management of fisheries, industrial sewage, and pollutants and conduct of sea rescues. Additionally, local governments have the primary role in developing plans for beach and foreshore development and zoning.

Other coastal and ocean management measures adopted by the Spanish government include a program to monitor pollution of the Mediterranean Sea. This and a related program monitoring hydrocarbon pollution in the Mediterranean are included in a national pollution control plan implemented in the early 1990s in relation to Spain's UNCED agenda (MOPT 1992).

Other marine-related initiatives include the establishment of a system of marine protected areas (MPAs). Following the creation of the first MPA in 1982 (a fishery preservation zone), numerous other zones have been designated as hunting refuges, marine reserves, national marine terrestrial parks, and conservation areas. Many of these MPAs have been zoned for multiple use and have operated with some success, although there is a disproportionate number of sites on the Mediterranean coast as compared with the Atlantic coast (Ramos-Espla and McNeil 1994).

Assessment of Government Responses

Although the Shores Act creates a framework for coastal management at the national level, its focus is relatively narrow. It works to regulate coastal development and tourism, manage the physical aspects of the coastline, and ensure public access to the coast. The evolution of a more integrated approach will require greater attention to integrating marine and coastal management, increased efforts at marine and coastal conservation, and coordination of the various sectoral activities in the coastal zone under the umbrella of a formal coastal program.

A major problem in Spain's coastal management scheme is boundary delimitation. The national government, by means of the Shores Act, generously defines the public coastal area. Termed the *seashore,* it includes the foreshore behind the dunes, the dune line itself, the berm of the beach, and submerged areas in the coastal waters (MOPU 1988b). The regions, however, claim jurisdiction to a similar strip of land (the foreshore and beach area) down to hydrographic zero (the low-tide line).

The courts have attempted to reconcile these boundary disputes, but the solution remains unclear. What results from the regime is local government extending urban land use planning to the foreshore and beach, for which the national government has the power to regulate uses and set guidelines for development. The regional governments, by means of the constitution, have the power to act over municipal districts in the foreshore and beach area but not in the submerged lands area, despite their jurisdiction over inland fisheries.

It appears that Spain is not effectively integrating terrestrial and marine management, nor is it integrating sectors effectively. Ocean management is administered by a separate ministry with its own policies, priorities, and resource levels. Sectoral activities in the coastal zone such as mariculture, mining, and forestry are managed under separate laws, out of the regulatory reach of the coastal program.

Intergovernmental integration is perhaps the weakest dimension of integration in the Spanish context. Much of this has to do with the respective governments trying to identify and develop their roles in coastal management given the nebulous framework of coastal jurisdiction. By passing the Shores Act, the national government reinstated a role for itself in coastal management that it had yielded to the regional governments ten years before. The result is a duplication of coastal management efforts and adversarial relations among the three levels of government. Interdisciplinary integration in ocean and coastal management in Spain appears to be satisfactory to moderately successful, based on the opinions expressed by several respondents to our cross-national survey.

Sources Consulted

CIA (Central Intelligence Agency). 1995. *The World Factbook.* Washington, D.C.: Government Printing Office.

Meltzoff, S. K., and E. Lipuma. 1986. The troubled seas of Spanish fishermen: Marine policy and the economy of change. *American Ethnologist* 13 (4): 681–699.

Montoya, F. 1996. Head of Coastal Service, Ministry of Public Works, Transport, and Environment, Tarragona, Spain. Response to ICM cross-national survey. Newark: University of Delaware, Center for the Study of Marine Policy.

MOPT (Ministerio de Obras Públicas y Transportes). 1992. *Informe Nacional*

UNCED, Brazil '92 (National report for UNCED 1992). Madrid: Ministerio de Obras Públicas y Urbanismo.

MOPU (Ministerio de Obras Públicas y Urbanismo). 1988a. *Actuaciónes en la costa (Coastal action).* Madrid: Ministerio de Obras Públicas y Urbanismo.

—-. 1988b. *Ley de Costas* (The Shores Act). Madrid: Ministerio de Obras Públicas y Urbanismo.

Ramos-Espla, A. A., and S. E. McNeil. 1994. The status of marine conservation in Spain. *Ocean & Coastal Management* 24 (2): 125–138.

Rivas, V., F. E. Diaz de Teran, Jr., A. Cendrero, J. Hidalgo, A. Serrano, M. Villalobos, I. Benito, and M. Herrera. 1994. Conservation and restoration of endangered coastal areas: The case of small estuaries in northern Spain. *Ocean & Coastal Management.* 23 (2): 129–148.

Suarez de Vivero, J. L. 1992. The Spanish Shores Act and its implications for regional coastal management. *Ocean & Coastal Management* 18 (4): 307–317.

——. 1996. Professor, Facultad de Geografía e Historia, Universidad de Sevilla, Sevilla, Spain. Response to ICM cross-national survey. Newark: University of Delaware, Center for the Study of Marine Policy.

The Republic of Korea

Background: General Country Context

The Republic of Korea (South Korea) lies on the southern half of the Korea Peninsula, situated in northeastern Asia. It is bordered to the east by the Sea of Japan, to the west by the Yellow Sea, to the south by the Korea Strait, and to the north by the Democratic People's Republic of Korea (North Korea), demarcated by a Demilitarized Zone (DMZ) roughly along the 38th parallel north. The climate is temperate: very cold, dry winters, with an average temperature of –6°C, and hot, humid summers, with an average temperature of 25°C (*Europa* 1995). About 40 percent of the population of 45 million lives in the coastal areas. The overwhelming proportion of the population is Korean, with the exception of about 20,000 Chinese; 48.6 percent of the country's population is Christian and 47.4 percent is Buddhist. The population growth rate was 1.04 percent in 1995 (CIA 1995). South Korea has a 2,413-kilometer (1,499-mile) coastline. The country claimed a 12-nautical-mile territorial sea (3-nautical-mile in the Korea Strait) in 1978 (CIA 1995; Churchill and Lowe 1988) and a 200-nautical-mile Exclusive Economic Zone in 1996.

Since the early 1960s, South Korea has experienced rapid economic growth under a succession of export-oriented Five-Year Economic Development Plans. By the mid-1980s, it became one of the world's largest trading nations, with an average annual growth rate of more than 10 percent (CIA 1995). In 1994, South Korea's per capita GNP was $8,260 (World Bank 1996). South Korea is not a country richly endowed with natural resources. For that reason, the mining and quarrying sector contributed only 0.3 percent of the GDP in 1993. Agriculture, including forestry, provided only 7.0 percent of the GDP in 1993, employing 13.6 percent of the labor force. The principal crop is rice. Fishing is also important, providing an annual catch of about 2.9 million metric tons (3.2 million short tons) in the early 1990s. In contrast, industry and manufacturing contributed 43.3 percent and 26.8 percent of the GDP, respectively, in 1993. As of 1993, South Korea had the world's largest shipbuilding capacity. Since his inauguration in 1992, South Korea's president, Kim Young Sam, has provided major economic reforms with the aim of eliminating corruption. The country achieved an average annual GDP growth rate of 6.9 percent during the latest Five-Year Economic Development Plan (1993–1997) (*Europa* 1995).

Since the retreat of Japanese forces at the end of World War II, the Korea Peninsula, formerly called Chosun, has been divided into two Koreas, the Republic of Korea to the south and the Democratic People's Republic of Korea to the north. A United Nations–supervised election of a new legislature took place in the southern division of the peninsula in May 1948, establishing the Kuk Hoe (National Assembly). The Republic of Korea was established on August 15, 1948, under the first constitution adopted by the National Assembly. This constitution was suspended by a military coup led by General Park Chung Hee in May 1961. After a period of rule by military juntas from 1961 to 1987, a new constitution for the Sixth Republic was adopted on October 27, 1987 by a national referendum. According to this constitution, the president is directly elected for one term of five years by universal suffrage. The president, as chief of state, appoints the prime minister and cabinet members composing the State Council, led by the prime minister, who is head of government. Legislative power is vested in the unicameral National Assembly, members of which are elected for four-year terms. South Korea has nine provinces and six special cities (*Europa* 1995). It is a democratic republic that has given its national government a great deal of power. Recently, more local autonomy was granted under an administrative and legislative decentralization. Each province has its own legislature, and each provincial government is led by a governor who is elected by the citizens. Municipalities follow the same system.

Description of Coastal and Ocean Problems

During the First and Second Five-Year Economic Development Plans (1962–1971), South Korea emphasized labor-intensive and export-oriented light industry. During this period, the major coastal economic activities were reclamation of the Youngsan River area and establishment of the Free Export Zone in Masan, both along the southern coast of the country. No significant level of concern with coastal and ocean issues was apparent during this phase (Hong 1991; Lee et al. 1993; Hong and Lee 1995). However, after completion of the Third and Fourth Five-Year Economic Development Plans (1972–1981), South Korea began to recognize serious conflicts among coastal users and environmental deterioration of wetlands and tidal lands due to a high rate of coastal area reclamation, establishment of heavy industrial complexes along the southern coast, and an increase of the coastal population.

In accordance with the national economic and developmental policy promoting heavy and chemical industry, a large number of heavy industrial complexes were established along the southeastern coast, at Ulsan (shipbuilding industry) and P'ohang (steel industry), and along the southern coast, at Ch'ang-won (integrated machinery), Yosu (petrochemical industry), Kwangyang (steel industry), and Okpo and Pusan (shipyards) (Hong 1991). The rapid urbanization of South Korea's coasts closely paralleled the area's economic modernization and concentration of industrial complexes. This rapid economic growth and increasing density of coastal population demanded more space for development.

Coastal areas, especially along the western coast, which is marked with many indentations, attracted land developers and government decision makers to reclaim the bays and estuaries. Following the Reclamation Act of Public Waters of 1962 (amended in 1986), a total area of 1,100 square kilometers, or 420 square miles—15 percent of the total potential area of 7,305 square kilometers (2,821 square miles)—was reclaimed for several purposes, including agriculture (65.1 percent), urbanization (20.3 percent), industry (11.3 percent), power plants (1.7 percent), and waste disposal (1.6 percent) (Lee et al. 1993; Hong 1995). More recently, the South Korean government planned to invest approximately $16 million in reclamation of 261 other areas, totaling 1,235 square kilometers (477 square miles), by 2001. Most of these areas (98 percent of the total planning areas) are located on the western coast (Lee et al. 1993). The majority of reclamation activities on the western coast has been accelerated by the West Coast Development Policy of 1989, which aimed internally to achieve more balanced regional development for the western coastal region and internationally to better connect with China in commerce (Hong 1991).

Although the reclamation of coastal areas contributed more space for agricultural, industrial, and urban expansion, it also resulted in a number of adverse environmental consequences, such as destruction of traditional fishing and mariculture farming grounds. Most of South Korea's mariculture farming activities are concentrated on the southern coast (74,185 hectares, or 183,311 acres) and the western coast (21,578 hectares, or 53,319 acres) where massive amounts of land have been reclaimed and coastal industrial complexes established since the early 1970s (Hong 1991). Conflicts arose between traditional fishermen and reclaimers. Coastal reclamation in Seosan A and B Districts on the western coast carried out by Hyundai Construction Company during 1982–1989 damaged traditional fish habitats and mariculture farms, incensing local fishermen. Coastal reclamation at Siwha on the western coast undertaken in 1986 also resulted in negative environmental effects on traditional fishing activities and mariculture of seaweeds, oysters, and shellfish. Massive reclamation of wetlands is ongoing. Although there are no immediate adverse environmental effects, loss of traditional fish hatchery and nursery habitats will result in irreversible long-term cumulative effects on the coastal fisheries.

Since the late 1970s, South Korea's distant-water fisheries have suffered from the oil crisis and the declaration of extended jurisdictional zones by coastal states (Hong 1991). In contrast, however, mariculture farming has grown rapidly. In 1989, the country's fish production was 3.6 million metric tons (4 million short tons): 0.9 million metric tons from distant waters, 1.7 million metric tons from coastal waters, and about 1 million metric tons from mariculture farms (Hong 1991). At present, more than 98 percent of the mariculture areas are concentrated on the southern and western coasts, in the same areas where massive coastal wetlands reclamation has taken place and a number of heavy industrial complexes have been built. Mariculture areas have been threatened by increasing coastal water pollution, red tides, and oil spills. On July 23, 1995, the Cypriot tanker *Sea Prince* grounded in Typhoon Faye near Sori Island, off the city of Yosu on the southern coast. Approximately 9,810 barrels (412,000 gallons) of oil spilled, seriously damaging a number of mariculture farms on the southern coast (*OSIR* 1995); the damage was worsened by coincidental red tides.

Evolution of Government Responses

Since the early 1970s, some major socioeconomic and environmental changes—rapid economic modernization, increased demand for land space, growing coastal population density, and increased shipping capacity—have prompted the creation of new government structures and new laws dealing with coastal resources and environments. After completion of the Third Five-Year Economic Development Plan (1972-1976), which achieved rapid development of heavy industry, the South Korean government began to recognize that this economic achievement was causing environmental degradation (Hong and Lee 1995). In 1977, the government enacted two important marine environmental laws: the Environmental Preservation Act (EPA) and the Marine Pollution Prevention Act (MPPA). The EPA set forth environmental management tools and principles such as the environmental impact statement, special management zones, and the "polluter pays" principle.

In 1980, the Environment Administration was created under the Ministry of Public Health and Social Affairs; in 1990, it was upgraded to the Ministry of Environment (MOE). The MOE is responsible for implementing the EPA and the MPPA. In 1982, the MOE designated approximately 930 square kilometers (359 square miles) of the southern coast connecting Ulsan, Pusan, Chinhae, and Kwangyang Bay, as "Special Sites of Coastal Pollution" in accordance with the EPA. Within these special zones, industrial development activities, which cause the pollution of coastal waters, are strictly limited and regulated through effluent standards (Hong 1991).

In order to maximize the efficiency of coastal and ocean environmental protection in response to "the new international oceanic order," the South Korean government enacted the Marine Development Basic Act of 1987 and created the Marine Development Council (MDC), which is headed by the prime minister (Hong and Lee 1995). The Office of the Prime Minister formulated a marine policy statement, "New Marine Policy Direction toward the Twenty-First Century," through the Working Group for the Integrated Marine Policy, the membership of which is almost identical to that of the MDC (Hong 1995). The statement emphasizes major ocean policy issues such as the Law of the Sea, integrated coastal management, development of ocean industries, and promotion of public awareness of coastal environments (Hong and Lee 1995). Under the direction of this national marine policy, the Ministry of Science and Technology played a leading and coordinating role in preparation of the Marine Development Basic Plan (MDBP), completed in 1996.

The MDBP was prepared with four agendas: (1) establishment of a national ocean management system, (2) maximization of the socioeconomic value of ocean resources, (3) enhancement of marine environment quality, and (4) improvement of ocean technologies, oceanographic surveys, and national ocean services (Hong and Lee 1995). It also proposed to enact a Coastal Management Act by 1997 and prepare a National Coastal Management Plan by 1998 (Lee 1996b). To support and provide basic information for these two efforts, the Ministry of Construction and Transportation led the National Coastal Zone Assessment Project in 1996–1997. The Korea Ocean Research and Development Institute (KORDI), part of the Ministry of Science and Technology, has also been carrying out an integrated coastal management pilot study since 1994 on the southern coast in the region of Chinhae Bay, one of the country's most severely polluted bays. The main objectives of this pilot study are to test the applicability of the ICM model before applying it nationwide under the new Coastal Management Act (Lee 1996a).

In order to centralize the country's fragmented national marine programs, the government of South Korea recently set up the Ministry of Maritime Affairs and Fisheries, which integrates the Maritime and Port Administration, the National Fisheries Administration, the Hydrographic Affairs Office, and the National Maritime Police Administration into one institution (Hong 1996). The new ministry was approved by the National Assembly on August 12, 1996. This action is significant because it combines in one ministry the major functions related to both oceans and coasts.

Assessment of Government Responses

South Korea's coastal zone management has been under the sectoral jurisdiction of many different ministries. The country has created more than twelve institutions, with their subordinate agencies, to set policies and priorities, neither conforming to nor being responsible for creating an efficient national coastal zone management program (Hong 1991). Of these ministries, the Ministry of Construction and Transportation (MCT), has played the most powerful role in managing activities in coastal areas (Lee et al. 1993). Using the slogan "the Era of West Coast," the MCT amended the Reclamation Act of Public Waters in 1986 to give construction firms (returning from the Middle East) incentives to participate in reclamation projects on the western coast. This mid-1980s reclamation boom had adverse consequences on coastal fisheries, mariculture farming, and coastal ecosystems, but no mechanism existed to deal with interagency conflicts such as those involving the Ministry of Construction and Transportation and the Ministry of Agriculture, Forestry, and Fisheries on one side and the Ministry of Environment and the National Fisheries Administration on the other side.

Another characteristic of South Korea's coastal zone management is a strong centralized management system. All major coastal and ocean policy decisions are formulated at the national level and flow down to the subnational governments. This process tended to be reactive rather than proactive, especially regarding preservation of the coastal environment (Lee et al. 1993), since the South Korean government tended to prefer development-oriented policy to conservation-oriented decision making.

Although ICM efforts in South Korea began only recently—after UNCED, in 1992—the country promptly put the prescriptions of UNCED's Chapter 17 of Agenda 21 into practice. In terms of ICM application in South Korea, the "New Marine Policy Direction toward the Twenty-First Century" readily acknowledged ICM as holding high priority on the national government's agenda and required the enactment of integrated coastal management legislation by 1997.

The ICM pilot study in Chinhae Bay represents an important preparatory government action to test the ICM concept before ultimately applying it nationwide. The pilot study is backed by national government policy makers, regional government officials, and local stakeholders. In addition, one of the most difficult challenges of any ICM effort—institutional integration—is being achieved in Korea through the successful establishment of the new Ministry of Maritime Affairs and Fisheries (MAF). The MAF and the Ministry of Construction and Transportation will be lead agencies under the proposed Coastal Management Act and National Coastal Management Plan.

In terms of capacity building in ICM, the Oceanographic Society of Korea, established in the mid-1960s, has annually published data and information related to marine sciences in the *Journal of the Oceanographic Society of Korea*. Since Seoul National University first created a program in oceanography in 1968, more than ten universities have developed programs in oceanography (Hong 1995), but as yet, there is no program in marine affairs and policy. In terms of marine policy, the Korea Ocean Research and Development Institute (KORDI), an important national marine scientific research institution established in 1973, investigates coastal zone management issues, conducts the Chinhae Bay ICM pilot study, and is establishing a coastal resource database. In 1997, the Korea Maritime Institute

(KMI), a government-sponsored research institution, was established through the integration of several institutions related to marine policy, maritime affairs, and fishery issues. KMI conducts research and related activities in the fields of ocean policy, fisheries, shipping, and ports.

As part of the process of ICM in Chinhae Bay, a workshop on integrated coastal management titled "Developing Strategies for the Sustainable Development of the Masan-Chinhae Bay" was held on February 8–9, 1995. A number of officials from ministries and agencies, local government officials, university professors, research scientists, fishermen, and representatives of local NGOs participated in the workshop, suggesting management strategies and research and monitoring techniques. Other capacity-building efforts in the Chinhae Bay pilot study include creation of the Local Advisory Committee for Chinhae Bay Management, which enhances communication among local participants, KORDI researchers, and local government officials (Lee 1996a).

Sources Consulted

Churchill, R. R., and A. V. Lowe. 1988. *The Law of the Sea*. Manchester, England: Manchester University Press.

CIA (Central Intelligence Agency). 1995. *The World Factbook*. Washington, D.C.: Government Printing Office.

The Europa World Year Book. 1995. 36th ed. London: Europa.

Hong, S. Y. 1991. Assessment of coastal zone issues in the Republic of Korea. *Coastal Management* 19: 391–415.

———. 1995. A framework for emerging new marine policy: The Korean experience. *Ocean & Coastal Management* 25 (2): 77–101.

———. 1996. Letter to Professors Biliana Cicin-Sain and Robert W. Knecht, June 25.

Hong, S. Y., and J. H. Lee. 1995. National level implementation of Chapter 17: The Korean example. *Ocean & Coastal Management* 29 (1–3): 231–249.

Lee, J. H. 1996a. Policy issues and management framework of Chinhae Bay in the Republic of Korea. Paper presented at international workshop, Integrated Coastal Management in Tropical Developing Countries: Lessons Learned from Successes and Failures, May 24–28, Xiamen, People's Republic of China.

———. 1996b. Research Scientist, Korea Ocean Research and Development Institute, Marine Policy Center, Ansan, Republic of Korea. Response to ICM cross-national survey. Newark: University of Delaware, Center for the Study of Marine Policy.

Lee, J. H., Y. Lee, S.-Y. Hong, and Moonsang Kwon. 1993. Coastal zone of Korea: Status and prospects. In *International Perspectives on Coastal Ocean Space Utilization*, ed. P. M. Grifman and J. A. Fawcett, 99–118. Los Angeles: University of Southern California, Sea Grant Program.

OSIR (Oil Spill International Report). 1995. Korea officials report widespread aquaculture damage from tanker spill. *OSIR* 18 (28) (August 3).

Australia

Background: General Country Context

The Commonwealth of Australia, in stark contrast to the countries around it, has a relatively small population (approximately 18 million) inhabiting an enormous amount of land space (7,686,850 square kilometers, or 2,967,890 square miles). The population is more than 93 percent Caucasian and 75 percent Christian. Although still somewhat dependent on the traditional primary resource-related economic activities that propelled it to its current developed socioeconomic status, the Australian economy has become increasingly dependent on tourism and other tertiary economic activities (CIA 1995). In 1994, Australia's per capita GNP was $18,000, indicative of a developed nation (World Bank 1996).

With more than three-quarters of the population living within 50 kilometers (31 miles) of the coast, it is not surprising that the coastal area and ocean area are highly salient features of Australia's society and economy. The coastal area is one of the largest in the world (more than 36,700 kilometers, or 22,800 miles) as is the Australian marine jurisdiction with a 200-nautical-mile Exclusive Economic Zone and an expansive continental shelf (14.8 million square kilometers, or 5.7 million square miles). Australia also claims a 12-nautical-mile territorial sea (Zann 1995).

Australia's political system is a hybrid system—based on the British Westminster model of government, with elements of federalism. The Australian constitution (1901) clearly outlines separation of powers between the federal government and the states and territories (six and two, respectively). The commonwealth (federal) government has authority to legislate for and manage issues of defense, coinage, customs and excise, taxation, and communications, whereas the states retain jurisdiction over all other activities, including coastal and ocean management. Local governments have no formal mention in the constitution; all their powers are provided by the states.

Description of Coastal and Ocean Problems

Coastal management in Australia started with the creation of the Port Phillip Authority in 1966 to manage problems associated with coastal erosion, land use conflicts, and a lack of coordination among public agencies in the Melbourne area (Cullen 1982). Similar problems emerged in other coastal areas shortly thereafter, prompting the creation of similar bodies or coastal planning legislation in the other states.

Problems in Australia's coastal zone worsened in the 1980s and 1990s with the emergence of more serious water pollution, habitat destruction, and resource depletion as a result of population growth and development in the coastal areas. In most cases, developers did not take into account changes in the profile of the coastline and its ecological character.

The construction of "canal estates" in southern Queensland and northern New South Wales has been particularly damaging to the coastal environment. These residential developments replace natural wetland barriers to erosion, and the waterways created in the estates tend to be stagnant and highly polluted. In addition, maintenance and upkeep of canal walls and revetments represent a major cost to local governments (Australia, RAC 1993, 55).

Coastal development has also had a profound effect on nearshore coral reefs and sea grass beds. Australia has the largest area of coral reefs of any country, made up primarily of the Great Barrier Reef and reefs in the Torres Strait area, the Coral Sea Islands Territory, and central and northern Western Australia. Coastal construction affects the reefs primarily through sedimentation and nutrient loading, smothering reef filter feeders and increasing algal growth. Other negative effects on coral reefs include introduction of exotic species such as the crown of thorns starfish and *Drupella* snail, effects from fishing and tourism, and threats of marine pollution from marine transportation operations (Zann 1995, 14). Although nearly all of the Great Barrier Reef is included in a marine protected area, few of the other reef areas in Australia, particularly the temperate reef areas, have the same level of protection.

Declining water quality around urban areas remains a major concern in Australia. The majority of this pollution is from point sources, discharges from sewage outfalls, and run-off from urban, industrial, and agricultural activities. Although most of Australia's marine environment is remote and isolated from these problems, a number of regional "hot spots" are forming around industrial areas including the Derwent River estuary in Tasmania and Port Pirie in South Australia (Australia, RAC 1993, 29).

With respect to fisheries, many stocks have been overexploited. Based on 100 major fishery groups, catch data show a major decline since the 1980s. No marine fish has been listed as endangered to date, although several stocks, such as whale sharks, white sharks, grey nurse sharks, black cod, and southern bluefin tuna, have been considered for inclusion (Zann 1995, 18).

To date, Australia has not experienced a major oil spill associated with offshore petroleum development or marine transportation. However, ship traffic has increased in recent years as has the frequency of slicks and small spills. Offshore oil and gas development continues to take place in the Bass Strait between Tasmania and Victoria and along the northwestern coast of Western Australia (Australia, RAC 1993, 33).

With respect to management problems, conflicts between state governments and the commonwealth government over resource management stem from the constitutional division of powers. States have jurisdiction over all resources within their territorial boundaries and thus take the lead role in coastal management. The commonwealth has full authority over income taxation and state borrowing, which it distributes as surplus revenues to the state governments. However, these federal subsidies are insufficient to cover state budgetary outlays, leaving the states in a perpetual search for revenue. As a result, the states have historically sought revenues from intensive natural resource development projects, which have had serious environmental effects, as well as some questionable economic results, such as inflation and underpricing of natural resource goods (Walker 1992). When the commonwealth sought to intervene in state environmental management (in some cases as a result of questionable state management practices), the move was seen as an intrusion into the traditional state jurisdiction set out in the constitution. These intergovernmental tensions remain a major barrier to integrated coastal management.

Evolution of Government Responses

The commonwealth government conducted three major inquiries in the period from 1980 to 1993, resulting in the 1980 *Australian Coastal Zone Management Report*, a 1991 report titled *Injured Coastline Protection of the Coastal Environment*, and the 1993 *Resource Assessment Commission Coastal Zone Inquiry— Final Report*. The inquiries were carried out to determine the extent of the coastal problems and identify the role the commonwealth should play in coastal management. Some of the major management problems outlined in the reports were as follows (Kenchington and Crawford 1993):

- Fragmentation of government responsibilities for planning and use of coastal and marine resources.
- Competing and/or conflicting multiple uses.
- Complex multiagency approval processes.
- Limited public participation in decision making.
- Lack of strong national planning and policy coordination among levels of government.

In 1995, two years after the RAC's report, the long-awaited commonwealth presence in coastal matters was articulated in the form of the Commonwealth Coastal Policy (CCP) (Australia, DEST 1995). Although states retain primary responsibility for coastal management, the Commonwealth Coastal Policy puts forth a Coastal Action Plan (CAP) composed of commonwealth initiatives relevant to coastal management. The flagship of the CAP is the Coastcare program, modeled after Australia's successful Landcare program, providing funding to coastal projects that involve state, local community, and industry collaboration. Other commonwealth initiatives include the Strategic Planning Program (originally recommended by the RAC), which assists state and local governments when coastal planning affects issues of national importance and which contains several initiatives to increase local and community capacity to manage coastal issues. The CCP has an approved budget of $A53 million over a four-year period.

States have also undertaken major modifications of their coastal programs, including introduction of new coastal legislation (Queensland and Victoria) or new coastal planning policies or strategies (Tasmania, Western Australia, and South Australia). The Northern Territory has also begun a plan to reshape its entire environmental management capacity. The new programs address a wider range of policy issues, including protection of coastal resources, public access, management of coastal development, coastal erosion, and coastal hazards (Australia, RAC 1993).

With respect to marine issues, the 1980 Offshore Constitutional Settlement (OCS) and the Great Barrier Reef Marine Park (GBRMP) also represent integrated approaches. The OCS defines jurisdictional boundaries in the offshore area, with states controlling waters and submerged lands out to 3 nautical miles and the commonwealth controlling these areas from 3 to 200 nautical miles. Agreements have been reached between states and the commonwealth for joint management (and revenue sharing) of nonliving and living resources in the adjacent offshore area (Haward 1989, 337–338).

The Great Barrier Reef Marine Park was created as a result of the 1975 Great Barrier Reef Marine Park Act for the conservation and management of the Great Barrier Reef. Under the guidance of the Great Barrier Reef Marine Park Authority, the reef is managed jointly by the state of Queensland and the commonwealth government through a system of multiple-use zoning plans and permitting schemes.

Assessment of Government Responses

With respect to the commonwealth's role in coastal management, the 1995 Coastal Action Plan provides critically needed financial and technical aid to local government and community groups, which lack local management plans and are generally overburdened with other duties and hindered by an uncoordinated approach (Fisk 1996).

Whereas intergovernmental integration through the OCS and in the Great Barrier Reef Marine Park have been somewhat successful, integration of coastal management between states and the commonwealth has been slow in evolving. Negotiations between states and the commonwealth (which have resulted in memoranda of understanding for cooperative management of coastal resources) are crucial first steps in achieving greater integration of activities in the coastal area.

Spatial integration (i.e., integration of terrestrial, coastal, and marine management) probably remains the most problematic. Three commonwealth agencies administer three separate programs: Landcare, Coastcare, and the Offshore regime. Presently, a commonwealth Oceans Policy is being formulated to enhance integration of activities. Modified or new approaches to coastal management at the state level have strengthened state programs, allowing greater integration of coastal sectors. However, traditional sectoral management structures still predominate within most state bureaucracies.

Representation of a broad range of perspectives and disciplines on the new Commonwealth Committees indicates a strong commitment by the Commonwealth to involve a wide range of parties and expertise in coastal decision making. In terms of integrating natural and social sciences, the Ocean Rescue 2000 program, a ten-year commonwealth program promoting conservation and understanding of the marine environment, has encouraged more effective integration of natural science reports and data into the policy stream.

In terms of capacity building, roughly one-third of the Coastal Action Plan is dedicated to training and capacity building of local and community bodies involved with coastal management. A nationwide coastal information network called CoastNet has recently been set up by the commonwealth's Department of the Environment, Sport, and Territories (DEST) to facilitate greater exchange of expertise among coastal managers. New, specialized staff members are being hired in each state to implement the CAP by serving as liaison between government and the community. New tertiary programs and funding are also to be created as a result of the CAP.

Sources Consulted

Anutha, K. 1996. Program Manager, Coastal and Marine Planning Division, Department of Environment and Land Management, Hobart, Tasmania, Australia. 1996. Response to ICM cross-national survey. Newark: University of Delaware, Center for the Study of Marine Policy.

Australia. DEST (Department of the Environment, Sport, and Territories). 1995. *Living on the Coast: The Commonwealth Coastal Policy.* Canberra, Australia: Department of the Environment, Sport, and Territories.

Australia. RAC (Resource Assessment Commission). 1993. *Resource Assessment Commission Coastal Zone Inquiry—Final Report.* Canberra: Australian Government Publishing Service.

CIA (Central Intelligence Agency). 1995. *The World Factbook.* Washington, D.C.: Government Printing Office.

Cullen, P. 1982. Coastal zone management in Australia. *Coastal Zone Management Journal* 10 (3): 183–212.

Fisk, G. W. 1996. Integrated coastal management in developed countries: The case of Australia. Master's thesis, Graduate College of Marine Studies, University of Delaware.

Haward, M. 1989. The Australian offshore constitutional settlement. *Marine Policy* 13 (4): 334–348.

———. 1996. Senior Lecturer, Department of Political Science, University of Tasmania, Hobart, Australia. Response to ICM cross-national survey. Newark: University of Delaware, Center for the Study of Marine Policy.

Kay, R. 1996. Coastal Planning Coordinator, State Government of Western Australia, Perth, Australia. Response to ICM cross-national survey. Newark: University of Center for the Study of Marine Policy.

Kenchington, R., and D. Crawford. 1993. On the meaning of integration in coastal zone management. *Ocean & Coastal Management* 21 (1–3): 109–128.

Walker, K. J. 1992. *Australian Environmental Policy: Ten Case Studies.* Sydney: University of New South Wales Press.

Zann, L. P. 1995. *Our Sea, Our Future: Major Findings of the State of the Marine Environment Report (SOMER) for Australia.* Canberra, Australia: Department of the Environment, Sport, and Territories, Great Barrier Reef Marine Park Authority.

Part II: Middle Developing Countries

••

Brazil

Background: General Country Context
The Federative Republic of Brazil lies in central and northeastern South America. The country borders Venezuela, Colombia, Guyana, Suriname, and French Guiana to the north; Peru and Bolivia to the west; Paraguay, Argentina, and Uruguay to the south; and the Atlantic Ocean to the east. The climate varies from hot and wet in the tropical rain forests of the Amazon basin to temperate in the central and southern part of the country, with an annual average temperature between 17°C and 29°C (*Europa* 1995). More than 38 percent of the country's population of 160 million lives in coastal areas. Almost all Brazilian people are Christian, and 90 percent of them are Roman Catholic. The population growth rate was 1.22 percent in 1995 (CIA 1995). Brazil has a long coastline, stretching for 7,491 kilometers (4,655 miles) on the Atlantic Ocean. In 1970, Brazil claimed a 200-nautical-mile territorial sea, but in accordance with the United Nations Convention on the Law of the Sea it currently claims a 12-nautical-mile territorial sea, a 24-nautical-mile contiguous zone, a 200-nautical-mile continental shelf, and a 200-mile Exclusive Economic Zone (CIA 1995; Churchill and Lowe 1988).

Brazil's economy prospered during the 1960s and early 1970s but gradually declined in the 1980s and 1990s, incurring the world's largest debt (more than $122 billion) and developing an unbelievably high inflation rate of 1,094 percent (1994 estimate) (CIA 1995). In 1994, Brazil's per capita GNP was $2,970 (World Bank 1996). Agriculture, employing 22.6 percent of the labor force, contributed 10.0 percent of GDP in 1991. The principal cash crops are soybeans, coffee, tobacco, sugarcane, and cocoa beans. Brazil dominated the world's coffee market in the 1980s, but its coffee exports have suffered since 1991 with competition from Colombia and other Central American countries. Industry contributed 34.6 percent of the GDP in 1991, employing 22.7 percent of the labor force. In spite of Brazil's rich endowment of natural resources, the country's economic condition remains very unstable due to a lack of steady political direction. In 1993, in response to sky-rocketing inflation, Brazil's finance minister, Fernando Henrique Cardoso, launched the Plano de Verdade to stabilize prices. Inflation subsequently dropped through the end of 1994. Cardoso, subsequently elected as president, has called for implementation of the market-oriented reform the previous president, Fernando Collor de Mello, had attempted.

Brazil became independent from Portugal in 1822 and was declared a republic in 1889. A federal constitution for the United States of Brazil was adopted in 1891. On October 5, 1988, after political turmoil and military coups from the 1960s through the mid-1980s, a new constitution was adopted, restoring direct presidential election by universal suffrage and a five-year presidential term (amended in 1994 to a four-year term). Brazil is a federal republic that grants numerous autonomous powers to the subnational levels of government. Legislative power is vested in a bicameral National Congress composed of the Federal Senate, with members elected by majority rule in rotation for eight years, and the Chamber of Deputies, with members elected by a proportional representation rule for four years. Executive power is vested in the president, who is chief of state and head of government. The president appoints and leads the cabinet. Brazil is divided into twenty-six states and the Federal District (Brasília). Each state is governed by a directly elected governor and legislature. The local governments are constituted by municipalities (*Europa* 1995).

Description of Coastal and Ocean Problems

Brazil's coasts and ocean have been utilized for port and shipyard development, shipping business, offshore oil production, tourism, urbanization, industrialization, scientific research, fisheries operations, and military objectives (Pires-Filho and Cycon 1987). Heavy use of Brazilian coastal resources has placed enormous stress on the coastal environment, posing a number of challenges to coastal planning, management, and implementation. In general, coastal and ocean issues in Brazil can be reviewed according to the country's three major biophysical regions: (1) the Amazonian and northeastern coast, (2) the eastern coast, and (3) the southern coast.

The Amazonian and northeastern coast lies from Cape Orange to Baía de Todos os Santos (All Saints) Bay. The Amazonian area is characterized by a low shoreline, mangrove swamps, large estuarine areas, and an extremely unstable coast due to tidal waves and river runoff, whereas the northeastern coast is shaped by high dunes and rocky cliffs, with mangrove swamps along the rivers. The Amazonian coast is the site of the most important Brazilian shrimp, snapper, and lobster fisheries; the abundant mangrove swamps along the coast are vital to shrimp and lobster reproduction. Fishery resources have been threatened by economic activities. In Pará, for example, the coastal mangrove swamps are threatened by industrial effluent from various mineral mining processes taking place upstream. The northeastern segment is stressed by contamination of freshwater and coastal water from sugarcane residues, and an erosion problem near Recife resulted from the removal of cashew and coconut trees (Pires-Filho and Cycon 1987).

The eastern coast, from Salvador to Vitória, is characterized by low shoreline and beach ridges. This coast is endowed with rich coral reefs such as the Abrolhos Reef. Vitória, one of Brazil's busiest seaports, is located on the southern part of the coast. The majority of Brazilian iron exports are shipped from Vitória. Here lies a potential risk for severe water pollution, but the southern equatorial current quickly disperses potential water pollutants. Of the three coastal regions, the eastern coast has been the least degraded, thereby offering great offshore sportfishing (Pires-Filho and Cycon 1987).

The southern coast, from Cape Frio to Chuí, comprises well-developed sandy beaches and some rocky shores. The states in this region, such as Rio de Janeiro, São Paulo, Paraná, Santa Catarina, and Rio Grande do Sul, are the site of extensive economic activities. Consequently, competition for coastal and ocean resources and space for urbanization and economic development is more prevalent here. Rio de Janeiro in particular faces pollution problems from municipal sewage, industrial effluent, and oil discharged by ships. One of the country's largest shipyards is also located in Rio de Janeiro. In São Paulo and Santa Catarina, a boom in coastal tourism has brought urban infrastructure and road building to the coast, leading to destruction of mangrove swamps and direct discharge of sewage into coastal waters. In Rio Grande do Sul, several estuaries have been contaminated by fertilizers and runoff from upstream agricultural activities, and by heavy metals from industry (Pires-Filho and Cycon 1987).

Evolution of Government Responses

In 1970, the Brazilian Government claimed a 200-nautical-mile territorial sea for the rational utilization and development of coastal and ocean resources. However, neither a regulatory nor a coordinating agency existed for laws governing Brazil's coastal resources and environment. Because of the inactivity of existing institutions, the Interministerial Commission for Marine Resources (CIRM) was created in 1974 to develop a national ocean resources policy, promote and manage research and rational development of marine resources; and coordinate intergovernmental and interministerial conflicts (Herz 1989; Pires-Filho and Cycon 1987).

The CIRM, headed by the Ministry of the Navy, included representatives from seven other ministries—Foreign Relations, Transportation, Agriculture, Education and Culture, Industry and Commerce, Mines and Energy, and Interior—and two federal organizations: the Planning Office of the Presidency and the National Council for Scientific and Technological Development. In 1979, the CIRM's secretariat was established. It has coordinated three major projects: (1) a sectoral plan for ocean resources, (2) Antarctic research, and (3) coastal zone management (Pires-Filho and Cycon 1987).

The CIRM developed guidelines for a National Ocean Resources Policy (approved in 1980) aimed at rational utilization of ocean resources within the 200-nautical-mile territorial sea. The first stage pursuant to the guidelines was formulation of the Sectoral Plan for Ocean Resources (PSRM). It was the first practical, sector-oriented tool for Brazilian coastal resources management. The PSRM's five project areas were: (1) ocean systems, (2) coastal systems, (3) ocean resources, (4) human resources, and (5) oceanographic support (Andriguetto-Filho 1993; Pires-Filho and Cycon 1987). The second phase of Brazilian coastal and ocean planning dates back to 1983, when the CIRM's secretariat initiated coordination of the Coastal Zone Management Program (PROGERCO), which provided grants for research on coastal resources (Pires-Filho and Cycon 1987). In order to attain legal clout for effective implementation of the PROGERCO, the CIRM sponsored a draft National Coastal Zone Management Act (PNGC), which was passed by the National Congress in 1989 (Andriguetto-Filho 1993).

The PNGC is characterized by its strong environmental inclination; its major provisions include the National Environmental Policy (PNMA) and the National Ocean Resources Policy (PNRM) (Herz 1989). The PNGC aims to guide rational utilization of coastal resources with regard to environmental protection, conservation of special areas, and maintenance of quality of the human environment. The PNGC provides bases for state and municipal planning of coastal zone management policies within the framework of the National Environmental System. One PNGC initiative was an attempt to reorder institutional priorities toward more balanced, coordinated, and integrated coastal resource management and environmental protection (Pires-Filho and Cycon 1987). The CIRM was in charge as its central agency until 1990, when administration of the PNGC was transferred from the CIRM to the Secretary of the Environment—Presidency of Republic (SENAM—PR), which is coordinated by the Brazilian Institute for the Environment and Renewable Natural Resources (IBAMA) (Andriguetto-Filho 1993).

As part of UNCED follow-up, in February 1997 Brazil established the Commission for Sustainable Development and Agenda 21 under direct supervision of the General Staff of the Presidency of Republic. The commission coordinates government policies and aims to design national strategies and action for sustainable development. Since the Earth Summit in 1992, the Fishing Sector Executive Group (GESPE), or Grupo Executivo do Setor Pesqueiro, was created to implement the National Fishing and Fish Farming Policy. The GESPE, with the participation of the Ministry for the Environment, Water Resources, and the Amazon and the IBAMA, is drawing up a draft fishing bill and also the master plan for Brazil's fishing industry (Brazil 1997).

Eighteen of Brazil's twenty-three states are coastal states, and some of them have their own management structures. For example, the government of São Paulo has developed the State Council for Environment (CONSEMA). In 1985, the

CONSEMA designated a 2,000-square-kilometer (772-square-mile) area in the Iguape/Paranaguá region for a special coastal area management program (CAMP) because of increasing conflicts among resource users and increasing coastal pollution (Pires-Filho and Cycon 1987). Another example is a case in the state of Paraná in which the state government has created the Environmental Institute of Paraná (IAP) under the State Bureau of Environment. The IAP, similar to the Brazilian Institute for the Environment and Natural Resources (IBAMA) at the national level, carries out an ICM program in the state's coastal zone that is significant for the Atlantic rain forest and one of the largest estuaries in southern Brazil.

Assessment of Government Responses

The most remarkable institutional innovation in Brazil's coastal and ocean resources management strategies was creation of the CIRM as an interministerial coordinating commission. It has provided, developed, and sponsored numerous coastal and ocean policies, programs, and guidelines. Many other coastal issues are addressed from a sectoral viewpoint, but the PNGC, coordinated by the CIRM, was designed to call for intergovernmental consistency and interdisciplinary approaches to conflicts among coastal resource users.

In accordance with institutional reform in the late 1980s, the roles of the CIRM and the national government's Environmental Protection Agency (SEMA) in the PNGC were transferred to the IBAMA. This inefficient institutional reform has resulted in management discontinuities and institutional conflicts in coastal zone management (Herz and Mascarenhas 1993). The PNGC has not been fully implemented, due to lack of infrastructure and inappropriate designation of the IBAMA as the central coordinating agency for PNGC implementation (Herz and Mascarenhas 1993; Andriguetto-Filho 1993). One lesson to be learned from this situation is that institutional integration should be cautiously executed, with consideration of its capacity for implementation of ICM projects. Otherwise, the newly integrated institution may be overloaded and thus suboptimally implement the programs. The IBAMA has expanded its functions by absorbing the fisheries development function and the role of the CIRM, but its outcomes in coastal resources management are not as evident.

Brazil's PNGC emphasizes decentralization of environmental coastal resource management strategy with a federal consistency provision. This has resulted in potential constitutional controversies among different government levels, since the new constitution of 1988 mandates that states can legislate on environmental problems, competing with the federal government (Herz and Mascarenhas 1993). In Brazil, state and municipal governments have played significant autonomous roles in planning environmental matters within their jurisdictional boundaries (Pires-Filho and Cycon 1987). An example is the CAMP program implemented in the Iguape/Paranaguá region, designed by CONSEMA, an autonomous state council of São Paulo.

Sources Consulted

Andriguetto-Filho, A. M. 1993. Institutional prospects in managing coastal environmental conservation units in Paraná State, Brazil. In *Coastal Zone '93,* ed. O. T. Magoon, W. S. Wilson, and H. Converse, 2354–2368. New York: American Society of Civil Engineers.

Brazil. 1997. *Towards Brazil's Agenda 21: Principle and Actions, 1992–97.* Prepared by the Ministry for the Environment, Water Resources, and the Amazon, with contributions from the other ministries and state agencies. Brasília: Ministry for the Environment, Water Resources, and the Amazon.

Churchill, R. R. and A. V. Lowe. 1988. *The Law of the Sea.* Manchester, England: Manchester University Press.

CIA (Central Intelligence Agency). 1995. *The World Factbook.* Washington, D.C.: Government Printing Office.

The Europa World Year Book. 1995. 36th ed. London: Europa.

Herz, R. 1989. Coastal ocean space management in Brazil. In *Coastal Ocean Space Utilization,* ed. S. D. Halsey and R. B. Abel, 29–52. New York: Elsevier Applied Science.

Herz, R., and A. S. Mascarenhas Jr. 1993. Political and planning actions on the Brazilian Coastal Management Program. In *Coastal Zone '93,* ed. O. T. Magoon, W. S. Wilson, and H. Converse, 1084–1091. New York: American Society of Civil Engineers.

Pires-Filho, I. A., and D. E. Cycon. 1987. Planning and managing Brazil's coastal resources. *Coastal Management* 15 (1): 61–74.

Turkey

Background: General Country Context

The republic of Turkey is located between southeastern Europe and western Asia. It borders the Black Sea to the north, the Mediterranean Sea to the south, and the Aegean Sea to the west. It is also adjacent to many other countries: to the east, Georgia, Armenia, the Nakhichevan Autonomous Soviet Socialist Republic, and Iran; to the south, Iraq and Syria; and to the west, Greece and Bulgaria. The climate is hot and dry in summer and cold and snowy in the winter in the Asian portion and mild and warm on the Mediterranean coast (*Europa* 1995). More than 47 percent of the population of 63.4 million lives in coastal areas. More than 99.8 percent of the population is Muslim, and 0.2 percent of the people Christian and Jewish. The population growth rate was 1.97 percent in 1995 (CIA 1995; OECD 1992). Turkey has a coastline of 8,300 kilometers (5,160 miles). In 1964, it claimed a 6-nautical-mile territorial sea in the Aegean Sea and 12-nautical-mile territorial seas in the Black and Mediterranean Seas, and a 200-nautical-mile Exclusive Economic Zone was claimed in the Black Sea (CIA 1995; Churchill and Lowe 1988).

In spite of extensive economic reform adopted by Turkey's late prime minister Turgut Ozal, who held office during 1983–1989, Turkey continues to suffer through economic crises such as governmental-budget deficit, devaluation of the Turkish lira, and high inflation. In 1994, Turkey's GDP was $305.2 billion, with a GDP growth rate of –5 percent (CIA 1995). Agriculture contributed 16.3 percent of the GDP in 1994, employing about 43.6 percent of the labor force. The country is self-sufficient in most food items; principal agricultural exports are cotton, tobacco, wheat, fruits, and nuts. Industry contributed 31.3 percent of the GDP in 1994, providing 22.2 percent of employment (CIA 1995). In 1994, Turkey's per capita GNP was $2,500 (World Bank 1996). Tourism is one of Turkey's fastest-growing sources of foreign exchange revenue. In 1994, Turkey earned $4.321 million revenue from 6.7 million foreign tourists in spite of attacks on tourists by radical members of the Kurdistan Workers Party. During the 1980s and 1990s, Turkey also experienced violent ethnic tension between the Turkish (80 percent of the population) and Kurdish (less than 20 percent) peoples as the Kurdish people fought to create a national homeland in Turkey (*Europa* 1995).

Turkey was formally ruled by a monarch, called the sultan, from the time of the Ottoman Empire until the end of World War I. In 1922, Turkey abolished the sultanate, and the country was declared a republic on October 29, 1923. Mustafa Kemal, a former army officer, became the first president and ruled until his death in November 1938. After World War II, a long period of political instability was fostered by unsuccessful political coalitions and a succession of military coups. In November 1982, under pressure from the European Community and NATO, which urged Turkey to return to democratic rule, a new constitution was approved. Turkey is now a republican parliamentary democratic country in which the national government has a great deal of power. Legislative power is vested in the unicameral National Assembly, composed of 550 deputies (1995 estimate) elected for five-year terms. Executive power is held by the prime minister, who is appointed by the president. The president, elected by the National Assembly for a seven-year term, is empowered to dissolve the National Assembly but only under very special circumstances, which are very clearly described in the constitution. Turkey comprises seventy-nine provinces and 2,074 municipalities. Each province is ruled by a governor (*vali*), who is appointed by the national government. Municipal mayors and council members are elected by town residents (*Europa* 1995; Ozhan 1996a).

Description of Coastal and Ocean Problems

Turkey's coastal areas are richly endowed with natural beauty, cultural attractions, and bays, estuaries, and wetlands replete with resources (OECD 1992). These resources have been degraded, polluted, and threatened by a sharp increase in coastal population density and economic activities such as agriculture, industry, tourism, fishing, aquaculture, and urban development. Turkey's population growth rate (1.97 percent in 1995) is one of the highest in Europe (CIA 1995; METAP 1991). As mentioned earlier, almost half of the national population resides in the coastal areas. At present, there is a rapid shift of population toward the coast, particularly with the migration of Turks from central Anatolia in search of better living conditions (METAP 1991). In addition, rapid growth of the tourism industry along the coastal areas has doubled the population pressure on the coastal zone, resulting in many environmental and socioeconomic effects, for example, pollution of coastal waters threatens swimming, public health, fisheries, and biodiversity (OECD 1992; METAP 1991).

Turkey's agricultural production constitutes a major economic activity in the coastal areas. Remarkably, 90 percent of tobacco, 80 percent of cotton, and 70 percent of rice production of the country take place in coastal provinces. Consequently, a great challenge facing the country is to reduce agricultural pollution runoff resulting from intensive use of fertilizers and pesticides (OECD 1992).

Industrial waste is one of the country's most serious sources of marine pollution. Most of Turkey's industrialization has also taken place in the coastal provinces, including Istanbul, Izmir, Izmit, Adana, Mersin, Iskenderun, and two Black Sea provinces. Although such industrial development is economically important, its rapid expansion along the coast has caused severe coastal water pollution and deterioration (OECD 1992). For example, Izmir, on the Aegean Sea, is an important industrial site encompassing 6,000 factories. Most of these are tanneries, textiles, and paint manufacturers located along the coast of the Bay of Izmir, causing serious pollution of bay water (OECD 1992; UNEP 1994).

Tourism is Turkey's largest single earner of foreign exchange revenue, As a result of the Tourism Incentives Law of 1982, an average of 10 million tourists, both foreign and domestic, have visited the coast annually in the 1990s, requiring substantial investment in tourism infrastructure (OECD 1992). Construction of tourist accommodations and vacation houses along the coast, especially the southwestern coast, has contributed significantly to sewage and solid waste problems and degradation of water quality. As tourism infrastructure and service facilities were rapidly expanding along the coastal zone, industrial and agricultural developments competed with tourism for coastal land use in some areas, such as Adana and Iskenderun on the eastern Mediterranean (OECD 1992; METAP 1991).

Commercial fishing is one of Turkey's traditional marine economic activities. The country's long coastline provides good natural conditions for fishing and aquaculture, producing 581,000 tons of fish in 1988. The Black Sea is exceptionally productive, providing more than 80 percent of Turkey's total fishery catch. However, since the late 1980s, the total value of the catch and the numbers of fish species have declined progressively, as a result of pollution of coastal waters (OECD 1992).

Evolution of Governmental Responses

In response to Turkey's emerging coastal and ocean problems, the national government, with the cooperation of a number of international organizations, such as the Regional Activity Center of the UNEP-MAP for Priority Actions Programme, the OECD, the World Bank, and the Global Environment Facility (GEF), has played a major role in the country's coastal zone management. The national government's involvement in management of coastal resources and environment is mandated by numerous laws and regulations on a sector-by-sector basis; the laws were passed primarily during the 1980s and 1990s.

The major laws and bylaws which relate to various issues of coastal zone management are well described in articles by Erdal Ozhan (Ozhan et al. 1993; Ozhan 1996a). Major marine-oriented laws are selected and summarized as follows:

Fisheries Law (22.3.1971, Amendments 15.5.1986). For protection, production, and control of living resources, Turkey's Ministry of Agriculture and Rural Affairs was authorized to regulate fisheries and mariculture. This law prohibits dumping of harmful substances into inland waters and prohibits bottom-trawling in inland waters, the Sea of Marmara, and Bosphorus and Dardanelles Straits.

Tourism Incentives Law (12.3.1982). In order to urge, guide, and regulate tourism development, tourism areas were declared by a decree of the Council of Ministers following a proposal by the Ministry of Tourism, which is responsible for national tourism development. The result was a boom of investment in tourism development, mainly in the coastal zone (especially along the coasts of the Aegean and Mediterranean Seas) during the mid-to late 1980s.

Environmental Law (9.8.1983). The bylaw on water pollution control provides water quality criteria for lakes and seawater. Another bylaw requires an environmental impact assessment (EIA) to be prepared by organizations, companies, and establishments that have the potential to cause environmental problems through planned activities. The metropolitan municipalities are authorized to permit sea outfalls within their borders on approval of the Ministry of Environment. The Council of Ministers is authorized to designate areas that have ecological significance and are sensitive to degradation as specially protected areas (SPAs).

National Parks Law (9.8.1983). National parks are identified by a decree of the Council of Ministers following a proposal by the Ministry of Forestry, which is responsible for management of national parks, including coastal parks. As of 1996, three coastal national parks have been designated: Olympus-Bey Daglari, Dilek Peninsula, and Gelibolu. A fourth coastal park was recently declared near the town of Marmaris, one of the major tourism resorts on the southern Aegean Sea.

Shore Law (4.4.1990. Amendment 1.7.1992). Turkey's Shore Law sets out principles for protection of the country's coastal and ocean areas. It defines the coastal landward boundary as an area at least 100 meters (328 feet) wide horizontally, starting from the shore edge line, which is defined as the natural limit of the sand beach, wetland, and similar areas, created by seawater motion. All construction is prohibited within the first 50 meters (160 feet) from the shore edge line; in the remaining landward part, only public facilities and recreational and tourism facilities may be built.

An important national institution established in 1989 in connection with Turkey's ICM effort is the Agency for Specially Protected Areas (SPAs), which considers the use of all kinds of measures in solving environmental problems. The agency was initially part of the Office of the Prime Minister but is now part of the Ministry of Environment. The main body of the agency is located in Ankara; two other major departments, the Department of Planning and Project Implementation and the Department of Environmental Protection, Research, and Investigation, are located in Köyceğiz and Silifke, respectively. At present, nine SPAs are designated within the coastal zone (Ozhan 1996a; METAP 1991).

Other institutions newly established for integration of coastal zone management are two ad hoc bodies: the Interministerial Executive Council (IEC), created in 1994, and the Ad Hoc Working Group (AHWG), created in 1995. The IEC, headed by a senior advisor to the prime minister, and comprising representatives from nine ministries related to coastal area management, coordinates intergovernmental and interdepartmental sea-related affairs. The AHWG, created under the Office of the Prime Minister, and composed of representatives from the Ministries of Public Works and Settlement, Culture, Tourism, and Environment, implements of the Shore Law and the Settlement Law (discussed in the next paragraph) (Ozhan 1996a).

In terms of Turkey's local, site-specific integrated coastal zone management efforts, the Bosphorus Law (18.11.1983) is a comprehensive, integrated coastal area specific management project in the Bosporus area of Istanbul. It was designed to protect and enhance the natural and cultural endowment within this area. After enactment of the Bosporus Law, two coordinating institutions, the Bosporus Supreme Coordination Council for Land Use and Development and the Bosporus Executive Council for Land Use and Development were in charge of implementing the Bosporus area management project. In 1985, however, the Settlement Law (3.5.1985) removed these institutions, reauthorizing the mayor of metropolitan Istanbul to facilitate implementation of the Bosporus Law and handle CZM planning (Ozhan 1996a; METAP 1991).

A Priority Action Project concentrating mainly on water quality had been sponsored and carried out in the Izmir Bay on the Aegean Sea during 1988–1989 by the UNEP-MAP Priority Actions Programme's Regional Activity Center, located in Split, Croatia. In 1989, it was decided to carry out an Integrated Management Study for the area of Izmir. The study was carried out in 1991–1993 by a team of Turkish and UNEP-MAP/ Priority Action Programme experts (Ozhan 1996a; OECD, 1992).

In 1990, Turkey's coastal zone management project was undertaken by the Undersecretariat of Environment through a World Bank METAP grant (METAP 1991). Around the same time, the OECD conducted a review of Turkey's environmental issues and, in 1992, published *Environmental Policies in Turkey* (OECD 1992). Currently, there are two ongoing CZM pilot studies in Turkey: (1) the Bodrum Peninsula Coastal Zone Management Project, carried out by Middle East Technical University, Ankara, and (2) the Mersin Coastal Zone Management Project, conducted by a private company through a contract with the Ministry of Environment (Ozhan 1996a).

Assessment of Governmental Responses

The emerging conflict between development in the coastal provinces (e.g., tourism, urbanization, agriculture, and industrialization) and environmental conservation and protection (e.g., of water quality, natural beauty and aesthetics, endangered species, and historical wealth) added a sense of urgency to the creation of an integrated coastal zone management strategy in Turkey (Ozhan 1996b). Turkey's government responded by creating two broad classes of government agencies: developmental agencies, such as the Ministry of Tourism and the Ministry of Public Works and Settlement; and the conservation and environmental agencies, such as the Ministries of Environment, Forestry, and Culture. Because of the lack of an overall institutional and regulatory integrative coordinating scheme among these ministries, Turkey has, in some cases, suffered from overlaps and gaps in its administration and implementation of projects in coastal areas (OECD 1992).

For example, pertinent to the Tourism Incentives Law (TIL) (12.3.1982), the Ministry of Tourism (MOT) provided many economic incentives such as tax exemptions and low-interest loans, for the development of tourism facilities. Tourism grew successfully, especially along the coasts of the Aegean and Mediterranean Seas. However, the MOT initially provided no consideration for the range and severity of the adverse environmental effects of tourism. In the 1990s, the MOT has discontinued several administrative incentives provided by the TIL because of secondary housing problems and degradation of water quality and the aesthetic value of natural coastal features, which conflicts with the Environmental Law (9.8.1983) (OECD 1992; METAP 1991; Ozhan 1996a).

One of Turkey's most noteworthy institutional arrangements is the Agency for

Specially Protected Areas. The agency's authority extends to land use management as well as environmental management within SPAs; therefore, the agency takes on duties and prerogatives of all Ministries and municipalities relevant to SPA management. A successful result of the SPA program is illustrated by the compromise between tourism development and nature conservation in Dalyan, on the eastern edge of the Mediterranean Sea (OECD 1992). Tourism development along the entire coastline of Dalyan Beach, an important hatchery for sea turtles, posed serious threats to this natural ecosystem. With the support of the World Wide Fund for Nature and IUCN, Dalyan Beach was designated an SPA in 1988. The result of this designation was that 70 percent of sea turtles could be protected while sustainable tourism development was practiced on this coast.

Another of Turkey's successes appears to be the multidisciplinary effort involved in the Integrated Management Study for the Area of Izmir, carried out by a team of Turkish and UNEP-MAP Priority Action Programme experts since 1988. Through research of the water treatment system in Izmir performed by the Institute of Marine Sciences at Dokuz Eylul University, an integrated management plan for the inner bay was adopted as the most strategic approach to provide a bay-wide management perspective (UNEP 1994). There was cooperation between national and local authorities through the joint UNEP-MAP program and the Metropolitan Municipality of Izmir. Other effective and essential aspects of this project were political commitment, funding, staff training, public involvement in the bay-wide planning process, EIA preparation, and generation of basic information (OECD 1992; UNEP 1994).

In terms of capacity building in ICM, Turkey introduced various measures to further meet ICM goals of protecting natural environments and achieving sustainable development in the coastal zone. As mentioned previously, Turkey's CZM efforts have been given financial sponsorship and technical assistance by international organizations such as the UNEP-MAP Priority Action Programme's Regional Activity Center, the World Bank's METAP, and the OECD.

At the national level, Turkey has also built CZM capacity through its 1993 establishment of the Turkish National Committee on Coastal Zone Management (KAY), which functions as a national network in connection with international organizations. It held the First International Conference on the Mediterranean Coastal Environment, MEDCOAST '93, on November 2–5, 1993, in Antalya, Turkey, and the follow-up MEDCOAST '95 in Tarragona, Spain, in November 1995, and MEDCOAST '97 in Malta. The KAY is located at the Middle East Technical University in Ankara. Its executive board includes eleven members representing universities, ministries, municipalities, the private sector, NGOs, and individuals. The KAY provides a medium for information exchange, develops scientific research projects, serves as a central storage facility for coastal resource data and information, and sponsors the development of national educational programs in coastal zone management (Ozhan 1996a).

Another example of CZM capacity building is the Global Environment Facility's Black Sea Environmental Programme (GEF-BSEP), administered by a coordinating unit established in Istanbul, Turkey, in 1995. The GEF-BSEP, coordinated by Turkey's Ministry of Environment, emphasizes creation of a national network for CZM, preparation of a national report, and execution of at least one pilot project by each participating country. These action plans will help reinforce Turkey's ICM efforts in the future (Ozhan 1996a).

Sources Consulted

Churchill, R. R., and A. V. Lowe. 1988. *The Law of the Sea*. Manchester, England: Manchester University Press.

CIA (Central Intelligence Agency). 1995. *The World Factbook*. Washington, D.C.: Government Printing Office.

The Europa World Year Book. 1995. 36th ed. London: Europa.

METAP (Mediterranean Environment Technical Assistance Program). 1991. *Coastal Zone Management (CZM) in Turkey*. Consultant report submitted to the undersecretariat of Environment and the World Bank.

OECD (Organization for Economic Co-operation and Development). 1992. *Environmental Policies in Turkey*. Paris: Organization for Economic Co-operation and Development.

Ozhan, E. 1996a. Coastal zone management in Turkey. *Ocean & Coastal Management* 30(2–3): 153–176.

———. 1996b. Professor, Civil Engineering Department, Middle East Technical University, Ankara, Turkey. Response to ICM cross-national survey. Newark: University of Delaware, Center for the Study of Marine Policy.

Ozhan, E., A. Uras, and E. Aktas. 1993. Turkish legislation pertinent to coastal zone management. Paper presented at MEDCOAST '93, the First International Conference on the Mediterranean Coastal Environment, November 2–5, Antalya, Turkey.

UNEP (United Nations Environment Programme). 1994. *Integrated Management Study for the Area of Izmir. MAP Technical Reports Series No. 84*. Split, Croatia: Regional Activity Center for Priority Action Programme.

Thailand

Background: General Country Context

The kingdom of Thailand lies in Southeast Asia, bordering Myanmar to the west and north, Laos to the northeast, and Cambodia to the southeast. Thailand also extends southward along the Malay Peninsula, bordering the Andaman Sea to the west and the Gulf of Thailand to the east and reaching Malaysia to the south. Thailand has a tropical climate: rainy and warm during the southwestern monsoon season (May–September); dry and cool during the northeastern monsoon season (November–March); and hot and humid year-round on the southern isthmus (*Europa* 1995). About 70 percent of the total population of 60 million lives in the coastal areas. Approximately 95 percent of the population is Buddhist; 3.5 percent is Muslim; 0.5 percent is Christian; and 0.7 percent is Hindu or affiliated with other religions. The population growth rate was 1.24 percent in 1995 (CIA 1995; Thailand 1992). Thailand's coastline is 3,219 kilometers (2,000 miles) long, including a short coastline on the Andaman Sea and a long coastline on the Gulf of Thailand. Thailand claimed a 12-nautical-mile territorial sea in 1966, a 200-nautical-mile Exclusive Economic Zone in 1981; and claimed jurisdiction over the continental shelf to a depth of 200 meters (656 feet) or the depth of exploitation (CIA 1995; Churchill and Lowe 1988).

Thailand's economy has grown rapidly since the late 1980s in spite of political turmoil. In 1994, Thailand's GDP was $355.2 billion, with a GDP growth rate of 8 percent. Agriculture contributed 10 percent of the GDP in 1993, employing 56.7 percent of the labor force. The principal cash crop, rice, is an export agricultural commodity. Timber was formerly a major source of export revenue, but uncontrolled logging was banned in 1989 after deforestation caused severe flooding. Industry provided 39.2 percent of the GDP in 1993, employing 17.5 percent of the labor force. Thailand's manufacturers contributed to the country's major export earnings in 1993. Tourism is the principal source of foreign exchange; more than 5 million foreign tourists visited Thailand in 1993. As a result of rapid economic growth, the income disparity between rural agricultural and urban industrial areas has caused serious population concentration in Bangkok in spite of the government's attempts to disperse industry away from Bangkok and into other provinces (*Europa* 1995). In 1994, Thailand's per capita GNP was $2,410 (World Bank 1996).

Thailand, formerly known as Siam, was founded in 1238 and has never been colonized. Since Phibun Songkhram took political power after a military coup in 1938, the country has undergone a long and bitter political struggle against the military regimes. A new constitution was approved by the National Legislative Assembly on December 7, 1991. The major criticism of the new constitution involved its provision that 360 senators and the prime minister were to be appointed by the National Peace-Keeping Council (NPC), which is dominated by the head of the armed forces. Following strong antigovernment demonstrations, the constitution was amended on June 10, 1992, to require that the prime minister be an elected member of the House of Representatives. The number of senators was reduced to 270, two-thirds of the membership of the House of Representatives. Thailand's legislative power is vested in a bicameral National Assembly composed of the House of Representatives (elected) and the Senate (appointed). Thailand is a constitutional monarchy over which the royal Thai government holds great power. The king is chief of state. Executive power is vested in the prime minister, who is head of government and is elected by the National Assembly. The Council of Ministers is headed by the prime minister. Thailand has seventy-two provinces (*Europa* 1995), each headed by a governor who is appointed by the national government. All provincial- and district-level institutions are strongly influenced by the national government.

Description of Coastal and Ocean Problems

Thailand's coastal areas are richly endowed with a variety of coastal resources and its extensive beaches are suitable for tourism development. The abundant mangrove forests and coral reef communities provide fishery habitats, wildlife refuges, beach protection, and recreational sites. However, during periods of rapid economic growth, particularly during the term of the Sixth National Economic and Social Development Plan (1987–1991), various conflicts arose among competing coastal resource users. The major coastal management issues that gave rise to the need for ICM in Thailand are coral reef destruction, mangrove deforestation, fishery stock declines, and various adverse environmental consequences from rapid economic development and tourism growth.

About 55 percent of Thailand's major coral reef lie in the Andaman Sea, and the remainder are located in the western and eastern portions of the Gulf of Thailand (Lemay, Ausavajitanon, and Hale 1991). Not only are these coral reefs important as a source of livelihood for local village people, but they also serve as shelters for fish and crustaceans and as recreational sites for tourists. However, the condition of Thailand's coral reefs has degraded at an accelerated rate, with more than 60 percent of all major coral reef groups damaged or degraded (Lemay, Ausavajitanon, and Hale 1991). This coral reef destruction has resulted from human-related activities such as dynamite and trawl fishing methods, boat anchoring, land-based sources of pollution, and tin mining in the coastal areas.

The mangrove forests of Thailand are ranked third among the Southeast Asian countries. They covered 287,308 hectares (709,940 acres) in 1979, mostly in the coastal areas adjacent to the Andaman Sea, although quite sparse and poor mangrove communities had formed along the coast of the Gulf of Thailand. In particular, from 1975 to 1979, about 25,392 hectares (62,744 acres) of mangrove trees were destroyed and the land converted for shrimp aquaculture, agricultural farming, and industrial sites (Kongsangchai 1986). More recently, about 5,300 hectares (13,100 acres) of mangrove forest in Ban Don Bay has been converted to shrimp ponds because of the high profitability of shrimp farming (Thailand 1992).

The use of dynamite fishing in Thailand's small-scale fisheries has degraded coral reefs and sea grass beds. Coastal shrimp aquaculture represents one of Thailand's most important fisheries; national consumption and foreign export increased from 22,000 metric tons (24,000 tons) in 1987 to 146,000 metric tons (162,000 tons) in 1991 (Clark 1996). As described earlier, this rapid expansion of shrimp cultivation encroached on mangrove communities. In 1989, only 180,560 hectares (446,160 acres) of mangrove forests in Thailand were left—that is, about 64 percent of the original mangrove cover had been converted, primarily for shrimp aquaculture (Clark 1996).

The tourism industry in Thailand has grown significantly as a result of the country's natural coastal attractions and the government's interest in promoting international tourism. However, tourism in Thailand, although crucial for foreign exchange earnings, has generated a number of adverse environmental and socio-cultural effects. The most serious environmental effects have occurred at three major beach resorts at Pattaya, Phuket, and Ko Samui. Pattaya, located on the eastern seaboard along the Gulf of Thailand, was a small, quiet coastal village in the early 1960s. It attracted fewer than 400,000 tourists by the mid-1970s but more than 900,000 a decade later. This rapid increase in tourism was accompanied by unplanned and unrestricted construction of hotels and infrastructure, resulting in the deterioration of beach scenery and coastal water quality (Dobias 1989).

Phuket, the country's largest island located between the Andaman Sea and Phangnga Bay, is replete with rich natural resources, including thirty kilometers (19 miles) of white sand beaches, minerals, fisheries, rubber, and orchards (Dobias 1989). Phuket's rapidly growing tourism industry hosted more than 726,000 tourists in 1988 (Bunpapong and Ausavajitanon 1991). The rapid expansion of accommodations and infrastructure resulted in beach degradation, nearshore water pollution from liquid effluents, and beach erosion (Dobias 1989; Bunpapong and Ausavajitanon 1991).

Ko Samui, an island in the Gulf of Thailand, houses a newly developed resort; the island's beautiful natural white sand beaches attracted 300,000 tourists in 1987. Ko Samui is also popular as a domestic vacation destination. Following the increased influx of visitors to Ko Samui, extensive building construction on its shoreline and pressure from the increasing tourist population began to produce adverse effects in the coastal environment similar to those occurring in the other two resorts (Thailand 1992; Dobias 1989).

Evolution of Government Responses

Prior to the ICM work of the ASEAN-USAID Coastal Resources Management Project (CRMP), Thailand's coastal area management methods were largely oriented toward sectoral issues. The Department of Fisheries (DOF) had formulated a fisheries development plan for small-scale and commercial fisheries in accordance with the Fisheries Act of 1947. However, the royal Thai government recognized that the national fisheries program alone could not manage and control open-access fisheries resources. In 1993, the DOF, with assistance from Kasetsart University, established a community-based fisheries management program to advocate the involvement of local fishermen in planning, managing, and implementing processes (Pomeroy 1995).

In the coral reef management sector, efforts were obviously fragmented before the National Coral Reef Management Strategy was approved in March 1991 (Pintukanok and Borothanarat 1993). There are three different sectoral national laws and major agencies related to coral reef protection. The Fisheries Act of 1949, with the DOF as lead agency, classified all coral reefs as protected areas prohibiting any environmentally adverse activity in coral reef fish habitats. Nine of the fifteen marine national parks in Thailand contain major coral reef groups. Under the National Park Act of 1961, the Royal Forestry Department was authorized to plan and manage the marine parks. Under the Enhancement and Conservation of National Environment Quality Act (NEQA) of 1975, the Office of the National Environment Board (ONEB) published *Coastal Water Quality Guidelines* for the island of Phuket, including a coral reef preservation framework (Lemay, Ausavajitanon, and Hale 1991). The strategy classified coral reefs according to three zoning categories: local management zones (coral reefs in good condition and in rural areas), tourism and recreation zones (coral reefs used intensively for tourism), and national ecological and scientific benefits zones (coral reefs of scientific interest) (Pintukanok and Borothanarat 1993).

In October 1989, a National Coral Reef Management Workshop was held in Bangkok, inspired by technical data on coral reef protection gathered from an ASEAN-Australian baseline study and by lessons on community-based coral reef protection learned from the Phuket pilot study. The workshop findings eventually led to cabinet-level consideration of urgent coral reef protection measures. By mid-1990, the Thai Cabinet approved four urgent measures: (1) enforcement authority for fisheries regulations to the navy and the Harbor Department; (2) patrol of offshore coral reefs; (3) installation of mooring buoys and educational activities; and (4) allocation of funds (Lemay, Ausavajitanon, and Hale 1991). Then, in 1991, the Office of Environmental Policy and Planning prepared the National Coral Reef Management Strategy, which offered a more active, harmonious, and integrative approach to managing Thailand's coral reefs through partnership with local communities and the private sector (Lemay, Ausavajitanon, and Hale 1991).

No single comprehensive plan or core agency can solve the conflict between mangrove deforestation and shrimp farming expansion. For example, the DOF has authority over fisheries resources and aquaculture farming, whereas the Royal Forestry Department (RFD) manages and controls mangrove forests. With no efficient interdepartmental cooperation achievable between the DOF and the RFD, mangrove forests will continue to disappear.

Thailand's Fifth and Sixth National Economic and Social Development Plans (1982–1986 and 1987–1991, respectively) emphasized promotion of international tourism. The Tourism Authority of Thailand (TAT) is the key agency in the Tourism Development Plan, and the Ministry of Science, Technology, and Energy, the ONEB, the Ministry of Industry, and other agencies are also involved. During Thailand's tourism boom, the country's practical tourism promotion efforts began to be decentralized. For example, a provincial tourism development plan for Surat Thani Province, including the island of Ko Samui, was designed in 1984 (Dobias 1989). The goal was to pursue a comprehensive development plan with greater local government control over administration and management of tourism development.

The Eastern Seaboard Project was the first comprehensive and detailed coastal management project in Thailand. Detailed guidelines called for environmentally healthy economic development along the coast of Pattaya. The Songkhla Lake Basin Project, undertaken because of concern that projected urban and industrial development could damage water quality in the basin, linked three subplans on natural resources, socioeconomics, and environment (Tabucanon 1991).

Among other ICM efforts in Thailand, the Phuket Island Action Plan of 1986–1989 and the 1992 Integrated Management Plan for Ban Don Bay and Phangnga Bay (IMPBP) are technically sound and comprehensive attempts at ICM. The University of Rhode Island and USAID's International Coastal Resources Management Program, which works in Sri Lanka, Thailand, and Ecuador to help set up national ICM programs and build local ICM capacity, provided technical assistance to the Phuket Island pilot study (Crawford, Cobb, and Friedman 1993). The study investigated tourism-related issues such as land use conflicts, wastewater treatment, and degradation of coral reefs. Based on this investigation, the Phuket Island Action Plan was developed with a budget of B111.37 million, and was funded for implementation in 1992 (Pintukanok and Borothanarat 1993).

After the Phuket Island pilot study, Ban Don Bay and Phangnga Bay in the upper southern region were chosen by the ASEAN-USAID Coastal Resources Management Project as broader pilot study areas because of serious resource use conflicts and socioeconomic considerations there (Thailand 1992). For the launching of this multidisciplinary coastal management plan, Thailand produced, with support from the Japan International Cooperation Agency (JICA), the *Subregional Development Study of the Upper Southern Part of Thailand* (1985), with an outline of development addressing internationalization, industrialization, and decentralization (Thailand 1992). The IMPBP was to describe the condition of the region's natural resources, assess resource use conflicts, suggest solutions to resource use problems, and propose action plans and specific projects (Thailand 1992).

Assessment of Government Responses

Although many laws and a number of government agencies in Thailand address coastal resources and environmental management, each appears to act according to a narrowly defined mandate. The ONEB has played a significant role in the country's coastal resources and environmental management, but it is not a single coherent coordinating agency in charge of coastal area issues. The lack of interagency cooperation was evident in spatial use conflicts between preservation of mangrove forests and expansion of shrimp ponds (Clark 1996). Despite the urgent need for integrated management to harmonize this important issue, no single planning or regulatory authority has been effective in orchestrating sustainable expansion of shrimp ponds with minimal degradation of mangrove ecosystems.

However, Thailand did experience some success in the lessons learned from ICM application during the pilot study on Phuket Island. Community-based coral reef protection and waste management were successfully implemented there through a local community education and awareness campaign at Patong. Public awareness education activities included training of escort guides, education on the installation of mooring buoys, and establishment of informational signs (Bunpapong and Ausavajitanon 1991). Moreover, solid waste management is a crucial environmental problem at Patong. This problem could not be solved by the Patong Sanitary District (PSD) alone because of inadequate staffing, a weak financial base, and insufficient equipment. The ONEB initiated a Demonstration Project of Solid Waste Management Strategies in 1990 at Patong to stimulate public awareness and cooperation from tourists and residents. Hotel and bungalow owners built their own incinerators to dispose of waste. The ONEB's voluntary public involvement in the coastal resources conservation campaign was successful in this area (Bunpapong and Ausavajitanon 1991).

In terms of capacity building in ICM, Thailand established a public awareness campaign with local sewage treatment operators and private sector groups such as hotel managers, tour guides, and tour boat operators. Technical, administrative, and financial assistance was provided by international organizations such as ASEAN-USAID, the University of Rhode Island, the Asian Development Bank, and the JICA. The country also promoted training of high-level officials in the planning of national ICM programs (Crawford, Friedman, and Cobb 1993). In the academic area, a postgraduate program in coastal resources management (CRM) was set up at Prince of Songkla University (PSU) with funding assistance from USAID, and a master's degree program in marine affairs was established at Chulalongkorn University in 1992 (Piyakarnchana et al. 1991). PSU opened a new center called the Coastal Resources Institute (CORIN). The CORIN has implemented faculty training in the United States, ICM training courses for CORIN staff, and ICM strategy in a local bay, thus gaining experience in special area management (Crawford, Friedman, and Cobb 1993).

More strategically, the IMPBP requires strong capacity-building measures in several action plans. Some of the more noteworthy measures include the Water Quality Management Action Plan, which establishes environmental training courses, assesses in-country and overseas training capacities, and conducts a series of television and radio programs; the Fisheries Action Plan, which conducts educational seminars and incorporates basic concepts of fisheries management into school curricula; and the Tourism Action Plan, which conducts four seminars per year, conducts special monthly seminars in areas with problems, initiates studies involving local participants, incorporates conservation principles in local school curricula, and is establishing a tourist information and island conservation center on Ko Samui (Thailand 1992).

Sources Consulted

Bunpapong, S., and S. Ausavajitanon. 1991. Saving what's left of tourism development at Patong Beach, Phuket, Thailand. In *Coastal Zone '91*, ed. O. T. Magoon et al., 1688–1697. New York: American Society of Civil Engineers.

Churchill, R. R., and A. V. Lowe. 1988. *The Law of the Sea.* Manchester, England: Manchester University Press.

CIA (Central Intelligence Agency). 1995. *The World Factbook.* Washington, D.C.: Government Printing Office.

Clark, J. R. 1996. *Coastal Zone Management Handbook.* New York: Lewis.

Crawford, B. R., A. Friedman, and J. S. Cobb. 1993. Building capacity for integrated coastal management in developing countries. *Ocean & Coastal Management* 21 (1–3): 311–338.

Dobias, R. J. 1989. Beaches and tourism in Thailand. In *Coastal Area Management in Southeast Asia: Policies, Management Strategies and Case Studies*, ed. T. E. Chua and D. Pauly, 43–55. ASEAN-USAID Coastal Resources Management Project. ICLARM Proceedings No. 2. Manila, Philippines: International Centre for Living Coastal Resources Management.

The Europa World Year Book. 1995. 36th ed. London: Europa.

Kongsangchai, J. 1987. The conflicting interests of mangrove resources use in Thailand. Paper presented at UNDP/UNESCO Regional Project RAS/79/002, Workshop for Mangrove Zone Managers, September 9–10, Phuket, Thailand. New Delhi, India: Vijayalakshmi Printing Works.

Lemay, M. H., S. Ausavajitanon, and L. Z. Hale. 1991. A national coral reef management strategy for Thailand. In *Coastal Zone '91*, ed. O. T. Magoon et al., 1698–1712. New York: American Society of Civil Engineers.

Pintukanok, A., and S. Borothanarat. 1993. National coastal resources management in Thailand. In *World Coast Conference 1993: Proceedings,* vols. 1 and 2. CZM-Centre Publication No. 4. The Hague: Ministry of Transport, Public Works, and Water Management, National Institute for Coastal and Marine Management, Coastal Zone Management Centre.

Piyakarnchana, T., et al. 1991. Environmental education curricula at the tertiary levels in Thailand: Case study of marine science and marine affairs programs. In *Coastal Area Management Education in the ASEAN Region,* ed. T. E. Chua, 55–63. ASEAN-USAID Coastal Resources Management Project. ICLARM Conference Proceedings No. 8. Manila, Philippines: International Centre for Living Aquatic Resources Management.

Pomeroy, R. S. 1995. Community-based and co-management institutions for sustainable coastal fisheries management in Southeast Asia. *Ocean & Coastal Management* 27 (3): 143–162.

Tabucanon, M. S. 1991. State of coastal resource management strategy in Thailand. *Marine Pollution Bulletin* 23: 579–586.

Thailand. Office of the National Environment Board and Ministry of Science, Technology, and Environment. 1992. *The Integrated Management Plan for Ban Don Bay and Phangnga Bay, Thailand.* ICLARM Technical Report No. 30. Manila, Philippines: International Centre for Living Aquatic Resources Management.

Malaysia

Background: General Country Context

The independent country of Malaysia lies on the Malay Peninsula (West Malaysia) in Southeast Asia, bordering Thailand to the north, the Strait of Malacca to the west, the South China Sea to the east, and the island of Singapore to the south. The country also occupies the northern one-third of the island of Borneo (East Malaysia), bordering Indonesia to the south, the South China Sea to the north, and the Sulu Sea and Celebes Sea to the east. The climate is tropical. From April to October, the country experiences the southwestern monsoon season, and from October until February, the northeastern monsoon season. The annual average precipitation is 200 centimeters (78.7 inches), and the average temperature is about 27°C, with little variation throughout the year (*Europa* 1995). About 70 percent of the population of 19.7 million lives in coastal areas. Nearly 53 percent of the country's population is Muslim, 19 percent is Buddhist, and 28 percent is Christian, Hindu, and of other religions. The population growth rate was 2.24 percent in 1995 (CIA 1995; Malaysia 1992). Malaysia has a coastline of 4,675 kilometers, or 2,905 miles (West Malaysia, 2,068 kilometers, or 1,285 miles; East Malaysia, 2,607 kilometers, or 1,620 miles). The country claimed a 12-nautical-mile territorial sea in 1969, a 200-nautical-mile Exclusive Economic Zone and Exclusive Fishery Zone in 1980, and the continental shelf to a depth of 200 meters (approximately 656 feet) or the depth of exploitation in 1966 (CIA 1995; Churchill and Lowe 1988).

From 1988 until 1994, Malaysia experienced remarkable economic growth, with a 9 percent average annual growth rate. This economic prosperity substantially reduced poverty but also resulted in a high level of resource depletion and environmental degradation. In 1994, Malaysia's GDP was $166.8 billion, with a GDP growth rate of 8.7 percent (CIA 1995). Malaysia is rich in natural resources, including natural rubber, timber, tin, and oil and gas. Agriculture contributed 14.6 percent of the GDP in 1994, employing 21.4 percent of the labor force. The principal crop is rice. Malaysia is the world's leading producer of palm oil, natural rubber, and timber. Industry provided 44.2 percent of the GDP in 1994, employing 31.9 percent of the labor force. Malaysia is also one of the world's largest producers of tin. Export of manufactured goods expanded rapidly by an annual average of 10.0 percent from 1980 to 1993. Tourism significantly contributes to foreign exchange revenues. In 1994, about 7.2 million tourists visited Malaysia, contributing about $3.6 billion to the country's revenues (*Europa* 1995). In 1994, Malaysia's per capita GNP was $3,480 (World Bank 1996).

On August 31, 1957, Malaysia, then composed of eleven states, obtained limited independence from Great Britain. The independent Federation of Malaysia was established by constitution on September 16, 1963, and included three new states: Singapore, Sarawak, and Sabah. However, Singapore left the federation to become an independent nation in August 1965. Currently, Malaysia is an independent country consisting of thirteen states and two federal territories. It grants substantial autonomous power to subnational levels of government. Malaysia is a constitutional monarchy; the monarch, as chief of state, is elected for a five-year term by and from the hereditary rulers of nine of the states. The monarch appoints the prime minister and cabinet members from Parliament. Legislative power is vested in a bicameral Parliament: the Dewan Negara (Senate), with members appointed by the monarch, the state legislatures, and others for six-year terms, and the Dewan Rakyat (House of Representatives), with members elected by the general population for five-year terms. Most of the thirteen Malaysian states are governed by hereditary rulers, with the exception of Melaka (ruled by a governor appointed by the Malaysian Pulua Pinang government) and two self-governing states, Sarawak and Sabah, on the island of Borneo (*Europa* 1995).

Description of Coastal and Ocean Problems

The coastal areas exposed to the South China Sea—the eastern coast of West Malaysia and the western coast of East Malaysia—are characterized by well-developed sandy beaches with clay-composed soils and mudflats. Mangrove trees are found along the west coast of West Malaysia, generally associated with mudflats and clay swamps (McAlister and Nathan 1987). Increased coastal population, rapid urbanization, oil and gas production, tourism development, heavy rainfall throughout the year, and various economic activities have created numerous environmental and ecological problems in Malaysia's coastal areas, including beach erosion, resource depletion and environmental degradation, and destruction of natural habitats.

In the late 1980s, serious, widespread coastal erosion along more than 1,300 kilometers (810 miles) of coastline—about 27 percent of the country's total coastline—drew government attention to Malaysian coastal zone management (McAlister and Nathan 1987). Sixty percent of those eroding sites (most of them in West Malaysia)—totaling 196 kilometers, or about 122 miles of coastline—were identified as serious erosion areas requiring engineering works on an urgent basis (Loi 1993). Natural coastal processes such as tidal waves, currents, storm surges, and changes in coastal sediment transport also erode Malaysia's coasts. Typical heavy rainfalls cause rivers to deposit large volumes of sediment in coastal areas. In recent years, dam construction and other upriver projects have decreased the amounts of sediment rivers carry downstream, thus disturbing the natural coastal sediment replacement process. With fewer sediment deposits, natural coastal erosion is augmented, resulting in damage to agriculture, housing, transportation, and recreational sites in the coastal areas (McAlister and Nathan 1987). As coastal areas became more popular in the economic and social life of Malaysians, these adverse consequences of coastal erosion caused serious concern among members of the general public and government officials.

The mangrove forests of Malaysia originally covered a total area of about 600,000 hectares, or 1.5 million acres. Sixty percent of these were designated as forest reserves (Ch'ng 1987). Most of the country's mangrove forests are found along the western coast and southern tip of West Malaysia, in the states of Kedah, Perak, Selangor, and Johore, although there are small patches of mangroves along the coasts of the South China Sea. Many mangrove areas have been destroyed by reclamation for agriculture, housing, and shrimp mariculture. For example, in 1960, southern Johore, on the southern tip of West Malaysia, had almost 23,000 hectares (57,000 acres) of mangroves, but by 1986 this area had been reduced by more than 20 percent, mostly by agricultural development and shrimp mariculture (Darus and Haron 1988; ICLARM 1991). In addition, in Sabah and Sarawak on the island of Borneo, 128,950 hectares (318,630 acres) of mangrove forests were slated to be harvested for the export of wood chips (Ch'ng 1987).

Mangrove forests are important to Malaysian coastal fisheries, especially as nursery and feeding grounds for commercially valuable species such as prawns. In 1984, the total prawn catch on the western coast of West Malaysia, the location of most of the mangrove forests, was 38,800 metric tons (42,700 short tons), whereas landings on the eastern coast, with poorly developed mangrove forests, were 2,600 metric tons or 2,900 short tons (Ch'ng 1987). Continuous removal of mangrove forest areas is a critical reason for the decline of nearshore fisheries in Malaysia.

Coral reefs are essential breeding and nursery grounds for many types of tropical fish. Most of Malaysia's coral reefs, located off the eastern and western coasts of West Malaysia, are rich and well developed (ICLARM 1991). However, destructive fishing techniques such as the use of trawls and fish traps in and around coral reef areas damage the corals (Malaysia 1992). Increasing tourism also has stressed coral communities through boat anchoring, illegal coral collection, and water pollution.

Although tourism is an important source of foreign revenue in Malaysia, it has generated a number of adverse environmental effects that have highlighted the need for some level of coastal area management. In 1990, 7.48 million foreign tourists visited Malaysia, an increase of 232.4 percent from a decade earlier (Smith 1992). Many beach resorts have been established to accommodate rapidly increasing tourist activity along the western and eastern coasts of the Malay

Peninsula. Among others, the town of Batu Feringgi, on the island of Penang, off the northwestern coast of West Malaysia, houses a well-established and rapidly developing resort. Marine pollution resulting from the resort's sewage has increased. Another adverse environmental consequence of tourism is economic dislocation in the fishing villages as many fishermen seek new jobs in the tourism industry (Smith 1992).

Evolution of Government Responses

Because Malaysia's severe coastal erosion was caused by a variety of processes, it has been a subject of national concern. Accordingly, during 1984–1985, the Malaysian government launched the National Coastal Erosion Study. The Environment and Natural Resources Division of the Economic Planning Unit (EPU) in the Office of the Prime Minister was the lead agency for this study (McAlister and Nathan 1987; Basiron 1996; Loi 1993). The study reported that about 1,390 kilometers (864 miles) of coastline were subject to erosion and emphasized the need to implement proper long-term planning to prevent coastal erosion (Loi 1993).

Following the recommendations of the coastal erosion study, two important institutions related to coastal zone management were established in 1987: the Coastal Engineering Technical Center (CETC) and the National Coastal Erosion Control Council (NCECC). The CETC is presently known as the Coastal Engineering Control Unit (CECU), under the Department of Drainage and Irrigation (DDI) in the Ministry of Agriculture (Malaysia 1992; Ann 1996). The DDI's CECU is responsible for implementing coastal erosion control, engineering works for critical erosion areas, providing technical support to the NCECC, providing technical advisory services to other government agencies, and collecting coastal engineering data (Ann 1996).

The NCECC is composed of representatives from the Economic Planning Unit; the Ministry of Finance; the Ministry of Science, Technology, and Environment; the Department of Drainage and Irrigation; the Public Works Department; the Town and Country Planning Department; the Forestry Department; the governments of Sabah, Sarawak, and two other states, on a rotating basis; and professional institutions and universities. The NCECC is headed by the Director General of the Implementation Coordination Unit in the Office of the Prime Minister (Ann 1996). Administrative guidelines created by the NCECC in 1987, the General Administrative Circular No. 5 (Loi 1993), require all government agencies to submit any development proposal along the coast to the DDI's CECU for comment (Malaysia 1992).

Another major national-level coastal area management effort involved activities prescribed by the Environmental Quality Act. For example, the Environmental Impact Assessment (EIA) Order of 1987 required all large-scale development projects (e.g., conversion of mangrove swamps, port expansion, coastal reclamation, and construction of resort facilities) to prepare EIAs prior to project approval by the Department of Environment under the Ministry of Science, Technology and the Environment (Loi 1993). To further national interests in environmental protection and sustainable coastal development, the EPU led the preparation of the National Coastal Resources Management Policy in 1992, creating an Inter-Agency Planning Group (IAPG) to complete the policy. The IAPG was divided into three working groups: (1) Coastal Resources, (2) Coastal Planning Processes, and (3) Legislative and Institutional Aspects (Loi 1993). These working groups focused on, respectively, development of a National Coastal Resources Programme in an integrated, systematic, and scientifically sound manner; establishment of effective, coordinated institutional mechanisms at the federal and state levels; and enhancement of the staffing and expertise of relevant agencies (Loi 1993).

Many states had ad hoc working groups or committees established by relevant ministries, and a number of state Economic Planning Units (the principal policy and development arm of state governments) now have a desk officer in charge of coastal area management. However, the desk officer assigned to such duties is not necessarily a specialist and may also have other tasks. A number of states have established committees that include a member of the State Executive Council and officials of other relevant agencies to examine coastal management and development (Basiron 1996).

At the national level, coastal management activities are coordinated on a project or program basis by the Office of the Prime Minister's Environment and Natural Resources Division rather than being a permanent responsibility of the Economic Planning Unit. Various agencies also take the lead concerning particular aspects of coastal management; for example, the Coastal Engineering Division of the Department of Drainage and Irrigation is responsible for coastal protection.

A pilot CZM study in southern Johore is considered to be the most comprehensively integrated and multidisciplinary regional coastal resources management effort in Malaysia. The Coastal Resources Management Plan for South Johore, Malaysia (CRMPSJ), was sponsored by the ASEAN-USAID Coastal Resources Management Project in cooperation with the Ministry of Science, Technology, and Environment (MOSTE); the Office of the Prime Minister's Implementation Coordination Unit; and the Department of Fisheries (DOF) (Malaysia 1992).

In the coastal resources management planning process, collaboration was facilitated through the establishment of two committees, the National Steering Committee and the Johore State Consultative Committee. With numerous collaborative efforts by resource managers and university-based research scientists, the MOSTE completed the CRMPSJ in 1992. The CRMPSJ's goal was to develop a rational coastal zone management plan for a coastal belt covering 300 kilometers (190 miles) that was undergoing massive economic development through rapid industrialization, urbanization, and expansion of the tourism industry. The multisectoral management issues addressed included concerns about coastal forests, mangrove forests, mariculture, sand mining, coastal erosion, water quality, tourism, and fisheries. (Malaysia 1992).

Assessment of Government Responses

In contrast with other Southeast Asian countries that have emphasized community-ty-based approaches to coastal resources management (e.g., the Philippines, Indonesia, Vietnam, and Thailand), Malaysia has not stressed community-based self-management systems for nearshore fisheries and coral reef management. The National Fishery Act (NFA) of 1985 is a comprehensive law that provides for the management, development, and conservation of marine, estuarine, and riverine fisheries (Malaysia 1992), but it does not favor decentralization of marine fisheries management (Pomeroy 1995). Only riverine fisheries are within the purview of the states. For achieving the goals of monitoring, control, and surveillance in fisheries management, Malaysia's NFA calls for shared responsibility with local communities, fishermen's organizations, and NGOs (Pomeroy 1995).

The lack of public awareness of the need for shared responsibility in coastal resources management has resulted in the continuous destruction of coral reefs. State governments play a more comprehensive role than does the federal government in coral reef management in Malaysia. However, this causes a jurisdictional problem between the federal and state governments because a coastal state manages and controls coastal resources within its three-nautical-mile territorial waters, whereas the federal government is responsible for waters beyond the three-mile mark (Loi 1993). For example, the Department of Fisheries is the national core agency for the designation of marine parks. However, the state government of Johore promulgated the National Parks–Johore Enactment of 1989 for designation of parks on islands within its jurisdiction, even though it had agreed to delegate the administration of island parks to the DOF. The state recently prepared development proposals for the islands without prior consultation with the DOF (Malaysia 1992). As a result of this lack of intergovernmental coordination, destruction and degradation of coral reefs have continued due to trawl fishing and illegal coral collection.

According to the National Forestry Act of 1984, the Department of Forestry (DOFor) is responsible for management of mangrove forests as part of the national forest resources. There is an interdepartmental conflict of jurisdiction between the DOFor and the DOF because the DOF has a policy for preserving mangrove forests as reserves for fish habitat under the NFA of 1985 (Malaysia 1992).

In general, Malaysia has practiced ICM through problem-oriented sectoral approaches, especially in the field of coastal erosion and mangrove community management. However, based on its experiences in sectoral ICM, Malaysia has proceeded, since the National Coastal Erosion Study in 1984–1985, toward an integrated coastal management strategy. In particular, establishment of the National Coastal Erosion Control Council (NCECC), composed of all the relevant ministries, departments, and agencies, represents an important step in national ICM efforts. Other manifestations of growth in ICM capacity building include the formulation in 1992 of the National Coastal Resources Management Policy and the Coastal Resources Management Plan for South Johore, both of which aim to apply a multisectoral, holistic approach to coastal zone planning. Continuing problems include the need for greater coordination and integration among governments at the national, state, and local levels when attempting to resolve resource use conflicts and when addressing lack of capacity or expertise at the provincial and local levels to implement ICM plans.

In terms of capacity building in ICM, as mentioned earlier, Malaysia began to recognize the need for coastal resources management through the National Coastal Erosion Study during 1984–1985. Because of the traditional absence of community involvement in coastal resources management in Malaysia, the new ICM policies and plans emphasize public education and participation in the coastal resources decision-making process. The National Coastal Resources Management Policy of 1992 focused on capacity-building efforts such as enhancement of staffing and expertise in relevant federal and state agencies, collection of scientific research data for managing coastal resources, and in particular, development of public education and awareness of coastal resource issues.

Malaysia is one of the countries that has received technical and administrative assistance from the ASEAN-USAID Coastal Resources Management Project. As discussed earlier, the southern Johore area was selected as an ICM pilot study site. The experience and knowledge gained from this pilot study is expected to be shared and exchanged with other coastal area managers in the country.

One of the most important agencies in Malaysia's ICM efforts, the Coastal Engineering Control Unit (CECU) of the Department of Drainage and Irrigation, contributes much to the country's ICM capacity building. It not only provides technical information on coastal erosion for the NCECC but also supports other agencies with its technical engineering comments on coastal erosion issues. The CECU maintains a coastal engineering database to assist in the planning of coastal engineering works in the country (Ann 1996).

Sources Consulted

Abdullah, D. 1996. Director, State Economic Planning Unit, Bangunan Sultan Ibrahim, Johor Bahru, Malaysia. Response to ICM cross-national survey. Newark: University of Delaware, Center for the Study of Marine Policy.

Ann, O. C., Jr. 1996. Director General, Department of Irrigation and Drainage, Kuala Lumpur, Malaysia. Response to ICM cross-national survey. Newark: University of Delaware, Center for the Study of Marine Policy.

Basiron, M. N. 1996. Senior Analyst, Malaysian Institute of Maritime Affairs, Kuala Lumpur, Malaysia. Response to ICM cross-national survey. Newark: University of Delaware, Center for the Study of Marine Policy.

Ch'ng, L. K. 1987. Coastal zone management plan development in Malaysia: Issues and possible solutions. In *Coastal Zone '87*, ed. O. T. Magoon, H. Converse, D. Miner, and L. T. Tobin, 4601–4615. New York: American Society of Civil Engineers.

Churchill, R. R., and A. V. Lowe. 1988. *The Law of the Sea.* Manchester, England: Manchester University Press.

CIA (Central Intelligence Agency). 1995. *The World Factbook.* Washington, D.C.: Government Printing Office.

Darus, M., and H. Haron. 1989. The management of Matang mangrove forest reserves in peninsular Malaysia. In *Coastal Area Management in Southeast Asia: Policies, Management Strategies, and Case Studies,* ed. T. E. Chua and D. Pauly, 77–84. ASEAN-USAID Coastal Resources Management Project. ICLARM Conference Proceedings No. 2. Manila: International Centre for Living Aquatic Resources Management.

The Europa World Year Book. 1995. 36th ed. London: Europa.

ICLARM (International Centre for Living Aquatic Resources Management). 1991. *The Coastal Environmental Profile of South Johore, Malaysia.* ASEAN-USAID Coastal Resources Management Project. ICLARM Technical Series No. 6. Manila: International Centre for Living Aquatic Resources Management.

Loi, H. K. 1993. Coastal zone management in Malaysia. In *World Coast Conference 1993: Proceedings,* vols. 1 and 2. CZM-Centre Publication No. 4. The Hague: Ministry of Transport, Public Works, and Water Management, National Institute for Coastal and Marine Management, Coastal Zone Management Centre.

Malaysia. Ministry of Science, Technology, and the Environment. Coastal Resources Study Team. 1992. *The Coastal Resources Management Plan for South Johore, Malaysia.* ASEAN-USAID Coastal Resources Management Project. ICLARM Technical Series No. 11. Manila: International Centre for Living Aquatic Resources Management.

McAlister, I. J., and R. A. Nathan. 1987. Malaysian national coastal erosion study. In *Coastal Zone '87,* ed. O. T. Magoon, H. Converse, D. Miner, and L. T. Tobin, 45–55. New York: American Society of Civil Engineers.

Pomeroy, R. S. 1995. Community-based and co-management institutions for sustainable coastal fisheries management in Southeast Asia. *Ocean & Coastal Management* 27 (3): 143–162.

Smith, R. A. 1992. Conflicting trends of beach resort development: A Malaysian case. *Coastal Management* 20: 167–187.

Fiji

Background: General Country Context

The independent republic of Fiji comprises more than 3,000 islands, 100 of which are inhabited, in the South Pacific Ocean. Of the four main islands, Viti Levu contains nearly 70 percent of the estimated population of 771,000. Fiji's coastal area (1,129 kilometers, or about 702 miles) surrounds a small land area (18,270 square kilometers, or 7,054 square miles) that is mainly volcanic and mountainous. The coast and nearshore areas exhibit a wide range of tropical coastal ecosystems including extensive coral reefs. Fiji's maritime claims, measured from archipelagic baselines, consist of a 12-nautical-mile territorial sea, a 200-nautical-mile Exclusive Economic Zone, and the continental shelf to the 200-meter (approximately 656-foot) isobath or to the depth of exploitation (CIA 1995, 140).

According to 1994 estimates, Fiji's GDP was approximately U.S.$4.3 billion. In 1994, Fiji's per capita GNP was $2,250 (World Bank 1996). The economy is primarily agricultural (23 percent of the GDP), with a large subsistence sector. The major sources of foreign exchange are sugarcane production and tourism. Other products include coconuts, cassava, rice, sweet potatoes, bananas, and livestock. The sugar industry supports one-third of the industrial sector, which contributes 13 percent of the GDP. Other industries include lumber, precious metals, and clothing. Power is generated mainly by hydroelectric methods (CIA 1995, 141).

Fiji became a British possession in 1874. Britain introduced a constitution in 1966, establishing a ministerial form of government and a Parliament elected by means of an electoral system. Shortly after gaining independence in 1970, Fiji established a second parliamentary chamber, the Senate. The head of state is the president. The population of Fiji is roughly 49 percent indigenous Fijian and 46 percent Indian; 52 percent of Fijians are Christian, 38 percent are Hindu, and 8 percent are Muslims.

Racial tensions—primarily between native Fijians and Indian immigrants and centered on issues of landownership and representation in the legislature—prompted a restructuring of government and the writing of a new constitution in 1990. The bicameral legislature, composed of a House of Representatives and a Senate, was maintained, with the majority of seats in the House to be elected by indigenous Fijians. Membership of the Senate was to be appointed by the newly designated president of the republic, the head of state. However, the head of government, the prime minister, remains a powerful figure in Fijian politics. In response to international concern regarding the continued existence of Fiji's racially based constitution, the prime minister established a commission in 1994 to complete a review of the constitution by 1997 (*Europa* 1995, 1138–1140).

Description of Coastal and Ocean Problems

In the past twenty years especially, adverse effects of coastal building (associated with tourism) and population increase have significantly increased Fiji's vulnerability to global environmental changes, including sea-level rise (SPREP 1995, 2). Even though Fiji's land areas are generally larger and higher than those of most of its South Pacific neighbors and hence more resistant to coastal storms and erosion, storms still can affect its coastal environment. Sea-level rise has the potential to flood critical coastal agricultural lands and to overwash critical resource-bearing fringing coral reefs. Pollution from coastal development, disposal of toxic materials, and agricultural runoff; nutrient loading of coastal waters resulting from deforestation; disposal of plastics; and spillage from inadequate sewage facilities are all having an effect on the coastal environment (SPREP 1995, 8). In addition, reliance on and resultant depletion of coastal resources represents a significant need for a more integrated framework for coastal management.

The success and economic viability of the tourism industry in Fiji have also been responsible for many of the country's present coastal and ocean problems. Loss of habitat (particularly mangrove forests and coral reefs), loss of nearshore fisheries, disturbance of shoreline processes, climate change, poor freshwater management and waste disposal practices, and displacement of recreational and traditional uses all stem from the development of large coastal resorts (UNDP, ADB, and SPREP 1992, 68).

With approximately 83 percent of the land area in Fiji owned by indigenous people who carry out traditional resource management practices, the majority of coastal management activities take place at the local or community level. Although the majority of traditional resource management in Fiji is inherently sustainable (e.g., tribal taboos are involved to prevent overfishing), not all practices are inclined toward conservation. Some unregulated practices, such as use of poisons in fishing, collection of coral, and general overfishing, have led to resource depletion. Two species in particular, the giant clam and the sea turtle, are severely threatened (Matthews, Veitayaki, and Ram-Bidesi, in press).

Coastal management activities in Fiji are also greatly dependent on financial aid from international and regional entities. There remains a great need to increase efforts in providing vocational training, apprenticeship programs, and "on-the-job skill-learning opportunities" (SPREP 1995, 6).

Evolution of Government Responses

Under the Native Land Trust Act of 1940, the control of all native lands is vested in the Native Land Trust Board (NLTB), which administers the lands for the benefit of all Fijian landowners. One of the NLTB's major purposes is to lease native land for tourism development; more than 78 percent of the planned tourism in Fiji will take place on these lands (Chape 1990). Tribal landownership also applies to fishing grounds and sacred areas, which are defined and owned by clans or chiefs who regulate their use and management (Matthews, Veitayaki, and Ram-Bidesi, in press).

This traditionally based management system is the foundation for many of Fiji's national programs, including national fisheries policy and regulatory fishing permit schemes. National patterns of land and marine resource use are strongly directed by traditional cultural values and viewpoints. However, no nationally or traditionally coordinated coastal management plan exists to date. At present, coastal management in Fiji is not a separate practice but a combination of efforts to control pollution, manage fisheries and marine industries, and regulate land uses. Specific environmental planning and management functions take place within the Department of Town and Country Planning, in the Ministry of Housing and Urban Development, although an interministerial Environmental Management Committee (EMC) was established, in 1980. Its purpose is to develop a coordinated, cross-sectoral approach to environmental planning issues (UNDP, ADB, and SPREP 1992, 109). The EMC is also charged with advising the director of town and country planning in the preparation of environmental impact assessments (EIAs) and with relating the activities of the South Pacific Regional Environment Programme (SPREP) to Fiji.

In 1982, the Department of Town and Country Planning established an Environmental Management Unit (EMU). Its purpose is to act as secretariat to the EMC, develop national environmental policy, coordinate the EIAs of development projects, and educate the public and build public awareness of environmental issues, although the unit was not staffed until 1989 (UNDP, ADB, and SPREP 1992, 110).

More recently, the government of Fiji, through the EMU and assisted by a grant from the Asian Development Bank, released the National Environmental Strategy (NES) in 1993. The EMU was upgraded to full departmental status, and combined with the Ministry of Housing and Urban Development to form the Ministry for Planning and the Environment. The overall recommendations of the NES include development of effective environmental management capacity, increased protection of the traditional culture and heritage, increased involvement of the private sector and the general public in environmental planning and management, and improvement regarding specific sectoral issues such as forestry, land degradation, and global warming (Chape and Watling 1993).

As a result of the NES, new coastal legislation and a new authority to administer it are under discussion. At the local level, conservation areas are being set up and projects related to ecotourism are being organized (Veitayaki 1996). A number of special planning boards have also been established to address coastal issues, including the Land Conservation Board, the Native Lands Conservation and Preservation Committee, and the National Sustainable Development Council (Nawadra 1996).

Assessment of Government Responses

The National Environmental Strategy is seen as a major step forward in Fijian environmental planning and management. It targets improvement of administrative and institutional structure, both between sister government agencies and between levels of government, through approval of environmental policies by all levels of government and passage of sound environmental and resource use legis-lation. The strategy also endorses sustainable development as a national goal. However, it is clear that the document has a pro-environmental focus. The major-ity of attention is given to regulation of development and enforcement of envi-ronmental statutes rather than balancing of development and conservation agen-das. It appears that coastal and ocean management in Fiji will continue to be handled within a larger framework of environmental management under the NES.

Whereas in the past, the strength of traditional management overpowered national initiatives, the NES, if implemented properly, could greatly improve communication links and coordination of activities between national and tribal leaders. However, the future success of the NES, and of ICM in Fiji, depends on how well the national government can garner the support of traditional entities.

Fiji has been moderately successful with respect to spatial integration. Accord-ing to the 1992 *Fiji National Report to UNCED*, the country's marine environ-ment "is better served by laws and regulations than any other sector" (Chape and Watling 1992). However, the coastal and ocean laws and regulations reported are administered by separate organizational units, do not operate with similar policy mandates and resource levels, and are not adequately coordinated.

As mentioned earlier, the NES targets specific sectoral issues in need of greater integration—forestry, land and soil degradation, and climate change—as a major action item. Fiji has also been involved in extensive vulnerability studies with the IPCC and several of its regional affiliates to develop an integrated coastal man-agement plan for coastal sea-level rise and climate change. A National Waste Management and Pollution Control Strategy also was recommended by the NES. As such—at the national level, at least—there appears to be some integration of sectors.

Fiji's affiliation with several prominent international and regional organiza-tions has greatly improved the integration of science into environmental and coastal policy. SPREP has recently completed a number of coastal vulnerability studies in Fiji related to sea-level rise and climate change (SPREP 1994), and the interdisciplinary skills and expertise of the World Conservation Union (IUCN) were incorporated into the NES.

Local capacity building is crucial to the success of the National Environmental Strategy. The NES has within its implementation schedule two capacity-building projects: an effort to upgrade environmental education in schools and a directed public awareness program (Chape and Watling 1993). Graduate-level university programs in ICM have recently been established. However, there remains a great need to improve educational and training efforts at the community and tradition-al levels.

Externally, Fiji has utilized specialized training overseas, participated in a large number of international and regional environment programs and training projects, and hired new staff with appropriate qualifications and training to build capacity in ICM.

Sources Consulted

Chape, S. P. 1990. Coastal tourism and environmental management in Fiji. In *Proceedings of the 1990 Congress on Coastal and Marine Tourism*, vol. 1, ed. M. Miller and J. Auyong, 67–77. Newport, Oreg.: National Coastal Resources Research and Development Institute.

Chape, S. P., and D. Watling, eds. 1992. *Fiji National Report to UNCED*. Apia, Western Samoa: South Pacific Regional Environment Programme.

———. 1993. *The National Environment Strategy: Fiji*. Suva, Fiji: Government of Fiji and World Conservation Union.

CIA (Central Intelligence Agency). 1995. *The World Factbook*. Washington, D.C.: Government Printing Office.

The Europa World Year Book. 1995. 36th ed. London: Europa.

Matthews, E., J. Veitayaki, and V. Ram-Bidesi. In press. Evolution of traditional and community-based management of marine resources in a Fijian village. *Ocean & Coastal Management*.

Nawadra, S. 1996. Project Coordinator, Coastal Management, Department of the Environment, Suva, Fiji. Response to ICM cross-national survey. Newark: University of Delaware, Center for the Study of Marine Policy.

SPREP (South Pacific Regional Environment Programme). 1994. *Assessment of Coastal Vulnerability and Resilience to Sea Level Rise and Climate Change: Case Study Yasawa Islands, Fiji. Phase 2: Development of Methodology*. Apia, Western Samoa: South Pacific Regional Environment Programme.

———. 1995. *Coastal Area Management in the Pacific Island Nations*. Apia, Western Samoa: South Pacific Regional Environment Programme.

UNDP, ADB, and SPREP (United Nations Development Programme, Asian Development Bank, and South Pacific Regional Environment Programme). 1992. *Country Report for UNCED*. Apia, Western Samoa: South Pacific Regional Environment Programme.

Veitayaki, J. 1996. Ocean Resources Management Programme, University of the South Pacific, Suva, Fiji. Response to ICM cross-national survey. Newark: University of Delaware, Center for the Study of Marine Policy.

Part III: Developing Countries

···

Ecuador

Background: General Country Context

The republic of Ecuador is a small country (283,561 square kilometers, or 109,483 square miles) on the tropical western coast of South America. The country's population is large relative to its land area, with an estimated 11 million in 1994, 45 percent of which inhabit the coastal area. Ethnic groupings are primarily mixed Indian and Spanish, or mestizo (55 percent) and indigenous Indian (25 percent) (CIA 1995). Ninety-five percent of the population is Roman Catholic. The 2,237-kilometer (1,390-mile) coastline, once dominated by coastal forests and mangrove swamps, has been extensively developed, first for agriculture and more recently as a result of the booming mariculture industry and related coastal development. In general, the southern coastline is drier and more arid than the tropical northern area. As with most of the western coast of South America, Ecuador's coastal waters exhibit upwellings of bottom water that support high surface productivity but are significantly affected by the periodic occurrence of the El Niño Southern Oscillation. Ecuador also owns the Galápagos Islands, about 966 kilometers (600 miles) offshore from the mainland, and claims the continental shelf between the mainland and the islands as well as a 200-nautical-mile territorial sea.

With a per capita GNP of $1,280 (World Bank 1996), Ecuador is considered a developing country. Oil resources and rich agricultural areas make up the bulk of the economy, although shrimp mariculture has risen to become the most profitable private sector activity. Bananas are the primary terrestrial agricultural product. Recent economic reform has inflated fuel prices and utility rates and eliminated subsidies in order to reduce both inflation and a high commercial debt (CIA 1995, 123).

Ecuador became independent from Spain in 1822. A republic, the country has a civil law system. The president is elected as chief of state and head of government, as set out in the 1979 constitution, and appoints the executive cabinet. The legislative branch is unicameral (Congreso Nacional), and the judicial branch is

387

headed by the Supreme Court (Corte Suprema). The country is divided into twenty-one provinces (including the Galápagos), each with a provincial governor.

Description of Coastal and Ocean Problems

Ecuador's rapid population growth and development in the past decades have been most intensive in the coastal area. Mangrove swamps and critical estuarine areas have been filled or dredged for mariculture development. The mariculture ponds interfere with circulation in coastal watersheds and reduce water quality through aquatic farming by-products such as toxins, excess wastes, and nutrient loads. Although the productivity of most of Ecuador's commercial pelagic fisheries is related to offshore upwellings, and fluctuations in the abundance of shrimp are related primarily to El Niño, the reduction of nursery areas in mangrove and estuarine systems degrades the quality and recruitment of existing offshore stocks.

This reduction of stocks has displaced many coastal fishermen. Although some have become involved in the mariculture industry, others have redeployed their boats to the Galápagos, where conflicts over the sea cucumber fishery, the black coral fishery, and the shark fishery have been occurring in recent years.

Notwithstanding the importance of tourism on the Galápagos, the mainland coast contains more than 100 beaches suitable for tourism and local recreation. Increased visitation has overpopulated many sites, and public facilities have become inadequate. In addition, the sites lack municipal or national planning systems to improve tourism development (Robadue and Arriaga 1993).

With respect to socioeconomic problems, weak compliance with coastal and environmental laws and decrees (such as a 1985 ban on mangrove cutting by the Ministry of Agriculture) has undermined government policies and strategies. Examples include noncompliance with wetland protection and fisheries closures, construction of ponds without proper licenses, and uncontrolled wastewater discharge. This weak compliance is said to result from a breakdown in implementation and inadequate public awareness of government regulations or the reasons for resource restriction (Robadue 1995, 8; Olsen 1993, 203).

Evolution of Government Responses

It was recognized as early as 1981, at a workshop conducted by the Ecuadorian navy, that there was a great need for coastal management in Ecuador. In March 1986, following some preliminary work by the Woods Hole Oceanographic Institution, the United States Agency for International Development (USAID) initiated an eight-year pilot coastal resources management project through a cooperative agreement with the University of Rhode Island's Coastal Resources Center (CRC). By 1988, a strategy had been formulated, and in 1989, the Ecuadorian Coastal Resources Management Program (PMRC) was adopted by presidential decree (Robadue 1995).

The PMRC, consisting of a national commission on coastal resources management and a technical secretariat, established special area management plans in six locations and a Ranger Corps for enforcement. The National Coastal Resources Commission (NMRC) is an interministerial council with representation from the six agencies in Ecuador most involved in ICM. The Office of the President serves as lead agency. The special area management zones are termed ZEMs (zonas especiales de manejo). Although they constitute only 8 percent of the total coastal area of Ecuador, the six sites were selected for the wide diversity of activities carried out within them, such as mangrove protection, tourism, water quality and environmental sanitation, mariculture management, and artisanal fishing (WCC 1994).

Actions within the ZEMs relate to four major themes: mangrove ecosystems, mariculture and artisanal fisheries, shoreline use management, and community sanitation and water supply. With respect to mangrove ecosystems, the main goals are to build public awareness of the need to conserve mangrove forests, to increase enforcement of mangrove protection by the Ranger Corps, and to facilitate agreements among users' groups. Demonstration projects in public outreach and improvement of fishing and mariculture techniques have also been carried out.

In relation to the shoreline use management theme, coastal access sites in the ZEMs have been identified and comprehensive plans for managing coastal development have been prepared. Activities covered in the comprehensive plan include zoning, shore construction, shore protection from erosion, and coastal tourism management. With respect to the community sanitation and water supply theme, hydraulic resource studies and small basin management have been undertaken, as has the building of local capacity to monitor water quality (Robadue and Arriaga 1993).

The ZEMs, implemented from 1989 to 1994, act as models or test sites whereby local users, in collaboration with government officials and the interministerial commission, create integrated plans for coastal resource management (Olsen 1993). In addition, the ZEMs increase exposure of coastal management initiatives to the general public while also building public awareness of the value of wetlands and support for their preservation (WCC 1993). The creation of a more effective management structure within the ZEMs, as mapped out during the USAID-supported phase of the program, would lead to a process by which the management philosophy behind their operation could be institutionalized along the entire Ecuadorian coast.

Assessment of Government Responses

A major success of the Ecuadorian approach to coastal management is its reliance on and cooperation between national and local levels of government in the ZEMs. It has been successful in building both local capacity and awareness of sustainable coastal resource use. However, with only 8 percent of the country's coastal area being managed, there is clearly a need to broaden the scope of the pilot program set out by the USAID-CRC program.

Integration among government levels has been moderately successful despite the fact that provincial governments (six of which are coastal) have little or no role in coastal management. The National Coastal Resources Commission sets overall policy for the program. However, the ZEM plans were developed at the community level and subsequently approved by the commission and incorporated into the National Development Plan.

Intersectoral integration has been moderately successful at the national level, but there is a great need to integrate and regulate sectoral activities occurring throughout the Ecuadorian coastal area. The economic boom associated with mariculture ponds provides little short-term incentive for users' groups to curtail activity, nor does it provide incentives for government officials to regulate mariculture activities, despite their adverse environmental effects. Thus, as discussed earlier, implementation of laws and policies for managing mariculture and protecting mangrove forests is weak and has no effective constituency (Olsen 1993, 204). Within the ZEMs, though, local integration of activities has been successful in resolving conflicts among the various uses.

With regard to spatial integration, ocean and coastal management is only loosely coordinated. Ocean and coastal management activities are not administered by the same unit, nor do they follow similar mandates or policies. Despite the fact that Ecuador claims a 200-nautical-mile territorial sea, there is poor management of adverse environmental effects on the coasts, such as pollution from mariculture and sewage outflows onto offshore areas. Moreover, management on the Galápagos Island is entirely separate from coastal and ocean management on Ecuador's mainland.

With respect to interdisciplinary integration, the USAID program brought an influx of coastal experts into the management and policy-making arena. As a result, the integration of natural science into national policies in Ecuador has been moderately successful.

In terms of capacity-building efforts, the primary goals of the USAID-CRC program were to build local capacity in coastal management and to garner public support for the evolution of a greater CZM program. Some approaches used in Ecuador to build capacity include specialized training of existing staff, specialized training of staff at overseas institutions, participation in various United Nations programs, and hiring of new staff with appropriate qualifications and training (Arriaga 1996).

Sources Consulted

Arriaga, L. M. 1996. Country Advisor, CRC-URI, Programa de Manejo de Recursos Costeros, Guayaquil, Ecuador. Response to ICM cross-national survey. Newark: University of Delaware, Center for the Study of Marine Policy.

CIA (Central Intelligence Agency). 1995. *The World Factbook*. Washington, D.C.: Government Printing Office.

Olsen, S. 1993. Will integrated coastal management programs be sustainable? The constituency problem. *Ocean & Coastal Management* 21 (1–3): 201–225.

——. Director, Coastal Resources Center, University of Rhode Island, Narragansett, Rhode Island, United States. Response to ICM cross-national survey. Newark: University of Delware, Center for the Study of Marine Policy.

Robadue, D. D., Jr., ed. 1995. *Eight Years in Ecuador: The Road to Integrated Coastal Management*. Kingston: University of Rhode Island, Coastal Resources Center.

Robadue, D. D., Jr., and L. Arriaga. 1993. Policies and programs toward sustainable coastal development in Ecuador's special area management zones: Creating vision, consensus, and capacity. In *Coastal Zone '93*, ed. O. T. Magoon, W. S. Wilson, and H. Converse, 2668–2682. New York: American Society of Civil Engineers.

WCC (World Coast Conference). 1993. *How to Account for Impacts of Climate Change in Coastal Zone Management: Concepts and Tools for Approach and Analysis*. Version 2. West Coast Conference 1993, November 1–5, Noordwijk, Netherlands. The Hague: Ministry of Transport, Public Works, and Water Management, National Institute for Coastal and Marine Management, Coastal Zone Management Centre.

——. 1994. *Preparing to Meet the Coastal Challenges of the 21st Century: Report of the World Coast Conference 1993*. November 1–5, Noordwijk, Netherlands. The Hague: Ministry of Transport, Public Works, and Water Management, National Institute for Coastal and Marine Management, Coastal Zone Management Centre.

The People's Republic of China

Background: General Country Context

The People's Republic of China occupies a vast area in the eastern part of the Eurasian continent. It borders Mongolia and Russia to the north; Tajikistan, Kyrgyzstan, and Kazakhstan to the northwest; Afghanistan and Pakistan to the west; and India, Nepal, Bhutan, Myanmar, Laos, and Vietnam to the south. It borders North Korea to the northeast and the Pacific Ocean to the east. The climate is subtropical in the far south; characterized by tropical monsoons in the east; temperate in the north, with an average temperature of less than 10°C; and arid in the northwest (*Europa* 1995). More than 40 percent of the population of 1.2 billion lives in the coastal areas. China is officially atheist, but 2–3 percent of the population is Daoist, Buddhist, and Muslim and 1 percent is Christian. The population growth rate was 1.04 percent in 1995 (Wang 1992; CIA 1995). The country has a coastline of 14,500 kilometers (9,010 miles) on the Pacific Ocean. In 1958, China claimed a 12-nautical-mile territorial sea, Exclusive Economic Zone, and all shallow areas of the continental shelf in the Yellow and East China Seas (Churchill and Lowe 1988; CIA 1995).

During the past decade, China has experienced one of the world's highest economic growth rates. In 1994, its GDP was $2.97 trillion, with a GDP growth rate of 11.8 percent (CIA 1995). Agriculture contributed 27 percent of the GDP in 1992, employing almost 59 percent of the labor force. The principal crop is rice, which accounted for 35.5 percent of the total world harvest in 1993. Industry contributed 34 percent of the GDP in 1992. Compared with the agricultural GDP, with an average annual growth rate of 5.4 percent in 1980–1992, the industrial GDP increased very rapidly—by 11.1 percent—during the same period. China is the world's largest producer of coal and natural graphite, raw cotton, and cloth (*Europa* 1995). Since the late 1970s, Chinese leadership has been steering the country's economy from socialist-styled centralization to a more flexible free-market system. As a result, China's economic growth rate has accelerated, particularly in the coastal areas, to more than 10 percent annually. However, one of the most serious threats associated with this continued rapid growth is deterioration of the coastal environment (CIA 1995). In 1994, China's per capita GNP was $530 (World Bank 1996).

The People's Republic of China was named on October 1, 1949, following the victory of communists led by Mao Tse-tung over Chiang Kai-shek's Kuomintang government. The leaders of the Kuomintang government fled to Taiwan. The new communist regime received widespread international recognition in 1971 when the People's Republic of China was admitted as a member to the United Nations. China is a unitary communist country that grants some autonomous authority to subnational levels of government. The highest authority in national power is held by the National People's Congress (NPC), with delegates indirectly elected for five-year terms by representatives from the People's Congresses of the provinces, autonomous regions, and municipalities and the People's Liberation Army. The NPC adopted a new constitution in 1982 that established the office of chief of state, represented by the president of China. Executive power is exercised by the State Council (cabinet), which is headed by the premier, who is the head of government. The State Council is appointed by the NPC rather than by the president of China. In real terms, political power in China is held by the Chinese Communist Party (CCP). Directly under the national government are twenty-two provinces; five autonomous regions, including Xizang (Tibet); and three municipalities (Beijing, Shanghai, and Tianjin) (*Europa* 1995).

Description of Coastal and Ocean Problems

China's economic modernization boom resulted in industrialization and a rapid population increase in the Pacific coastal areas. With the need for more natural resources, the Chinese leadership placed national coastal zone management priority on development and utilization of coastal resources rather than on protection of marine ecosystems (Degong 1989). This trend gave rise to environmental deterioration—in particular, serious marine pollution from land-based sources—and conflicts with use of coastal space.

The major categories of competition for coastal space are mariculture farming versus shipping and port operations; estuary reclamation versus wetland resource preservation; coral and sand mining versus beach erosion protection; and offshore oil and gas production versus fisheries operations. In the first of these categories, the mariculture industry was the fastest-growing element in China's fisheries production, producing 1.62 million metric tons (1.78 million short tons) in 1989— about 15 percent of China's total fish catch in that period (Yu 1991). During the 1980s and 1990s, however, the expansion of mariculture farming conflicted with rapid development of major seaports in China. Among other ports, Dalian Port in the Bohai Sea, one of the busiest vessel-traffic areas in China, faced conflicts with traditional mariculture farming operations, resulting in a number of lawsuits for damage recovery (Yu 1994).

In the second category, from 1950 through 1985, China reclaimed about 1.19 million hectares (2.94 million acres) of tidal areas (more than one-third of the country's total) for salt-making fields, agricultural farming areas, port development, and industrial sites (Yu 1994). Tidal land reclamation activities resulted in the removal of wide areas of mangrove communities in China, reducing these from 50,000 hectares (120,000 acres) to less than 20,000 hectares (49,420 acres) (Yu 1994). This mangrove destruction resulted in seawater intrusion and soil erosion, especially in Xiamen in Fujian Province (Yu 1994). Reclamation of bays and estuaries along China's Pacific coast conflicted with efforts to preserve wetlands for fish hatcheries, natural shellfish beds, reptile habitats, and bird feeding areas.

In the third category of use conflicts, coral and sand mining contributed to seawater invasion and shoreline erosion in Shandong Province, the Gulf of Liaodong, and Zhejiang Province. For example, during 1985–1990, a coastline 10 kilometers (about 6.2 miles) in length in Shandong Province retreated 60–80 meters (200–260 feet) as a result of sand mining, causing damage estimated at more than 2 million yuan (Yu 1994). The most serious seawater invasion due to coral mining was reported in Hainan Province, one of China's major coral reef areas. Coral miners had damaged or removed about 80 percent of coral reefs offshore from the province for construction materials and ornaments, causing the destruction of a local village by a 300-meter (approximately 984-foot) seawater invasion in 1992 (Yu 1994).

Finally, in 1990, China's offshore oil development activities produced 1.147 million metric tons (1.263 million short tons) of oil (WRI 1994), most of it concentrated in the shallow waters (6–32 meters, or 20–105 feet) of the semienclosed Bohai Sea (Yu 1994). Jinzhou Bay in the Bohai Sea has been recorded as the state of the country's most seriously polluted seawater, by oil and heavy metals (Zhijie and Cote 1991). The serious oil pollution in the Bohai Sea has destroyed China's largest fish spawning, breeding, and feeding grounds. Although conflicts between fisheries and the oil industry are settled by financial compensation for loss of fishing grounds (Yu 1994), long-term cumulative effects of oil pollution on the fisheries, especially in the Bohai Sea, challenge China's CZM planners.

One of the most urgent CZM issues in China is marine pollution, caused mainly by large coastal populations and rapid economic development along the coasts. More than 27 large cities, containing more than 100,000 people, and more than 40,000 factories lie along the country's long coastline (Zhijie and Cote 1990).

It is not only oil pollution that threatens fishery resources and human health; land-based marine pollution in China also represents a serious threat to estuarine and bay ecosystems. In 1988, 1.79 million tons of pesticide and 2.68 million metric tons (2.95 million short tons) of fertilizer were being used in agriculture in coastal areas each year; inevitably, some of those substances ran off into the estuaries and bays (Zhijie and Cote 1990). During the 1980s, China generated approximately 530 million metric tons (580 million short tons) of industrial waste per year, and half of this amount was indiscriminately discharged into the ocean (Zhijie and Cote 1991). Scientists demonstrated that fish deaths, species extinctions, and increased toxic residues in fish flesh were caused by these industrial pollutants. From 1964 to 1986 in the Jianozhou Bay, 154 species became extinct, reportedly due to marine pollution (Zhijie and Cote 1991). The treatment of marine pollution is currently the most critical problem in China, giving rise to the need for integrated coastal zone management strategies.

Evolution of Government Responses

Since the late 1970s, national ocean-related agencies in China have become aware of the need for coastal resources management (Yu 1996). In accordance with the national priority for development and utilization of the country's coastal resources, a Comprehensive Coastal Environment and Resources Survey was carried out from 1980 to 1986. The institution that was the most influential in obtaining data on coastal resources for scientific management of the coastal zone was the State Oceanic Administration (SOA), established in 1964 under the direct leadership of the State Council. With cooperative support from the Environmental Protection Administration and the Ministry of Agriculture, the SOA coordinates marine activities; conducts investigations, scientific research, and policy making; develops long-term coastal planning; takes steps to prevent oil pollution and ocean dumping; and pursues environmental protection in offshore oil exploitation (Zhijie and Cote 1990).

In order to provide more data for scientific coastal management policy making, after completion of the seven-year comprehensive coastal survey, the SOA initiated a second national program, the Comprehensive Environment and Resources Survey on Islands, during 1988–1994 (Yu 1996). These two projects have built national capacity for collection of data on coastal economic, geographical, and environmental phenomena. The data collected in these coastal surveys will be used in the upcoming China Coastal Zone Management Act and China Coastal Zone Management Regulations, both of which were being developed at the time of this writing.

In 1988–1990, under the leadership and coordination of the SOA and with the involvement of the Ministries of Forestry, Agriculture, Geology and Mineral Resources, and Water Utilization, China established a number of national marine natural reserves, including Changli Golden Beach Reserve (Hebei Province), Shankou Mangrove Ecosystem Reserve (Guangxi Province), Dazhou Island Marine Ecosystem, Sanya Coral Reef Reserve, Nanmi Island Chain Reserve, Tianjin Paleo-Wetland Reserve, and Sheahu Bay Paleo-Forest Reserve in Fujian Province (Li 1996).

At the national level, the SOA is the sole government organization that looks after China's marine affairs as a whole and in an integrated manner, formulating the country's overall strategy and policy in ocean development, protection, and planning. It has been responsible for drafting the most important ocean-related laws, such as the Law on the Territorial Sea, the Marine Environmental Protection Law, and other regulations concerning ocean dumping of wastes, marine environmental protection, and marine scientific research. Most of these laws and regulations were promulgated in the early and mid-1980s, not long after China's adoption of the Reform and Open Policy, which marked a cornerstone in the country's massive and rapid development of its coastal areas (Li 1996).

In 1991, the national government completed the National Marine Development Plan, the main objectives of which are to ensure that all coastal issues are considered in the rational development and utilization of coastal resources. The principles of the plan are integration, balance, and overall planning with due consideration of all relevant coastal resource management factors. The plan's time span is 1995–2020, with emphasis on marine resources for development, marine development strategy, structural arrangement of marine industries, oceanic land regulation and protection, construction of a marine service system, and policies and measures (Guochen 1993).

After completion of the national coastal survey projects, eleven provincial oceanic administrations were set up for managing, researching, planning, and exploiting each province's coastal and ocean resources. At the provincial level, CZM regulations were prepared for management of the coastal provinces' coastal waters (Yu 1996). The Jiangsu provincial government was the first to put the coastal management regulations into effect, issuing the Coastal Zone Management Provisional Regulations of Jiangsu Province of 1985 (Wang 1992).

In 1988, Hainan Province created China's first provincial ocean authority for comprehensive management of offshore activities, responsible for approval of development, construction, laying of submarine cables and pipelines, and discharge of waste into the ocean within the provincial jurisdiction (Wang 1992). The other provinces are following Jiangsu's lead in coastal zone management and Hainan's model of ocean management.

Among other subnational ICM efforts in China, the selection of Xiamen in late 1994 as an ICM demonstration site represents an important development at the local level. Xiamen, located on the southeastern coast of the mainland, is an international trade seaport. As one of the five special economic zones (the cities of Xiamen, Shantou, Shenzhen, and Zhuhai and the province of Hainan), it has experienced rapid economic growth in recent years, with an average annual growth rate of 20 percent and a population growth rate of about 18.6 percent in 1995 (Chua and Yu 1996).

With international technical support from the GEF, UNDP, and IMO's Regional Programme for Marine Pollution Prevention and Management (MPP-EAS), an Interagency Executive Committee, headed by the executive vice-mayor of the city of Xiamen, was established to provide policy advice and review the general progress of Xiamen's ICM activities. Strengthening the existing agency for ICM coordination, the Marine Management Division was assigned as lead implementing agency for the ICM project. Addressing the need for a mechanism to facilitate interaction between project managers and relevant scientists, an Integrated Management Task Force (IMTF) was established to ensure technical integration of scientific data into management programs. Through numerous meetings organized by the IMTF, a Strategic Environmental Management Plan (SEMP) was prepared. To achieve more effective project coordination, a Marine Management and Coordination Working Committee/Task Force (Leading Group) was created in late 1995, composed of members of the same agencies represented in the Executive Committee and four other vice-mayors in charge of agriculture, fisheries, ports, city planning and construction, and science and technology. The Leading Group can effectively address interagency issues in integrated coastal management (Chua and Yu 1996).

Assessment of Government Responses

During the 1970s and the early 1980s, the Chinese government undertook coastal zone management more for the purpose of economic development than for protection of the marine environment and preservation of coastal resources (Zhijie and Cote 1991). The need for CZM in China has been evident since the late 1970s, when efforts to achieve economic benefits from the development and utilization of coastal resources began. During the late 1980s and the 1990s after adoption of the Reform and Open Policy, the country's rapidly growing economic modernization, coastal development, and utilization of marine resources have resulted in a serious threat to the sustainability of natural resources, increased need for management of use conflicts, and environmental degradation and deterioration. During this period, China's policy in coastal and ocean management has been revised in an effort to balance economic development with protection of the marine environment, a goal that is clearly set forth in the Marine Environmental Protection Law (Li 1996).

China's initial CZM efforts emerged without technical or administrative aid from international projects offering assistance in ICM such as the ASEAN-USAID Coastal Resources Management Project and the program conducted by USAID and the University of Rhode Island. As is the case in many other coastal developing countries, China's coastal and ocean management initially has been exercised largely on a sectoral basis. Although the SOA has broad involvement in China's CZM matters, it was functionally unable to play an effective role in interagency coordination at least until the early 1990s (Zhijie and Cote 1990). A comprehensive legal CZM framework is not in effect, and too many different ministries are involved on a sectoral basis in China's coastal zone (Wang 1992). The sectoral orientation makes it difficult to harness the necessary political will for preventing and mitigating the adverse environmental and socioeconomic consequences of rapid coastal and ocean development. In most cases, intersectoral coordination is performed directly by the SOA; in some very controversial cases, it is carried out by the State Council. The SOA coordinating mechanism needs to be improved, with the SOA maintained as the chief coordinating agency and as a buffer for intergovernmental conflicts (Li 1996).

Since the early 1990s, China has employed a new ICM strategy for managing coastal resources and controlling environmental pollution. One of the noteworthy government ICM efforts is, with completion of the nationwide sea use zoning efforts led by the SOA, the selection of the Xiamen Demonstration Site supported by the MPP-EAS. This pilot project focuses on application of ICM to address local coastal environmental pollution arising from rapid economic development. The most successful practices in the Xiamen pilot project include the establishment of a coordinating mechanism (the Leading Group), prompt local government cooperation, practice-oriented ICM training courses, and a strong public awareness campaign involving publications, local television and radio broadcasts, and articles in local newspapers.

To promote more effective coastal management, eleven provincial oceanic administrations were given authority over all coastal activities within their jurisdictional boundaries. This bottom-up approach to ICM has made it easier for coastal managers to recognize specific local problems and then to respond to these problems in accordance with local conditions.

In terms of capacity building in ICM, three programs set up recently at Nanjing University—the Coastal Exploitation and Management Research Group and Marine Law Society, under the Chinese Society of Oceanography; the Special Committee of Marine Geography, under the Chinese Society of Geography; and the State Pilot Laboratory of Coast and Island Exploitation—work to collect and disseminate scientific data for ICM in China (Wang 1992). The Chinese Academy of Sciences not only provides scientific data for ICM in China but also has conducted a series of international symposia to improve coastal zone research and coastal management.

Although China obtained various coastal data and scientific research results through its two comprehensive coastal surveys, it has had some difficulty in implementing ICM due to a lack of core staff and experts in ICM. Nevertheless, Chinese government officials are aware of the importance of ICM and the need for improved management of coastal areas. This awareness has led decision makers and government agents to train existing staff and exchange ICM know-how through international ICM conferences. On May 24–28, 1996, China hosted an international workshop called "Integrated Coastal Management in Tropical Developing Countries: Lessons Learned from Successes and Failures" in Xiamen, Fujian Province. The workshop highlighted ICM case studies, indicators of success, ICM program formulation and implementation, information for management, human resource development, ICM principles and guidelines, and ICM initiatives.

Sources Consulted

Churchill, R. R., and A. V. Lowe. 1988. *The Law of the Sea*. Manchester, England: Manchester University Press.

CIA (Central Intelligence Agency). 1995. *The World Factbook*. Washington, D.C.: Government Printing Office.

Chua, T. E., and H. Yu. 1996. From sectoral to integrated coastal management: A case study in Xiamen, China. Paper presented at international workshop, Integrated Coastal Management in Tropical Developing Countries: Lessons Learned from Successes and Failures. May 24–28, Xiamen, People's Republic of China.

Degong, C. 1989. Coastal zone development, utilization, legislation, and management in China. *Coastal Management* 17 (1): 55–62.

The Europa World Year Book. 1995. 36th ed. London: Europa.

Guochen, Z. 1993. Coastal zone management in China. In *World Coast Conference 1993: Proceedings*, vols. 1 and 2. CZM-Centre Publication No. 4. The Hague: Ministry of Transport, Public Works, and Water Management, National Institute for Coastal and Marine Management, Coastal Zone Management Centre.

Li, H. 1996. Intergovernmental Oceanographic Commission Secretariat, United Nations Educational, Scientific, and Cultural Organization, Paris. Personal review of and comments on manuscript for *Integrated Coastal and Ocean Management: Concepts and Practices*, August 27.

Wang, Y. 1992. Coastal management in China. In *Ocean Management in Global Change*, ed. P. Fabbri, 460–469. New York: Elsevier Applied Science.

——. 1996. Director, State Pilot Laboratory of Coast and Inland Exploitation, Nanjing University, Nanjing, China. Response to ICM cross-national survey. Newark: University of Delaware, Center for the Study of Marine Policy.

WRI (World Resources Institute). 1994. *World Resources 1994–95*. New York: Oxford University Press.

Yu, H. 1991. Marine fishery management in the People's Republic of China. *Marine Policy* 15 (1): 23–32.

——. 1993. A new stage in China's marine management. *Ocean & Coastal Management* 19: 185–198.

——. 1994. China's coastal ocean uses: Conflicts and impacts. *Ocean & Coastal Management* 25 (3): 161–178.

——. 1996. Former Deputy Director General, Department of Integrated Marine Management, State Oceanic Administration, Beijing, People's Republic of China. Response to ICM cross-national survey. Newark: University of Delaware, Center for the Study of Marine Policy.

Zhijie, F., and R. P. Cote. 1990. Coastal zone of the People's Republic of China: Management approaches and institutions. *Marine Policy* 14 (4): 305–314.

——. 1991. Population, development, and marine pollution in China. *Marine Policy* 15(3): 210–219.

Indonesia

Background: General Country Context

The Republic of Indonesia includes a group of about 17,500 islands between the mainland of Southeast Asia and Australia. Indonesia is the largest archipelagic country in the world; its principal provinces include the islands of Java, Sumatra, and Sulawesi (Celebes); Kalimantan, occupying the southern two-thirds of the island of Borneo; and Irian Jaya, consisting of the Western half of the island of New Guinea and adjacent islands off New Guinea's northern and northwestern coasts. The climate is tropical—hot and humid, with heavy rainfall during most seasons—with an annual average temperature of 26°C (*Europa* 1995). More than 60 percent of the total population of 203.5 million lives in coastal areas. About 45 percent of the population is Javanese, 14 percent is Sundanese, 7.5 percent is Madurese, 7.5 percent is coastal Malay, and 26 percent is composed of other ethnic groups. About 87 percent of the population is Muslim; 9 percent is Christian (Roman Catholic, 3 percent; Protestant, 6 percent); 2 percent is Hindu and Buddhist; and 1 percent follows other religions. The population growth rate was 1.56 percent in 1995 (CIA 1995; Soegiarto 1996). The Indonesian archipelago has an enormously long coastline, stretching for 54,716 kilometers (about 34,000 miles). Indonesia claimed a 12-nautical-mile territorial sea in 1957, measured from archipelagic baselines; a 200-nautical-mile Exclusive Economic Zone in 1980; and the continental shelf to the depth of exploitation in 1969 (CIA 1995; Churchill and Lowe 1988).

Despite Indonesia's vast natural resources and average annual GDP growth rate of 6 percent, during the past decade, it remains in relative poverty. In 1994, Indonesia's GDP was $619.4 billion, with a GDP growth rate of 6.7 percent (CIA 1995). Agriculture contributed 18.5 percent of the GDP in 1993, employing 53.7

percent of the labor force in 1992. Principal crops for domestic consumption are rice, cassava, and maize. Having once been the world's largest rice importer, in 1992 Indonesia attained self-sufficiency in rice production. Indonesia is the world's second largest exporter of rubber and palm oil, after Malaysia. Industry provided about 40 percent of the country's GDP in 1993, employing 14.6 percent of the labor force. From 1980 until 1992, foreign investment boosted manufacturing output and exports, creating an average annual GDP growth rate of 12 percent. Monetary controls imposed from the mid-1990s, a high rate of foreign investment, and financial deregulation have resulted in an average annual GDP growth rate of 6–7 percent during the 1990s (*Europa* 1995). Indonesia's per capita GNP in 1994 was $880 (World Bank 1996).

Indonesia was known as the Netherlands East Indies in the seventeenth century, except for East Timor, which was a Portuguese colony. After Japan's occupation of the territory during World War II, a group of nationalists proclaimed Indonesia an independent republic on August 17, 1945. With the recognition of the Netherlands, Indonesia became legally independent on December 27, 1949. Sukarno, a leader of the nationalist movement since the 1920s, became the first President under a federal constitution. In August 1950, however, the federation was dissolved and the country became the unitary Republic of Indonesia. Indonesia is a republic that grants autonomous powers to subnational levels of government. Legislative power is vested in a unicameral legislature, the Dewan Perwakilan Rakyat (House of Representatives), or DPR. The DPR contains 500 seats: 100 appointed from the Indonesian Armed Forces (ABRI) and 400 directly elected. Theoretically, the highest authority of the country is the Majelis Permusyawaratan Rakyat (People's Consultative Assembly) or MPR, with 1,000 members who serve five-year terms. The MPR includes 500 DPR members, with the remaining 500 seats allocated to regional representatives, members of the ABRI belonging to the Joint Secretariat of Functional Groups, and representatives of other organizations. Executive power is granted to the president, who is elected for a five-year term by the MPR. The president serves as head of government and chief of state and appoints the cabinet. The country is divided into twenty-seven provinces, with each province headed by a governor, who is elected for a five-year term by the Provincial Assembly. However, the election of provincial governors must be confirmed by the president (*Europa* 1995).

Description of Coastal and Ocean Problems

Indonesia's archipelagic coastal zone is one of the world's richest areas in coastal resource. The region's long coastline contains highly productive coral reefs and mangrove ecosystems (Soegiarto 1996) as well as fishery resources, recreational sites for tourism, and an abundance of nonrenewable resources such as oil and gas, tin, bauxite, phosphate, and iron ore (Dahuri 1996a). However, the country's coastal areas have been severely stressed by overfishing, destructive fishing methods, habitat destruction, and marine pollution. These problems are attributed to the country's intensive economic development plan, a high population density along the coast, and poverty (Soegiarto 1996; Dahuri 1996a).

During its First Twenty-Five-Year Long-Term Development Plan (1969–1994), Indonesia achieved remarkable economic development. In 1969, when REPELITA I, the First Five-Year Development Plan, had just begun, the per capita income was $70. Twenty-five years later, per capita income has risen to $920 (Dahuri 1996a). During this rapid economic expansion, the country inevitably increased the utilization and development of its natural resources, including mangrove swamps, tidal wetlands, and fisheries. Thus, coastal pollution, physical degradation of natural habitats, and overfishing occurred (Soegiarto 1996).

More than 60 percent of the population of 203 million lives in Indonesia's coastal areas. The presence of approximately 120 million people on the coasts has resulted in severe depletion of coastal resources. The country's coastal population grew rapidly from the early 1970s to the mid-1980s as the result of a national transmigration program. To reduce population pressure in the crowded inner regions of the islands, particularly Java and Bali, the Basic Principles of Transmigration Act was enacted in 1972. By the end of 1984, approximately 641,000 people had moved to coastal wetlands or mangrove swamps that had been converted into habitable territory. About 4 million hectares, or 9.9 million acres, of coastal wetlands were converted into transmigration sites (Dahuri 1996a).

Apart from the high level of degradation and depletion of coastal resources by population pressure, about 27 million people (more than 13 percent of the total population) in Indonesia still live below the poverty line. This poverty forces them to exploit coastal and marine resources by employing dynamite fishing methods and mining coral just to meet their basic needs for subsistence (Dahuri 1996a). The mining of coral to build temples and human shelters has caused shoreline erosion in Indonesia. In southern Bali, for example, severe beach erosion resulted from heavy coral mining on the seafloor. About 975,000 cubic meters (1.275 million cubic yards) of coral was removed from the mid-1960s to 1980. As a result, stronger waves came through the lagoon, eroding the beachfront and rendering it unsuitable for tourist use (Clark 1996). Unisectoral overexploitation or severe destruction of some coastal zone resources, such as coral reefs, mangrove forests, and fisheries, and marine pollution from human-related economic activities have led to an urgent need for an innovative coastal zone management strategy.

Evolution of Government Responses

Prior to the Marine Resources Evaluation and Planning (MREP) project of 1993, coastal zone management in Indonesia operated under a single-sector approach. There are a number of regulations related to protection of coastal and marine resources: the Basic Provisions on Environmental Protection Act No. 4 of 1982;

the Act on Ecosystem Protection (1994); Greenbelt on Coastal Area (1972); Protection on Biodiversity (1995); and the National Fisheries Law (1985). However, they do not sufficiently integrate interdisciplinary and multisectoral issues related to coastal resources (Soegiarto 1996). Most government planners and decision makers, as well as private sector coastal users, viewed marine ecosystems and their resources as secondary, as waste dumping sites. Recognition of the importance of coastal resource management is a recent occurrence in Indonesia (Dahuri 1996a). The Environmental Protection Act of 1982 contains basic provisions for management of the living environment. The act authorized the Ministry of the Environment (LH), through its executive body, the Badan Pengendali Dampuk Lingungan (BAPZDAL), to implement controls over marine pollution, among other things. The LH is Indonesia's national coordinating agency for marine and coastal management and policy development. The National Agency for Environmental Impact Management (BAPEDAL), under the LH's oversight, manages the environmental impact assessment process (AMDAL) (Kusuma-Atmadja and Purwaka 1996; Sloan and Sugandhy 1994).

In 1993, as Indonesia embarked on its Second Twenty-Five-Year Long-Term Economic and Social Development Plan (1995–2020), the People's Consultative Assembly enacted the Guidelines of State Policy (Garis-Garis Besar Haluan Negara, or GBHN), which officially defined marine-related issues as a sector of Indonesian government interest. The basic principle of the GBHN is to pursue a policy of sustainable coastal resource development with socioeconomic harmony in a sound environment. The National Development Planning Board (BAPPE-NAS) controls coastal resource development funds for REPELITA I (the First Five-Year Development Plan) and oversees marine resource management (Sloan and Sugandhy 1994).

In REPELITA V, which ended in 1994, Indonesia recognized that marine protected areas (MPAs) could preserve and protect coastal and marine resources in a sustainable manner. With a target of 10 million hectares (24.71 million acres), Indonesia has continued to develop its system of MPAs over the past decade. Twenty-four MPAs have been declared since the first one was designated at Palau Dua, on the island of Java (Alder, Sloan, and Uktolseya 1994). There is legislation to direct the management of all protected areas in Indonesia—the Conservation of Living Natural Resources and Their Ecosystem Act (1990). The act reaffirms the Department of Forestry (DOF) as the lead agency for management of marine parks and reserves. In 1990, as part of a decentralization policy, responsibility for implementing management plans for marine parks and reserves was devolved from the Directorate-General for Forest Protection and Nature Conservation (PHPA) of the DOF to the Regional Forestry Staff (Alder, Sloan, Uktolseya 1994).

This environmentally sound, socioeconomically harmonized, and sustainable national policy for marine resource development was incorporated into the Marine Resources Evaluation and Planning (MREP) project of 1993, which exercises integrated and multisectoral disciplines at site-specific coastal areas in ten provinces in Indonesia. Partly supported by the Asian Development Bank (ADB), the MREP will foster the building of capacity to develop plans and to manage coastal resources in an integrated manner in those ten site-specific regions (Soegiarto 1996). Under the MREP project, ten marine and coastal management areas (MCMAs) and three Special Marine Areas (SMAs) have been selected. The MCMAs were set in the context of provincial responsibilities, with the provincial planning agencies (BAPPEDA) expanding their scope of responsibility for marine and coastal resources management planning. However, three SMAs—Makasar Strait, Lombok Strait, and Timor Gap—are of special interest and are managed primarily by the BAPEDAL (Rais 1993; Soendoro 1994).

Indonesia currently is formulating another multisectoral and multidisciplinary program, the Coral Reef Rehabilitation and Management Program (COREMAP). Supported by the World Bank, the Global Environment Facility (GEF), the Asian Development Bank (ADB), Australian Aid, and the Japan International Cooperation Agency (JICA), this program is designed to rehabilitate damaged coral reef habitats and encourage community-based coral reef management at thirty-five sites in Indonesia (Soegiarto 1996).

To promote the implementation of integrated and interdisciplinary coastal zone management in Indonesia, the Integrated Management Plan for Segara Anakan–Cilacap (IMPSA), Central Java, was approved in 1992, with cooperation and technical assistance from the ASEAN-USAID Coastal Resources Management Project. Segara Anakan's coastal area was highly endowed with natural resources such as large mangrove forests, enormous estuarine areas, and great biodiversity. Recently, however, these coastal resources have been rapidly depleted as a result of industry and economic development. The Indonesian government chose the 51,700 hectares (127,700 acres) of the Segara Anakan area as a pilot study site, situated west of Cilacap, Central Java (ASEAN-USAID CRMP and DGF 1992).

The management issues affecting Segara Anakan are pollution, siltation of lagoons, overutilization of fish and mangrove forests, and lack of institutional coordination. To determine the most practical strategies for solving these problems, an integrated management plan was undertaken. One strategy was the design of a zonation scheme to allow for orderly, planned utilization of resources. Zones include: protection zones, reserve zones, forest zones, development zones, agriculture zones, human settlements, aquatic zones, Ministry of Justice zones, and marine zones. For implementation of the ICM plan and interagency coordination, the Segara Anakan Task Force (SATF) was created, composed of representatives of all concerned parties (ASEAN-USAID CRMP and DGF 1992). At present, the IMPSA is being partially implemented by the SATF in cooperation with the BAPPEDA (ASEAN-USAID CRMP and DGF 1992; Chua and Paw 1996). The IMPSA is expected to serve as a model of ICM to be replicated in areas in need of other coastal resource management.

Assessment of Government Responses

Indonesia's remarkable achievement in designating twenty-four MPAs (2.8 million hectares, or 6.9 million acres), which few other nations have accomplished, demonstrates the country's will to protect its coral reef ecosystems. Yet in spite of this extraordinary effort, Indonesia is still lagging in completion of management plans for the MPAs, with plans completed for only three of them (Alder, Sloan, and Uktolseya 1994). The main reason for this lack of follow-through is jurisdictional boundary complexities among provincial governments, as most of the MPAs are developed and managed by Regional Forestry staff in cooperation with regional planning boards (Alder, Sloan, and Uktolseya 1994). Disagreement among the regional boards hinders the preparation and implementation of management plans. There is no single agency with ultimate responsibility for coordinating divergent perspectives among regional authorities.

Indonesia's National Fisheries Law (1985) does not address local community-based resource management systems in coastal fisheries. Fisheries administration and governance in Indonesia has been centralized (Pomeroy 1995). However, since the early 1990s, the Indonesian government has begun to recognize the importance of local community involvement in managing coastal resources. The Conservation of Living Natural Resources and Their Ecosystem Act of 1990 calls for community involvement in the management of marine conservation areas.

In 1994, a new program for alleviating poverty called Inpres Desa Tertinggal (Presidential Instruction on the Less Developed Village), or IDT, was created to promote economic development in fishing and farming villages through decentralization and active participation of the local community (Pomeroy 1995). Not only in the fisheries but also in the Coral Reef Rehabilitation and Management Programmes, a community-based management approach was adopted as a basic management strategy. During implementation of the IMPSA, active solicitation of public participation in project planning was mandatory and was performed with the assistance of NGOs. It is expected that as the government moves toward decentralization and recognizes the importance of managing coastal resources, ICM practices through pilot studies such as the IMPSA and improved ICM-oriented projects such as MREP and COREMAP will serve as building blocks for a future comprehensive integrated coastal management program.

Capacity building in Indonesia has been growing with the participation of NGOs and technical and administrative assistance from the ASEAN-USAID CRMP, the World Bank, the GEF, the ADB, the World Conservation Union (IUCN), the World Wide Fund for Nature (WWF), and JICA. The WWF and other NGOs, such as the Indonesian Forum for Environment, an association of eighty Indonesian environmental NGOs, are encouraging community participation and educating local residents. Their efforts support successful implementation of the MPA programs. WWF-Indonesia has bridged the communication gap in marine conservation issues over the long term, emphasizing community participation and awareness of the importance of coral reef habitats (Alder, Sloan, and Uktolseya 1994; Sloan and Sugandhy 1994).

Moreover, six universities and other higher learning centers are developing and strengthening their staff under the Marine Science Education Program. Six universities are funded, in part, by loans from the Asian Development Bank: Pattimura University, Sam Ratulangi University, Hasunudin University, Diponegoro University, the Bogor Agricultural University, and the University of Riau (Soendoro 1994). Under the MREP project, some 2,300 worker-months have been allocated for training in computer simulation, remote sensing and GIS, ICM strategy, mapping, hydrography, fish habitat, and fisheries management (Soendoro 1994).

Sources Consulted

Agoes, E. R. 1996. Law Lecturer, Center for Law and Archipelago Studies. Responses to ICM cross-national survey. Newark: University of Delaware, Center for the Study of Marine Policy.

Alder, J., N. Sloan, and H. Uktolseya. 1994. Advances in marine protected area management in Indonesia: 1988–1993. *Ocean & Coastal Management* 25: 63–75.

ASEAN-USAID CRMP and DGF (Association of Southeast Asian Nations–United States Agency for International Development Coastal Resources Management Project and Directorate-General of Fisheries). 1992. *The Integrated Management Plan for Segara Anakan–Cilacap, Central Java, Indonesia.* ICLARM Technical Report No. 34. Manila; International Centre for Living Aquatic Resources Management.

Chua, T. E., and J. N. Paw. 1996. A dichotomy of the planning process used in the ASEAN-US project on coastal resources management. Paper presented at international workshop, Integrated Coastal Management in Tropical Developing Countries: Lessons Learned from Successes and Failures, May 24–28, Xiamen, People's Republic of China.

Churchill, R. R., and A. V. Lowe. 1988. *The Law of the Sea.* Manchester, England: Manchester University Press.

CIA (Central Intelligence Agency). 1995. *The World Factbook.* Washington, D.C.: Government Printing Office.

Clark, J. R. 1996. *Coastal Zone Management Handbook.* New York: Lewis.

Dahuri, R. 1996a. Coastal zone management and transmigration in Indonesia. Paper presented at international workshop, Integrated Coastal Management in Tropical Developing Countries: Lessons Learned from Successes and Failures, May 24–28. Xiamen, People's Republic of China.

———. 1996b. Head of Coastal Zone Management, Bogor Agricultural University, Bogor, Indonesia. Response to ICM cross-national survey. Newark: University of Delaware, Center for the Study of Marine Policy.

The Europa World Yearbook. 1995. 36th ed. London: Europa.

Kusuma-Atmadja, M., and P. T. Purwaka. 1996. Legal and institutional aspects of coastal zone management in Indonesia. *Marine Policy* 20 (1): 63–86.

Pomeroy, R. S. 1995. Community-based and co-management institutions for sustainable coastal fisheries management in Southeast Asia. *Ocean & Coastal Management* 27 (3): 143–162.

Rais, J. 1993. Marine Resource Evaluation and Planning Project for integrated coastal zone planning and management in Indonesia. In *World Coast Conference 1993: Proceedings,* vols. 1 and 2. CZM-Centre Publication No. 4. The Hague: Ministry of Transport, Public Works, and Water Management, National Institute for Coastal and Marine Management, Coastal Zone Management Centre.

Sloan, N. A., and A. Sugandhy. 1994. An overview of Indonesian coastal environmental management. *Coastal Management* 22: 215–233.

Soegiarto, A. 1996. Integrated coastal zone management in Indonesia: Problems and plans of action. Paper presented at international workshop, Integrated Coastal Management in Tropical Developing Countries: Lessons Learned from Successes and Failures, May 24–28, Xiamen, People's Republic of China.

Soendoro, T. 1994. The Indonesian experience in planning and management of coastal zone through MREP project. Paper presented at IOC-WESTPAC Third International Scientific Symposium. November 22–26, Bali, Indonesia.

———. 1996. Bureau Chief, Marine Aerospace, Environment, Science, and Technology, Bappenas, Jakarta, Indonesia. Response to ICM cross-national survey. Newark: University of Delaware, Center for the Study of Marine Policy.

The Islamic Republic of Pakistan

Background: General Country Context

The Islamic Republic of Pakistan, situated in southern Asia, is bordered to the east and southeast by India, to the west by Afghanistan and Iran, to the northeast by the People's Republic of China, and to the southwest by the Arabian Sea. The climate is mostly hot and dry, with an average annual temperature of 27°C, except in the mountains, which experience cold winters (*Europa* 1995). About 10 percent of the total population of 131 million lives in coastal areas. Ninety-seven percent of the population is Muslim (Sunni, 77 percent; Shia, 20 percent), and 3 percent follows other religions, including Christianity, Hinduism, and others. The population growth rate was 1.28 percent in 1995 (CIA 1995; Rizvi 1993). Pakistan's coastline is 1,040 kilometers (646 miles) long. The country claimed a 12-nautical-mile territorial sea in 1966, a 200-nautical-mile Exclusive Economic Zone in 1976, the continental shelf to 200 nautical miles, or the outer edge of the continental margin in 1976, and a 24-nautical-mile contiguous zone in 1976 (CIA 1995; Churchill and Lowe 1988).

Pakistan undertook an extensive economic reform program in 1988 that included liberalization of foreign trade, financial reform, and revitalization of the private sector. These reforms resulted in an average annual GDP growth rate of 1.5 percent during 1985–1993. In 1994, Pakistan's GDP was $248.5 billion, with a GDP growth rate of 4 percent (CIA 1995). Agriculture provided 26.0 percent of the GDP in 1993–1994, employing 40.2 percent of the labor force. The principal cash crops are cotton, rice, wheat, sugarcane, and maize. Industry employed 19.8 percent of the labor force and provided 25.1 percent of the GDP in 1993–1994. In December 1992, following severe flooding in Pakistan, the International Monetary Fund (IMF) provided about $262 million in emergency aid. The national government, headed at that time by Benazir Bhutto, agreed to $1.3 billion loan package and a three-year austerity plan with the IMF. As a result, the public deficit decreased from more than 8 percent of the GDP in 1992–1993 to 5.8 percent in 1993–1994; foreign exchange rose to $3 billion in 1994; and foreign and private sector investments increased. However, the country's economy is still hampered by a high level of debt as well as a large population and increasing sectarian, ethnic, and tribal violence (*Europa* 1995). Pakistan's per capita GNP in 1994 was $430 (World Bank 1996).

Pakistan became independent on August 14, 1947, from British India in accordance with the Muslims' desire to establish an Islamic state. Prior to 1972, Pakistan consisted of two separate regions, West Pakistan (now Pakistan) and East Pakistan (now Bangladesh). East Pakistan was the more populated region, mostly by Bengalis, but dominant political and military power was concentrated in the west. The country became a republic on March 23, 1956, under its first constitution, with Major-General Iskander Mirza as its first president. After a succession of military rulers over two decades (and the 1972 conversion of East Pakistan to Bangladesh), on December 30, 1985, the constitution was restored with changes that introduced a strong executive presidency with the power to dissolve the National Assembly and to dismiss the prime minister and cabinet members. Pakistan is a republic in which the national government holds great administrative power. Legislative power is vested in a bicameral Parliament: the Senate, with members elected by the four provincial assemblies for six-year terms, and the National Assembly, with members elected for five-year terms. The president is chief of state, and the prime minister is head of government and appoints the cabinet. Pakistan comprises four provinces, each with an appointed governor and provincial government and federally administered tribal areas (*Europa* 1995).

Description of Coastal and Ocean Problems

For the most part, Pakistan's coast is sparsely inhabited, except for the metropolitan area of Karachi, which is considered one of the world's most populated cities, with nearly 10 million residents. Three major distinct coastal areas can be identified: the Sindh coast, the Karachi coast, and the Balochistan coast, each with markedly different management issues (Rizvi 1993; Amjad 1996).

The coastal zone of Sindh Province consists of the Indus Delta coast and semi-arid coastal areas, with about 104,000 square kilometers ($40,000 square miles) of mangrove forests, swamps, and wetlands containing numerous creeks and extensive mudflats (Rizvi 1993). The Indus Delta coast is the country's densest area of mangrove swamps, representing the fifth largest single strand of mangrove forest in the world (IOC 1994). This coast has recently suffered from reduced freshwater flow attributed to dams on the Indus River, affecting mangrove eco-systems, shrimp aquaculture sites, and biological productivity (Amjad 1996). Over the years, the reduction in river flow has resulted in reduced sediment bud-gets in the delta area. Seawater encroachment and salt intrusion in the sediment have contaminated freshwater aquifers (IOC 1994).

Another serious environmental issue on the Indus Delta coast is destruction of mangrove forests. In the 1970s the country's total valuable mangrove forest area was about 612,000 hectares (about 1.5 million acres), of which 605,000 hectares were on the Indus Delta and only 7,000 hectares were on the Sindh coast (Rizvi 1993). However, mangrove cover on the Indus Delta had decreased considerably during the 1980s (IOC 1994), threatening fishery resources and thus the liveli-hood of a large number of fishermen. Particularly affected are the shrimp fish-eries, which contribute a significant proportion of foreign exchange revenue.

The Karachi coast consists of about 60 kilometers (37 miles) of coastline, with rocky shores, sandy beaches, and estuaries of the Hab River (Rizvi 1993). This coast, the most populated area of Pakistan, has been greatly stressed by severe, uncontrolled marine pollution from municipal and industrial sources, threatening natural habitats, aesthetic values, and human health. In terms of marine pollution, the Karachi coast has the most serious problem in Pakistan. Among the major sources are industrial complexes, the ship breaking industry, power plants, steel mill complexes, and oil refineries. More than other areas, Karachi Harbor, Gizri Creek, Boat Basin, Korangi Creek, Clifton Beach, and China Creek have been stressed by municipal sewage and oil (IOC 1994).

Finally, the coastal zone of Balochistan Province, from near Karachi to the Hingol River to the west, is characterized primarily by rocky shores and sand dunes. The Balochistan coast has two distinct areas: the Las Bela coast, covering Sonmiani Bay and the Miani Hor area, and the Makran coast, containing most of the region's rocky shores, cliffs, and some sandy beaches (Rizvi 1993). The Las Bela coast suffers from erosion of sandy beaches by strong monsoon waves. The Makran coast is vulnerable to seasonal torrential rains as well as strong winds, causing sand dune migration (Rizvi 1993). In general, Balochistan Province's economy is less developed than those of other coastal areas, such as the metrop-olis of Karachi and Sindh Province. Therefore, the province's coastal resources have been relatively less utilized and degraded by coastal economic activities. However, the Pakistani government has recently approved a number of coastal development and industrial projects for Balochistan (IOC 1994). Balochistan Province thus faces the challenge of preparing a proactive coastal management plan.

Evolution of Government Responses

ICM planning and management efforts in Pakistan are relatively weak but are at an emerging stage. On October 10–14, 1994, the "International Workshop on Integrated Coastal Zone Management," organized by the Ministry of Science and Technology, was held in Karachi with the cooperation and financial support of the Intergovernmental Oceanographic Commission, an agency of the United Nations Educational, Scientific, and Cultural Organization (UNESCO). Prior to this, Pakistan's coastal area management strategy was clearly dominated by a single-sector approach, inherently ignoring the interests of other sectors in management projects.

There is recognition at the national level of the importance of the coastal zone for industrial development, the marine fisheries industry, and water management. Fisheries in Pakistan's EEZ, beyond its twelve-nautical-mile territorial waters, are governed by the Marine Fisheries Directorate of the national government, whereas provincial fisheries, located within territorial waters, are managed and controlled by the Department of Fisheries of the relevant provincial governments (IOC 1994).

In terms of marine pollution, the Ministry of Pollution and Environmental Control has management responsibility. In addition, National Environmental Quality Standards (NEQS) have been prepared, to which all new industries are to conform after July 1994, with existing industries to conform by July 1996. A Marine Pollution Control Board has been created to monitor implementation of these standards. One important mechanism to prevent marine pollution is an environmental impact assessment, which new industries must prepare before they are established (IOC 1994).

A key institution in Pakistan's CZM planning and development is the National Institute of Oceanography (NIO), under the Ministry of Science and Technology (MST). During the IOC workshop, the importance of high-level national government institution in ICM was recognized, and it was strongly recommended that the NIO play a key role at the national level in generating ICM efforts in Pakistan. Previously, the NIO had conducted basic research for the country's potential ICM program, creating a number of CZM publications and a review statement for ICM in Pakistan (IOC 1994).

The provincial governments have aimed coastal area management at improved socioeconomic conditions through exploitation of coastal resources such as fisheries and aquaculture development, but with due consideration for environmental health. With regard to formulation of an ICM plan in Pakistan, the Sindh Coastal Development Authority (CDA), established in 1994, has the potential to become a key player. A similar coastal authority is being considered for Balochistan Province. The CDA has been given legal authority to conduct comprehensive coastal development planning along the Sindh coast. The Governing Council of the CDA consists of agencies from various sectors in the province. The CDA encourages the involvement of the private sector, NGOs, and the general public, but it still needs to be extended to coordinate sectoral and environmental perspectives (IOC 1994).

Assessment of Government Responses

Pakistan's general awareness of the importance of ICM appears to have grown recently as a result of reports, including "Coastal Zone Management Issues in Pakistan—An Overview," by the NIO (Rizvi 1993). These reports and other "issue papers" stress the need for an ICM plan (IOC 1994) and for international cooperation and assistance in ICM initiation from such entities as the World Bank, IUCN, and the IOC.

The 1994 IOC workshop examined the influence of socioeconomic conditions in the coastal zone of Pakistan and drafted guidelines for the ICM planning process and for capacity building at the national level in ICM implementation. The workshop recommended four steps for moving toward ICM in Pakistan: (1) integration of policy and programs across government jurisdictions and economic sectors; (2) establishment of a government regulatory framework to guide public participation and involvement in decision making; (3) linkage of ICM at the local, provincial, and national levels of government; and (4) creation of an interagency entity to coordinate development projects and ensure that they are consistent with ICM policies (IOC 1994).

A key component of the workshop's ICM recommendation for Pakistan is strengthening of the country's interagency coordinating mechanism to institute both vertical (local, provincial, and national levels of government) and horizontal (different sectors) integration. For the provincial level, it was recommended that the CDA's Governing Council (GC) broaden its coverage to include all aspects of ICM in Sindh and create an advisory board representing resource users, managers, and the scientific community to address multisectoral issues.

On the other side of Pakistan's coast, Balochistan has lagged in economic development. But industrialization and urbanization along the Balochistan coast are imminent; therefore, immediate establishment of a Coastal Development Authority for Balochistan is needed for timely implementation of ICM in that province (IOC 1994).

In terms of capacity building in ICM, important steps have been taken to build ICM capacity in Pakistan—such as the IOC workshop and the NIO's publications—but at present, capacity for establishing new ICM programs remains weak. The IOC workshop recommended a capacity-building strategy in three main areas—at the professional level, in the area of public education, and at the institutional level. The most interesting proposals include (1) a basic training program in ICM for national coastal zone managers, (2) a training program for ICM "trainers," and (3) expansion of capacity-building efforts at universities and other institutions. Further capacity-building measures recommended include public awareness campaigns, involvement of NGOs and the private and commercial sectors, GIS applications, strengthening of existing institutions, and establishment of cooperative programs with international agencies and institutions (IOC 1994).

Sources Consulted

Amjad, S. 1996. Director General, National Institute of Oceanography, Karachi, Pakistan. Response to ICM cross-national survey. Newark: University of Delaware, Center for the Study of Marine Policy.

Churchill, R. R., and A. V. Lowe. 1988. *The Law of the Sea*. Manchester, England: Manchester University Press.

CIA (Central Intelligence Agency). 1995. *The World Factbook*. Washington, D.C.: Government Printing Office.

The Europa World Year Book. 1995. 36th ed. London: Europa.

Haq, S. M. 1996. Former Professor and Director, Center of Excellence in Marine Biology, University of Karachi, Pakistan. Response to ICM cross-national survey. Newark: University of Delaware, Center for the Study of Marine Policy.

IOC (International Oceanographic Commission). 1994. *International Workshop on Integrated Coastal Zone Management (ICZM)*. October 10–14, Karachi, Pakistan. Intergovernmental Oceanographic Commission Workshop Report No. 114. Paris: United Nations Educational, Scientific, and Cultural Organization.

Rizvi, S. H. N. 1993. Coastal zone management issues in Pakistan—An overview. Paper presented at the IPCC Eastern Hemisphere Workshop on the Vulnerability of Sea-Level Rise and Coastal Zone Management, August 3–6, Tsukuba, Japan.

Democratic Socialist Republic of Sri Lanka

Background: General Country Context

The Democratic Socialist Republic of Sri Lanka comprises one large island and several smaller ones in the Indian Ocean, situated about 80 kilometers (50 miles) east of the southern tip of India. The climate is tropical, with very little seasonal variation in temperature and an average annual temperature of about 27°C. The tropical monsoon season occurs from December to March in the northeastern part of Sri Lanka and from June to October in the southwestern area (*Europa* 1995). Nearly 34 percent of the population of 18 million lives in the coastal regions. About 69 percent of the population is Buddhist; 15 percent is Hindu; 8 percent is Christian; and 8 percent is Muslim. The population growth rate was 1.15 percent in 1995. Sri Lanka's coastline is 1,340 kilometers (830 miles) long (CIA 1995). The country claimed a 12-nautical-mile territorial sea in 1971; a 200-nautical-mile Exclusive Economic Zone in 1976; continental shelf within 200 nautical miles or to the outer edge of the continental margin in 1976; a 24-nautical-mile contiguous zone in 1976; and a 200-nautical-mile pollution prevention zone in 1976 (CIA 1995; Churchill and Lowe 1988).

In 1994, Sri Lanka's GDP was $57.6 billion, with a 5 percent GDP growth rate (CIA 1995). Agriculture contributed an estimated 21.7 percent of the GDP in 1993, employing about 38.5 percent of the labor force. The country's principal cash crops are tea, rubber, and coconuts. In 1990, Sri Lanka became the world's largest tea exporter. Industry has become Sri Lanka's main source of export earnings, contributing more than 28.1 percent of the GDP in 1993 (*Europa* 1995). Since the late 1970s, unemployment (13.6 percent, 1993 estimate), a persistent fiscal deficit ($7.2 billion, 1993 estimate), and inflation (12 percent, 1994 estimate) have been the country's main economic problems (CIA 1995). In 1994, Sri Lanka's per capita GNP was $640 (World Bank 1996).

Major economic dislocation in Sri Lanka resulted from ethnic conflicts between the Sinhalese majority and the Tamil minority in the north. Continued conflict between the government and the Liberation Tigers of Tamil Eelam (LTTE), which emerged as the dominant Tamil separatist group, has plagued the country for more than twenty years. However, since resumption of peace talks between the government and the LTTE in January 1995, Sri Lanka's economy has showed signs of recovery, including an attempt to begin a major privatization program, inflow of foreign investment, and liberalization of exchange controls (*Europa* 1995).

Sri Lanka, known as Ceylon until 1972, became independent from the United Kingdom in February 1948. A presidential system of government was adopted in 1977 and confirmed by the constitution of 1978. Under the new constitution, very strong executive powers are vested in one individual—the president, who is chief of state and head of government. The president is directly elected for a six-year term. He or she can appoint or dismiss the Prime Minster and members of the cabinet. Sri Lanka's supreme legislative body is a unicameral Parliament, which the president is also empowered to dismiss. Sri Lanka is divided into nine provinces and twenty-four administrative districts. As a result of continuing violence in the northern and eastern provinces, where the Tamils are a majority, Parliament adopted legislation creating nine provincial councils with some regional autonomy (*Europa* 1995).

Description of Coastal and Ocean Problems

Since Sri Lanka's independence in 1948, more and more people have migrated to coastal areas, especially southwestern urban coastal areas such as Colombo,

Galle, and Gampaha, to take advantage of economic, educational, industrial, and business opportunities there (Lowry and Wickremeratne 1988). As a result, development activities along Sri Lanka's coastline, including construction of housing facilities for residents and tourists, mining of coral and sand, dredging and filling, fishing with the use of dynamite, and commercial extraction of mangrove trees for firewood, have created many coastal problems, including coastal erosion, degradation of valuable coastal habitats, and resource use conflicts.

In Sri Lanka, a key coastal management problem is coastal erosion resulting both from the natural action of tidal waves and currents and from human causes such as ill-designed coastal erosion protection works and coral mining. Shoreline erosion has resulted in damage to or loss of hotels and other buildings near the shoreline, destruction of coastal vegetation, and disruption of fishing and recreation. The most severe effects of shoreline erosion have occurred in Sri Lanka's western and southwestern coastal areas. For example, from the Jaffna Peninsula in the north to Weligama Bay in the south, about 175,000–285,000 square meters (210,000–340,000 square yards) of coastal land is lost to erosion each year (Lowry and Wickremeratne 1988). During the June–October monsoon season in the southwest, loss of coastal roads and beaches occurs frequently.

For most coastal villagers and hotel owners, one way to prevent shoreline erosion is to construct groins and breakwaters. However, the groins reduce coastal erosion in one area but caused erosion and accretion elsewhere. One example is an ill-designed long groin constructed to protect the outlet of the Panadura River. The groin has prevented the periodic flooding of adjacent inland areas but caused a serious erosion problem north of the river by blocking littoral supplies of sand to the entire coastline (Lowry and Wickremeratne 1988). In addition, coral and sand mining in Sri Lanka have contributed to serious beach erosion and exposure to natural hazards. Coral mining, most notably along the southwestern coast, was stimulated by a rapidly growing demand for lime in the 1960s and 1970s (Clark 1996). Unlike coral mining, mining of sand for construction materials takes place at river mouths and dunes along the coast.

The concentration of population in Sri Lanka's coastal areas has contributed to the increased rate of degradation of valuable coastal habitats. Mangrove areas stretch in a narrow band one kilometer (about six-tenths of a mile) landward of the mean low-water line along the coastline's intertidal belt. Many mangrove forests are under threat of being cut down for firewood and housing material. Sri Lanka's large sea grass beds, situated from Dutch Bay to Jaffna Lagoon, are an example of coastal habitat being destroyed by bottom trawling and dragnet fisheries.

The most serious resource use conflict occurs between coral and sand miners and local fishermen. As discussed earlier, coral and sand mining causes shoreline erosion, which leads to a loss of mangrove forests, small lagoons, and sea grass beds and eventually causes collapse of local fisheries.

Evolution of Government Responses

The most important issue propelling Sri Lanka into a national discussion about ICM was the severe coastal erosion caused by coral and sand mining (Clark 1996). Coastal management conflicts and problems were discussed at the national level in 1977, and as a result the national government set up a lead agency to oversee a full-scale CZM program covering the country's entire coastline. In 1978, the Coast Conservation Division was established within the Ministry of Fisheries, with planning, development, and regulatory responsibilities for Sri Lanka's CZM. Later it became a government department, officially called the Coast Conservation Department (CCD) (Premaratne 1991).

In 1981, Sri Lanka enacted the Coast Conservation Act (CCA), which gave the CCD jurisdiction over Sri Lanka's coastal zone, defined by Parliament as the area from 2 kilometers (about 1.2 miles) seaward to 300 meters (about 328 yards) inland. The law required the CCD to develop a Coastal Zone Management Plan (CZMP) within three years. The CCA went into force in October 1983, and the CCD prepared the CZMP in 1988. Required to address the shoreline erosion problem and a variety of other coastal resource management issues, the CZMP addressed four significant issues: coastal erosion; degradation and depletion of natural habitats and resources; loss and degradation of historic, cultural, and archaeological sites and monuments of significance; and loss of physical and visual access to the ocean (Sri Lanka 1987).

In Sri Lanka's coastal zone management, two agencies, the CCD and the Urban Development Authority (UDA), are the most significant authorities. The CCD has played a leading role in efforts to reduce and prevent coastal erosion in particular. The CCD has conducted a number of studies, produced a Master Plan for Coast Erosion Management, and regulated development activities along the coast by issuing permits and conducting workshops and seminars to discuss effective coastal management strategies (Lowry and Wickremeratne 1988). The UDA, established under Urban Development Authority Act No. 41 of 1978, has jurisdiction over all areas within one kilometer of the coastline described as "urban areas." All construction of buildings, hotels, and other structures within coastal urban areas requires a permit from the UDA.

The primary mechanism for implementing Sri Lanka's CZMP has been a permit system whereby anyone engaging in development activities within the designated coastal areas must first acquire a permit from the CCD. Development activities include construction of buildings and works; deposit of wastes; removal of sand, coral, shells, natural vegetation, or sea grass; dredging and filling; land reclamation; and mining or drilling for minerals (Clark 1996). The CCD began issuing permits in 1983. Between 1983 and 1995, according to a table in the draft *Coastal Zone Management Plan, Sri Lanka: 1996–2000*, 2,668 permits were issued.

As of 1991, the most prohibitive measure adopted in the CZMP was a complete ban on coral mining except for research purposes. Enforcement of this measure and its effects on the livelihood of coral miners were two of the most difficult challenges faced by the CZMP. The CCD continues to make various efforts to resolve coastal management issues by pursuing local support. In 1990, in a provincial-level program titled "Implementation of CZM Plan," and in 1995, in a local-level program called "Preparation of Special Area Management Plan" (Samaranayake 1996), residents were encouraged to become actively involved in the design and implementation of the coastal zone management program. This bottom-up approach was designed to make the local community "fully aware of and integrated into the planning effort so that it is truly participatory" (Clark 1996).

Since 1990, the CCD has been reexamining Sri Lanka's comprehensive coastal management strategy, commissioning a series of technical papers on various issues in coastal areas. These papers were bound in a two-volume report, *Coastal 2000: Recommendations for a Resource Management Strategy for Sri Lanka's Coastal Region*. A central concern of these recommendations is creation of a second-generation coastal resources management program, to be implemented simultaneously at the national, provincial, and local levels (Kahawita 1993).

Assessment of Government Responses

As is common in most coastal countries, Sri Lanka's CZM approach was highly fragmented until the CCD was established. More than fifty different laws and thirty-two different national and provincial agencies had been involved in managing the primary activities affecting coastal areas (Lowry and Wickremeratne 1988). Sri Lanka's CZMP is a comprehensive plan for managing and regulating the country's coastal zone issues. With the technical assistance and know-how of the USAID–University of Rhode Island project in Sri Lanka, CZM efforts evolved rapidly and the CCD's program was broadened from an erosion management program into a more comprehensive resource management program (Olsen 1993).

Sri Lanka's CZM program initially employed a top-down approach involving strategies such as inventories of coastal habitats and ecosystems, regulation of large-scale activities affecting habitats, and long-range management of coastal uses by means of zoning (Kahawita 1993). A bottom-up approach was also undertaken in special area management strategies. In 1990, the CZMP (under the CCD staff and consultants' broader approach to coastal management strategies) referred to a second-generation coastal resources management program that must be implemented simultaneously at the national, provincial, district and local level (WCC 1994). Therefore, the CCD has begun transferring the authority for issuing "minor" permits for modest houses, including the removal of small quantities of sand from beaches and some other limited and practicable activities, to local authorities (Clark 1996). The decentralization effort has resulted in national and local levels of government successfully sharing authority, which has encouraged local government officials and earned local support.

This bottom-up approach was also applied to the designation of areas for special area management (SAM), whereby local communities become actively involved in both design and implementation of the coastal management program. This small-scale management strategy can be achieved through "mutual education, persuasion, and cooperation to minimize activities that deplete and degrade coastal habitats to the detriment of all residents" (Clark 1996). There are currently two SAM sites, at Hikkaduwa, a resort and fishing community, and Rekawa Lagoon, a valuable mangrove and fishery area. In these experimental areas, the key aspect of success was decentralization supported by the community, which then became integrated into the CZM process.

One major problem in implementing Sri Lanka's CZM involved a lack of consideration of cultural factors when imposing a complete ban on coral mining (Premaratne 1991). First, the coral-mining ban directly affected the livelihood of 1,434 coral miners who depended on that industry in the southwestern coastal belt. Second, many local officials charged with enforcing the ban were reluctant to act in the face of public opposition. Finally, the most serious policy mistake was attempting to find alternative employment for the coral miners without considering existing social and cultural barriers in Sri Lanka (Clark 1996). The CCD selected 261 families to be allocated agricultural land in dry areas. It appears that the coral miners had no experience in inland agriculture; thus, the resettlement effort created new social and political tensions between the CCD and coral miners. The CCD was successful in eliminating offshore coral-mining problems, but its resettlement initiative did not succeed because the broader social and cultural aspects of traditional subsistence activities were not addressed in CZM planning.

In the 1960s and 1970s, Sri Lanka's coastal zone management was exclusively concerned with shoreline erosion. However, since implementation of the CZMP in the 1980s, coastal managers have become concerned with much broader management questions. In Sri Lanka, as in most developing countries, a major implementation problem is securing the compliance of the many users of the coast. Given the economic conditions in the country, Sri Lankan coastal managers have been remarkably successful in their efforts to encourage compliance with management plans, mostly through educational programs (Lowry 1996). As mentioned earlier, Sri Lanka's CZM program has also been supported by USAID through the University of Rhode Island. Considerable experience and technical assistance have been gained over the course of the program (Olsen 1993). The CCD has organized an extensive training program for local officials to implement the minor permit system, as conferred by the national authority. Capacity building has also been achieved by the National Aquatic Resources Agency, which researches coastal and ocean issues and employs a sizable staff to plan and implement CZM programs. Moreover, as the CZM program became decentralized, more professional staff members were required in order to carry out the work.

Sources Consulted

Churchill, R. R., and A. V. Lowe. 1988. *The Law of the Sea.* Manchester, England: Manchester University Press.

CIA (Central Intelligence Agency). 1995. *The World Factbook.* Washington, D.C.: Government Printing Office.

Clark, J. R. 1996. *Coastal Zone Management Handbook.* New York: Lewis.

The Europa World Year Book. 1995. 36th ed. London: Europa.

Kahawita, B. S. 1993. Coastal zone management in Sri Lanka. In *World Coast Conference 1993: Proceedings,* 669–675. CZM-Centre Publication No. 4. The Hague: Ministry of Transport, Public Works, and Water Management, National Institute for Coastal and Marine Management, Coastal Zone Management Centre.

Lowry, K. 1996. Professor, Department of Urban and Regional Planning, University of Hawaii, Honolulu, Hawaii, United States. 1996. Response to ICM cross-national survey. Newark: University of Delaware, Center for the Study of Marine Policy.

Lowry, K., and H. J. M. Wickremeratne. 1988. Coastal area management in Sri Lanka. In *Ocean Yearbook* 7, ed. E. M. Borgese, N. Ginsberg, and J. R. Morgan, 263–293. Chicago: University of Chicago Press.

Olsen, S. B. 1993. Will integrated coastal management programs be sustainable? The constituency problem. *Ocean & Coastal Management* 21 (1–3): 201–225.

Premaratne, A. 1991. Difficulties of coastal resources management in developing countries: Sri Lanka experiences. In *Coastal Zone '91,* ed. O. T. Magoon, H. Converse, V. Tipple, L. Thomas Tobin, and D. Clark. 3026–3046. New York: American Society of Civil Engineers.

Samaranayake, R. A. D. B. 1996. Manager, Coastal Resources Development, Coast Conservation Department, Maligawatte, Colombo, Sri Lanka. Response to ICM cross-national survey. Newark: University of Delaware, Center for the Study of Marine Policy.

Sri Lanka. Coast Conservation Department. 1987. *Coastal Zone Management Plan.* Colombo, Sri Lanka: Coast Conservation Department.

WCC (World Coast Conference). 1994. *Preparing to Meet the Coastal Challenges of the 21st Century: Report of the World Coast Conference 1993.* November 1–5, Noordwijk, Netherlands. The Hague: Ministry of Transport, Public Works, and Water Management, National Institute for Coastal and Marine Management, Coastal Zone Management Centre.

Federated States of Micronesia

Background: General Country Context

The nation known as the Federated States of Micronesia (FSM) (as distinct from the larger island group called Micronesia) is part of the Caroline

Islands chain, about 800 kilometers (500 miles) east of the Philippines. The FSM includes the islands of Yap, Chuuk, Pohnpei, and Kosrae and numerous smaller islands, with the capital, Palikir, situated on Pohnpei. The FSM gained independence in 1979 and ratified a constitution in 1980. The country FSM is made up of roughly 500 islands, with 6,112 kilometers (3,798 miles) of coastline, and claims a 12-nautical-mile territorial sea and a 200-nautical-mile Exclusive Economic Zone. The islands are at the southern edge of the typhoon belt, making them susceptible to occasional but severe damage from coastal storms. Although several of the islands are high and mountainous or actively volcanic, many are low-lying atolls with extensive fringing and barrier reef areas offshore (CIA 1995).

The population of the FSM, estimated to be about 123,000 in 1995, is made up of nine ethnic Micronesian and Polynesian groups. Major religious preferences are Roman Catholic (50 percent), Protestant (47 percent), and others (3 percent) (CIA 1995).

Economic activities consist primarily of subsistence farming and fishing. High-grade phosphate is the major mineral export. There is potential for growth in the tourism industry, but a lack of adequate facilities and the islands' remoteness hinder development. Government employees make up more than two-thirds of the workforce. Financial aid from the United States is the primary source of revenue, representing nearly $1 billion during the decade of the 1990s (CIA 1995).

The Federated States of Micronesia was formerly a part of the U.S.-administered Trust Territory of the Pacific Islands. In 1980, following demands for self-government in the 1970s, a constitution was drafted and ratified by the four main island states. The United States agreed to terminate trusteeship over the islands shortly thereafter and signed the Compact of Free Association in 1982, leaving the newly formed federal government of the FSM in charge of the country's internal and foreign affairs. Following approval by the United Nations Trusteeship Council in 1986, U.S. administration of the FSM formally ended in November of that year.

The locally drafted constitution of 1979 established a federal legislature composed of fourteen elected senators. Each of the four states elects a senator-at-large, serving a four-year term, with the remaining ten senators elected for two-year terms; seats are allocated by island population. The president and vice-president are elected by the senators-at-large for four-year terms (*Europa* 1995, 2089).

Ownership and control of land in the FSM are traditionally based. With the exception of the larger towns and cities, which elect magistrates or town councils, all villages, farmsteads, and wilderness lands are controlled by indigenous bodies or chiefs (Dahl, in press). However, the states have jurisdiction over most natural resource management activities. The federal government has a lesser role in resource management; its most notable function in coastal and marine management is management of fisheries in the Exclusive Economic Zone (Dahl 1995, 354).

Description of Coastal and Ocean Problems

The marine environment is of enormous cultural and economic importance to the people of the FSM. Yet coral reefs continue to be degraded by dynamite blast fishing and general overfishing despite efforts to control such practices. Coastal fisheries in close proximity to urban centers have been seriously depleted as have nearshore reef habitats through coral mining and harvesting.

Offshore, tuna is the primary fishery resource. The FSM's enormous EEZ supports a large catch by Micronesian fisherman—approximately 135,000 metric tons, or 150,000 short tons, per year, according to a 1993 estimate. However, the increasing presence and catch of distant-water fleets and the FSM's inability to monitor and enforce its EEZ are negatively affecting tuna stocks. The various states have been working to develop their tuna industries through increased fishing efforts and technology and onshore canneries and refrigeration units, but they often must rely on exportation of the goods to Japan, Hawaii, and Guam to generate income. The application of aquaculture and mariculture have been discussed in the FSM but remain at the planning stage.

Other coastal management issues relate to construction and infrastructure development. Both road alignment and land clearing in coastal areas contribute to an increased rate of coastal erosion and siltation of nearshore reefs. Increased erosion has also exacerbated saltwater intrusion and contamination of freshwater supplies. Dredging for marine facilities and navigation have destroyed the recreational value of many reefs and wetland areas, as well as adversely affecting fisheries. Sand mining and mangrove forestry have also contributed to habitat loss (SPREP 1994b).

Coastal management in the FSM is uncoordinated because of the isolated nature of the islands, fragmentation of decision-making powers among tribal chiefs, and the conduct of few comprehensive planning efforts by the state governments. The management plans and legal frameworks for coastal management that do exist were set up during the period of U.S. influence and thus are poorly suited to the FSM's traditional land tenure system.

Evolution of Government Responses

Coastal resources management in the FSM has been undertaken largely as a result of aid from external agencies, primarily from the U.S. government during administration of the Trust Territory of the Pacific Islands in the 1980s. During this period, the FSM was eligible for inclusion in the U.S. federal Coastal Zone Management Program, but none of the states chose to participate.

In 1986, the U.S. Army Corps of Engineers produced coastal atlases and resource inventories for a number of major urban centers on the islands. In addition, draft coastal management plans were created for Pohnpei and other states. These activities stimulated interest in coastal resources management that led to CZM planning in two of the four states (Dahl 1995). Currently, coastal management programs are operating in Yap and Kosrae. Both programs emphasize bottom-up local management of coastal resources under the aegis of state CZM plans.

Kosrae's state program centers on coastal resources management legislation passed in 1992. Administered by the state's Development Review Commission, the program issues permits for coastal development, sets harvesting regulations and environmental quality standards, and ensures that projects include environmental impact assessments. Proposed activities include creation of land and water use plans, more comprehensive regulation of activities, and creation of performance standards (URI 1996).

Yap's state program is in the process of being implemented based on a coastal resources management plan drafted in 1994. The Marine Resources Office of Planning and Budget and the state Environment Protection Agency are the chief implementing agencies. The program will address tourism development, habitat loss, water quality, and fisheries issues through proposed land and water use plans, coastal permitting, and education programs (URI 1996).

At the national level, the federal government is working to implement the National Environmental Management Strategy, which was formulated in 1993 with technical assistance from the South Pacific Regional Environment Programme (SPREP) and the World Conservation Union (IUCN) and funding from the Asian Development Bank (ADB). Within it are a set of eighteen strategies and thirty-nine programs, many of which relate to marine and coastal management. Chief objectives include integration of economic growth with environmental considerations, improvement of environmental awareness and education, and management of natural resources (SPREP, IUCN, and ADB 1993).

Assessment of Government Responses

As with the other small-island developing states in the Pacific, the demand for development and economic growth in the Federated States of Micronesia is developing faster than is environmental management. Also looming large on the horizon is the termination of U.S. aid following the country's adoption of free association status with the United States. Still, the FSM has made strides to create ICM programs in its states and to develop the National Environmental Management Strategy (NEMS).

Because of the islands' isolation and the difficulties of integrating western styles of coastal management with more traditional indigenous forms of management, integration among the federal, state, and local governments has been problematic and largely unsuccessful. Evaluation of more comprehensive, legislatively bound coastal programs will enhance communication and cooperation among levels of government, but there remains the problem of monitoring and enforcing CZM policies and plans and building capacity at the local level.

Spatial integration has been moderately successful in the FSM. Coastal and ocean management are administered by the same organizational unit and operate with similar policy mandates and program priorities. However, only the management of offshore fisheries in the EEZ necessitates national control.

Integration of sectors should improve with implementation of the programs set out in the National Environmental Management Strategy. The state programs of both Kosrae and Yap address a host of coastal and ocean concerns, such as habitat protection, water quality, management of development, management of coastal fisheries, and public access.

Integration of science and policy has been greatly enhanced by the FSM's affiliation with SPREP and other international organizations. The United States still provides technical assistance, such as consultation with the U.S. Army Corps of Engineers, on coastal matters.

Capacity building in the FSM has been aided by the University of Hawaii's Sea Grant Program, which employs a Sea Grant extension agent who is charged with working with states to formulate ICM plans. The College of Micronesia has started a marine science program to enhance in-country expertise in marine issues occurring in the waters off the FSM. In addition, coastal managers have received specialized training at overseas institutions and new staff members with appropriate qualifications and training have been hired. Participation with regional bodies like SPREP and international bodies has enhanced both the financial and technical capacity of the FSM to implement new coastal programs (Edward 1996).

Sources Consulted

CIA (Central Intelligence Agency). 1995. *The World Factbook.* Washington, D.C.: Government Printing Office.

Dahl, C. 1995. Micronesia. In *Coastal Management in the Asia Pacific Region: Issues and Approaches*, ed. K. Hotta and I. Dutton, 337–358. Tokyo: Japan International Marine Science and Technology Federation.

———. 1996. Program Assistant, Sea Grant Extension Pacific Program, Honolulu, Hawaii, United States. Response to ICM cross-national survey. Newark: University of Delaware, Center for the Study of Marine Policy.

———. In press. Integrated coastal resources management and community participation in a small island setting. *Ocean & Coastal Management.*

Edward, A. 1996. Sea Grant Office, Kolonia, Pohnpei, Federated States of Micronesia. Response to ICM cross-national survey. Newark: University of Delaware, Center for the Study of Marine Policy.

The Europa World Year Book. 1995. 36th ed. London: Europa.

SPREP (South Pacific Regional Environment Programme). 1994b. *Coastal Protection in the Pacific Islands: Current Trends and Future Prospects.* Apia, Western Samoa: South Pacific Regional Environment Programme.

SPREP, IUCN, and ADB (South Pacific Regional Environment Programme, World

Conservation Union, and Asian Development Bank). 1993. *National Environmental Management Strategy: Federated States of Micronesia.* Apia, Western Samoa: South Pacific Regional Environment Programme.

URI (University of Rhode Island, Coastal Resources Center). 1996. Database of integrated coastal management efforts worldwide. Internet: http://brooktrout.gso.uri.edu/.

The United Republic of Tanzania

Background: General Country Context

The United Republic of Tanzania consists of Tanganyika, on the eastern coast of Africa, and the nearby island of Zanzibar. Bordered by Kenya and Uganda to the north; Rwanda, Burundi, and Zaire to the west; Zambia, Malawi, and Mozambique to the south; and the Indian Ocean to the east, Tanzania has 1,424 kilometers (885 miles) of coastline. The country has roughly 945,090 square kilometers (364,900 square miles) of total land area, with flat plains in the coastal area, a central plateau, and highlands in the north and south of the African mainland. The coastal region supports a diversity of ecological systems such as coral reefs, tropical mangrove forests, and sandy beaches. Tanzania claims a 200-nautical-mile Exclusive Economic Zone and a 12-nautical-mile territorial sea.

Tanzania is one of the poorest countries in the world, with a developing economy heavily dependent on agriculture. However, its arid terrain limits agricultural production to only 5 percent of the total land area. In 1994, Tanzania had a GDP of $21 billion (CIA 1995). Major cash crops on the mainland are coffee, tea, cotton, and tobacco. Crops on Zanzibar are mainly food items such as corn, wheat, and bananas. The estimated per capita GNP in 1994 was $140 (World Bank 1996).

Ninety-nine percent of the 28 million inhabitants of Tanzania are native Africans; 45 percent are Christian, 35 percent are Muslim, and 20 percent hold indigenous beliefs. Formerly separate political entities, Zanzibar and Tanganyika were joined in 1964 as an independent united republic. Provisions of the 1977 constitution assigned legislative power to the unicameral National Assembly, which is made up of elected and nominated members. Executive power is held by the president (elected every five years), who appoints two vice-presidents: one for Zanzibar and the other to serve as prime minister of the union (*Europa* 1995, 412–414). Below the national government are twenty-five administrative regions, including two in Pemba and three in Zanzibar, although the national government retains most of the power.

Description of Coastal and Ocean Problems

Specific coastal and ocean problems in East Africa stem from broader problems of poverty and rapid population growth. The countries lack the resources, tools, and equipment to manage resources effectively (SAREC 1994, 5). Even more crucial, they lack the requisite scientific knowledge, technological capacity, and management skills to be able to explore or exploit resources (Annex 1994).

A recent publication by the World Bank titled *Africa: A Framework for Integrated Coastal Zone Management* lists destructive fishing methods and associated habitat degradation, eutrophication and siltation of coastal waters, and marine oil pollution from tanker traffic and ballast discharge as the main coastal and marine issues in East Africa (World Bank 1995a, 23). In Tanzania specifically, the main coastal and ocean problems relate to habitat degradation as a result of improper mining, forestry, and fishing efforts.

With respect to coral reefs, deforestation and poor cultivation methods have led to siltation that has affected fishing grounds. Sand mining and construction of dams on rivers have also contributed to siltation in coastal waters. In addition, these dams and mining projects have altered the overall sand budget of the coast, causing erosion of beaches and flooding of wetland areas.

The demand for aggregate and cement substitutes in Tanzania has led to extensive mining of live coral, on the order of 5,500 cubic meters (7,000 cubic yards) per year, for use in construction projects ranging from traditional houses to hotels and roads. Live coral mining has been identified as one of the greatest threats to continued productivity of coastal waters through the destruction of reef habitat (Dulvy and Darwell 1995).

Runoff from agriculture, untreated industrial sewage, and industrial wastes continues to pollute coastal waters off Tanzania, and bioaccumulation of these substances in fisheries remains a significant problem. Other sources of pollution in coastal waters include high-pH discharges from textile mills, discharge of oil and tar from shoreline industries, and discharge of oil and ballast water by ocean-going vessels offloading oil at port facilities near Dar es Salaam (Msangi 1991, 230).

Coastal fisheries are perhaps the most affected by Tanzania's coastal and ocean problems. In addition to the loss of habitat associated with water pollution, siltation, and coral mining, overfishing and unsound fishing practices have left many coastal stocks overexploited. Dynamite blast fishing continues to be a widespread problem in East Africa, along with other unsustainable fishing practices. Little in the way of mariculture has been attempted in East Africa, although there has been some success with seaweed culture in Zanzibar (Ngoile and Linden, in press).

Evolution of Government Responses

Although there is no overarching national ICM program in place in Tanzania, there has been government activity in ocean and coastal management in selected local areas. This work is funded by international lending organizations such as the World Bank and bilateral mechanisms. Activities of the Swedish Agency for Research Cooperation with Developing Countries (SAREC) and the Swedish International Development Authority (SIDA) in Tanzanian coastal and ocean areas (ongoing since the early 1990s) consist mainly of profiling ecological resources and generating resource databases in partnership with other international entities such as the World Wide Fund for Nature (WWF), the U.S. Agency for International Development (USAID) and the University of Rhode Island (URI), the World Bank, the World Conservation Union (IUCN), and others (World Bank 1995a).

With 150 kilometers (90 miles) of coastline in the northern section of mainland Tanzania, the region of Tanga has set up the country's most sophisticated management plan to date. The Tanga Regional Authority, with technical assistance from IUCN, has set up the Tanga Coastal Zone Conservation and Development Programme. Short-term objectives include working with regional resource managers, planning authorities, and community leaders to create an ICM plan and empowering coastal communities to restore degraded mangrove and coral reef environments. Current activities include strengthening government institutions, training resource managers, increasing awareness of the need for coordination of management between government and the community, and engaging in community development (Tanzania 1996). Other initiatives related to ocean and coastal management in Tanzania include community-based management in Manai Bay on Zanzibar and development of ICM for Kunduchi in Dar es Salaam (Ngoile and Linden, in press).

With respect to ocean management in Tanzania, there exist several pieces of legislation related to sea utilization. The Merchant Shipping Act (1967) provides for the control, regulation, and orderly development of merchant shipping (Msangi 1991). Part IX of this act relates to pollution of the sea by oil and sets penalties for discharge of oil within 100 miles of the Tanzanian coast. The Fisheries Act of 1973 and subsequent amendments provide for the regulation and conservation of fisheries. Regulations promulgated in connection with the act provide for licensing and regulation of boats and gear and impose penalties for use of explosives and poisons. The act also provides for the establishment of marine parks, sanctuaries, and reserves. Two parks have been established: one off Kunduchi Beach in Dar es Salaam, to preserve productive coral fishing grounds, and one on Mafia Island, in part to control the mining of live coral. Legislation also exists to regulate mining and petroleum exploration and exploitation in the seabed and subsoil of the continental shelf offshore from Tanzania.

Assessment of Government Responses

Although it is still in an early stage, ICM has taken hold in Tanzania. As more technical expertise and international support are channeled to the nation, more coastal and ocean initiatives are likely. However, the lack of financial resources undermines Tanzania's capacity to develop a countrywide ICM program. Pilot projects such as that of Tanga, though, open the door for initiatives in other regions of the country.

Essential to the success of ICM programs in Tanzania, as outlined in the Tanga pilot program, is the building of public awareness of and support for the initiatives. In the case of coral mining on Mafia Island, development of alternative income sources and alternative construction materials has served to limit collection of live coral by coral miners, and the development of the marine park on Mafia Island was a direct attempt to further limit mining (Dulvy and Darwell 1995, 588–590).

In terms of capacity building in ICM, Tanzania hosted the first in a series of workshops to promote regional progress in integration of coastal management activities and sustainable development in East Africa. The "Workshop on Integrated Coastal Zone Management in Eastern Africa Including the Island States," held in Arusha in 1993, produced the Arusha Resolution, emphasizing policies, programs, sectoral integration, and agency integration for achieving ICM and sustainable development (World Bank 1995a, 119). The workshop was fundamental to capacity building in Africa, bringing together heads of state with international researchers and experts in the fields of coastal and ocean management (SAREC 1994, 10). Tanzania's role was especially important, as the country was charged with coordinating activities in the intersessional period leading up to a follow-up 1996 workshop. Tanzania also participates in a number of East African regional programs that build capacity, such as the East African Action Plan (through UNEP's Regional Seas Programme), the Marine Science Program (through SAREC), and others.

Capacity building at the local level, using the Tanga program as an example, focuses on increasing community awareness of coastal issues, fostering communication between communities and government, and providing alternative economic incentives to local resource users through trial projects (Tanzania 1996).

Sources Consulted

Annex. 1994. An outline on the implementation of strategy and programme of action adopted by regional leadership. Seminar on marine and ocean affairs in Africa, March 28–April 2, Addis Ababa, Ethiopia.

CIA (Central Intelligence Agency). 1995. *The World Factbook*. Washington, D.C.: Government Printing Office.

Dulvy, N., and W. R. T. Darwell. 1996. Tanzania, Mafia Island: Participatory control of coral mining. In *Coastal Zone Management Handbook*, ed. J. R. Clark, 587–590. New York: Lewis.

The Europa World Year Book, 1995. 36th ed. London: Europa.

Msangi, J. P. 1991. The integrated utilization of the sea off the coast of Tanzania. In *The Development of Integrated Sea-Use Management*, ed. H. D. Smith and A. Vallega, 230–237. London: Routledge.

Ngoile, M. A. K. 1996. Director, Institute of Marine Science, UDSM, Dar es Salaam, Tanzania; Director and Program Coordinator, Marine Affairs, World Conservation

Union, Gland, Switzerland. Response to ICM cross-national survey. Newark: University of Delaware, Center for the Study of Marine Policy.

Ngoile, M. A. K., and O. Linden. In press. Lessons learned from eastern Africa: The development policy on ICZM at national and regional levels. *Ocean & Coastal Management.*

SAREC (Swedish Agency for Research Cooperation with Developing Countries). Marine Science Program. 1994. *Technical Recommendations of the Workshop on Integrated Coastal Zone Management in Eastern Africa Including the Island States.* April 21–22. Arusha, Tanzania.

Tanzania. Region of Tanga. 1996. *Tanga Coastal Zone Conservation and Development Programme.* Pamphlet. Tanzania: Region of Tanga.

World Bank. 1995a. *Africa: A Framework for Integrated Coastal Zone Management.* Washington, D.C.: World Bank, Land, Water, and Natural Habitats Division, African Environmentally Sustainable Development Division.

Republic of India

Background: General Country Context

The Republic of India forms a cone-shaped subcontinent between the Arabian Sea on the west and the Bay of Bengal on the east, bordering, to the north, China, Bhutan, Nepal, Pakistan, Bangladesh, and Myanmar. About 80 kilometers (50 miles) off the southern tip of India is the island country of Sri Lanka. Two tropical island groups, the Andaman Islands and the Nicobar Islands, situated to the east of the mainland between the Bay of Bengal and the Andaman Sea, are territories of India. The climate varies from temperate in the north to tropical in the south, with an average summer temperature of 27°C and heavy rain during the summer monsoon season in June and July (*Europa* 1995). About 80 percent of the population of 936 million is Hindu; 14 percent is Muslim; 2.4 percent is Christian; 2 percent is Sikh; 0.7 percent Buddhist; and 0.9 percent follow other religions. The population growth rate was 1.77 percent in 1995 (CIA 1995). India's coastline is about 7,000 kilometers (4,000 miles) long. The country claimed a 12-nautical-mile territorial sea in 1967; a 200-nautical-mile Exclusive Economic Zone in 1976; the continental shelf to 200 nautical miles or to the edge of the continental margin in 1976; and a 24-nautical-mile contiguous zone in 1976 (CIA 1995; Churchill and Lowe 1988).

From 1985 to 1993, India enjoyed positive economic growth at an annual average rate of 3.0 percent. However, over the same period, the population grew by an annual average rate of 2.1 percent, leaving about 40 percent of the population in poverty. In 1994, India's GDP was $1.25 trillion, with a GDP growth rate of 5 percent (CIA 1995). Agriculture contributed 30.3 percent of the GDP in 1993–1994, employing about 66 percent of the labor force. India's agriculture is characterized by highly labor-intensive crops such as cotton, tea, rice, cashew nuts, and coffee. Industry provided 28.2 percent of the GDP in 1993–1994, employing 19.4 percent of the labor force. In terms of industrial output, India is one of the twelve leading industrial nations in the world, with an annual average growth rate of 6.4 percent during 1980–1992. The government of Narasimha Rao, who came to power in June 1991, launched the country's Eighth Five-Year Plan (1992–1997), in an attempt to solve the severe economic problems of enormous foreign debt, high inflation, and extreme shortage of foreign exchange reserves. By early 1994, this radical economic reform had brought about many positive changes, but the country still faces the challenge of alleviating the desperate poverty of millions of citizens and reducing the adverse environmental effects of the expanding population (*Europa* 1995). In 1994, India's per capita GNP was $320 (World Bank 1996).

After long-term peaceful protests against British colonial rule, India became independent on August 15, 1947, with the British monarch as the head of state represented by an appointed governor-general. On January 26, 1950, India became a republic under a new constitution. India is a constitutional federal republic that devolves some autonomous power to subnational levels of government. Legislative power is vested in the Parliament, which consists of the president and two Houses: the Council of States (Rajya Sabha), with members directly elected by the State Assemblies and nominated by the president for six-year terms, and the House of the People (Lok Sabha), with members directly elected by the people. The president, as chief of state, performs primarily ceremonial functions, and is elected for a five-year term by both houses of Parliament and the State Assemblies. Actual executive power is vested in the Council of Ministers. The Council of Ministers is headed by the prime minister, who is head of government and appoints council members. India consists of twenty-five self-governing states, each with a governor, a legislature, and a Council of Ministers headed by a chief minister. Seven union territories administered by lieutenant-governors, all of whom are appointed by the president (*Europa* 1995).

Description of Coastal and Ocean Problems

India's coastline is characterized by three different types of biophysical features. The western coast of the Indian subcontinent consists of a narrow coastal strip with rocky cliffs; thus it experiences very little flooding and sedimentation. But a few areas, as in the state of Gujarat on the Gulf of Kutch (Gulf of Kachchh), located in northwestern India, are marked by low, wide intertidal mudflats dominated by mangrove forests. The eastern coast is characterized primarily by low, wide, and flat areas. Thus, urbanization, port development, and major coastal economic activities have taken place more rapidly in this coastal zone. Finally, the two tropical island groups between the Bay of Bengal and the Andaman Sea, the Andaman and Nicobar Islands, are clad with dense forests from their shores to their hills. Extensive coral reef communities and coastal vegetation are well developed around these islands (Nayak et al. 1992).

India's huge population and rapid growth have overstrained coastal resources and coastal ecosystems. Coastal urbanization and rapid industrialization along the coastline have given rise to coastal environmental concerns. In particular, intensive development in the western Ghalis coastal area, such as fertilizer plants, nuclear power plants, refineries, and steel manufacturers, have threatened the area's ecology (Murthy 1996). The Central Water Power Research Station (CWPRS) undertakes research on coastal engineering and related research work in India's coastal zone.

The total area of mangrove forest in India has been estimated at 700,000 hectares (2 million acres), which is about 7 percent of the world's mangroves. More than 80 percent of India's mangrove forests exist on the eastern coast, with less than 20 percent of the mangroves on the western coast (Untawale 1993). Diverse human economic activities, such as gathering of fuelwood and fodder, have destroyed mangrove communities on the Gulf of Kutch along India's northwestern coast and the state of Karnataka.

Karnataka's coastal area has become increasingly industrialized and urbanized because of its favorable conditions for harbors and ports. Expansion of large-scale industries in this region, along with the new port development at Mangalore, inevitably caused destruction of coastal wetland ecosystems. Encroachment of agriculture into vast mangrove communities is also expanding rapidly as mangrove habitats are converted into paddy fields or coconut plantations. In addition, numerous mangrove forests have been cut down by coastal residents for firewood and building materials. One of the major factors leading to the destruction of mangrove forests is the public's attitude towards mangrove trees. Many people consider mangrove habitat as useless areas, and treat them as waste dumping sites or sewage treatment facilties (Elkington 1996).

Mangrove forests within tidal areas are important to India's coastal fisheries and mariculture farming. Roughly 60 percent of the country's coastal fish species depend on mangrove estuarine complexes (Untawale 1993). In India, mariculture is conducted according to traditional methods, as in brackish water fish farming. This traditional fish farming has developed along estuaries fringed with mangrove trees (Untawale 1993). Accordingly, maintenance of mangrove areas in India is critical to coastal fisheries production.

Evolution of Government Responses

Throughout India's history, coastal and ocean areas have played an important role in maintenance of the economy. For example, India has been among the top fifteen shipping countries in the world (Saigal and Fuse 1994). Nevertheless, there has been a lack of comprehensive and coherent coastal zone management in the country.

At the national level, India's National Institute of Oceanography (NIO), created in 1966, conducts multidisciplinary research in physical, chemical, biological, and geological oceanography (Saigal and Fuse 1994). In 1971, the national government established the Central Beach Erosion Board to coordinate the collection of data relating to coastal processes, to guide general coastal investigation and research, to design techniques to combat beach erosion, and to train coastal engineers (Nayak et al. 1992).

In 1974, The Water Prevention and Control of Pollution Act mandated creation of the Central and State Water Pollution Boards to establish standards for water quality and for waste effluent released into waters (Murthy 1996). This law controls and regulates the country's water quality, but it is based on a sectoral approach to coastal area management.

Because of the inefficiency and lack of coordination that resulted from a large number of ministries and departments dealing with coastal and ocean affairs on a sectoral basis, a new institutional structure was thought necessary to centralize responsibility for all matters relating to coastal and ocean affairs. To achieve policy coordination, decrease duplication and conflict, and increase the effectiveness of task performance, the Department of Ocean Development (DOD) was created in 1981 (Saigal and Fuse 1994).

The DOD was designed to coordinate regulatory measures relating to all coastal and ocean matters and to have authority over the Ocean Commission, the Pan-Indian Science Association, the Ocean Science and Technology Agency, and India's Antarctic issues. The centerpiece of the new department's creation was a document titled "Ocean Policy Statement," presented to Parliament in 1982. This statement contains a number of important policy guidelines, but its central themes are coordinated and centralized marine development in the context of sensitive marine environments; use of appropriate technologies and knowledge; building of effective management systems; and conservation of the marine environment. The DOD was required to set up a centralized data system with a mechanism for collecting, analyzing, and disseminating information (Saigal and Fuse 1994).

As discussed below, India has begun the process of formulating a national coastal zone management policy in conjunction with its nine coastal states. Furthermore, the Central Beach Erosion Board has proposed the establishment of a Coastal Zone Management Authority through a high-level policy-making body such as the DOD (Nayak et al. 1992).

India recognized the importance of mangrove forest conservation as the population of royal tigers dwindled in the Gangetic Sunderbans (Untawale 1993). The Wasteland Development Board and the Forest Department are responsible for mangrove forest management at the national level. The mangrove forests of the Gangetic Sunderbans were designated a mangrove reserve area as a result of the

Tiger Project. Following this preservation movement, mangrove forests in the state of Goa were reserved. Similarly, some of the Andaman and Nicobar Islands have been declared biosphere reserves, supported by the World Wildlife Fund for Nature (Untawale and Wafar 1990).

Assessment of Government Responses

The government of India's prompt response to the dynamic changes in international ocean affairs of the 1970s and early 1980s (e.g., the Law of the Sea negotiations) was the creation of the DOD in 1981. Since that time, the DOD has produced significant results: drafting of the "Ocean Policy Statement" in 1982; acceptance as a consultative party member of the Antarctic Treaty in 1983; acceptance as a pioneering investor in deep-sea exploration in 1982; and establishment of an information networking system (Saigal and Fuse 1994). As demonstrated by these achievements, India had traditionally focused primarily on the ocean side of the coastal zone rather than on the nearshore and land sides, except for mangrove-related issues. However, in 1991 a new coastal zoning law was enacted that established a series of zones of differing sensitivity immediately adjacent to the shoreline. India's coastal states were mandated to prepare coastal management plans consistent with the coastal zoning scheme contained in the legislation. Regulations guiding state coastal program development were subsequently prepared by the DOD and, in 1996, several states began the preparation of coastal mangement programs.

India does not yet have a comprehensive, integrated coastal zone management legislation or policy. B. U. Nayak and colleagues suggest that a Coastal Zone Management Authority (CZMA) be established as a central agency in each of the maritime states to ensure proper interaction regarding coastal zone management among various government agencies, voluntary organizations, developers, and the public (Nayak et al. 1992). Under the CZMA, state governments would create coastal zone management committees for each coastal district to hold meetings, seminars, workshops, and public hearings and to educate planners, administrators, and coastal residents.

Moreover, the proposed CZMA would take over all matters pertaining to coastal zone management, including functions of the existing Central Beach Erosion Board, the Anti-Sea Erosion Committees in the coastal states, and the central and state pollution control boards. Nayak and colleagues strongly recommend that district-level coastal zone management committees be given the necessary authority and statutory power to plan and implement coastal development programs in each coastal district in accordance with the policies of the national and state CZMAs (Nayak et al. 1992).

In terms of capacity building in ICM, India has been collecting and developing a large volume of basic scientific and technological ocean and coastal data through the DOD and the NIO since their early days. Particular attention has been focused on multidisciplinary conservation and management of mangrove forests.

The "Symposium on the Significance of Mangroves," held at the Maharashtra Association for the Cultivation of Science Research Institute in March 1990, resulted in the following recommendations: (1) promotion of awareness among the public and decision makers of the need to preserve mangrove forests, (2) establishment of programs encouraging research and training in conservation, (3) establishment of integrated multidisciplinary programs in basic research, and (4) prohibition of activities that damage mangrove forests (Agate et al. 1990). However, the country faces the challenge of creating an effective CZM program and getting political leaders interested in the ICM concept. India has difficulties in implementing the ICM concept of coastal management because of the lack of government conceptualization of the ICM strategy (Murthy 1996).

Sources Consulted

Agate, A. D., S. D. Bonde, and K. P. N. Kumaran, eds. 1991. *Proceedings of the Symposium on the Significance of Mangroves.* March 16, Pune, Maharashtra, India. Pune, Maharashtra, India: Maharashtra Association for the Cultivation of Science Research Institute.

Churchill, R. R., and A. V. Lowe. 1988. *The Law of the Sea.* Manchester, England: Manchester University Press.

CIA (Central Intelligence Agency). 1995. *The World Factbook.* Washington, D.C.: Government Printing Office.

Elkington, N. 1996. The effects of human impact on some mangrove ecosystems in Karnataka, India. Paper presented at international workshop, Integrated Coastal Management in Tropical Developing Countries: Lessons Learned from Successes and Failures, May 24–28, Xiamen, People's Republic of China.

The Europa World Year Book. 1995. 36th ed. London: Europa.

Murthy, R. 1996. Research Scientist, Consultant to the Indian Government, CCIW–Environment Canada. Response to ICM cross-national survey. Newark: University of Delaware, Center for the Study of Marine Policy.

Nayak, B. U., P. Chandramohan, and B. N. Desai. 1992. Planning and management of the coastal zone in India: A perspective. *Coastal Management* 20: 365–375.

Saigal, K., and T. Fuse. 1994. National case-studies: India and Japan. In *Ocean Governance: Sustainable Development of the Seas,* ed. P. B. Payoyo, 119–135. New York: United Nations University Press.

Untawale, A. G. 1993. Significance of mangroves for the coastal communities of India. In *Proceedings of the Seventh Pacific Science Inter-Congress Mangrove Session,* ed. July 1, Okinawa, Japan.

Untawale, A. G., and S. Wafar. 1990. Socio-economic significance of mangroves in India. In *Proceedings of the Symposium on the Significance of Mangroves,* ed. A. D. Agate et al. March 16, Pune, Maharashtra, India. Pune, Maharashtra, India: Maharashtra Association for the Cultivation of Science Research Institute.

Republic of the Philippines

Background: General Country Context

The Republic of the Philippines is an archipelagic island nation, situated in the eastern part of Southeast Asia and on the western Pacific Ocean. To the southwest of the Philippines is the island of Borneo and to the southeast, New Guinea. The main islands of the Philippines are Luzon, in the north, and Mindanao, in the south. More than 7,000 islands of the Visayas are positioned between these two main islands. The Philippines has a tropical marine climate characterized by abundant rainfall throughout the year—a northeastern monsoon season from November to April and a southwestern monsoon season from May to October (*Europa* 1995). Nearly 80 percent of the total population of 73 million lives in the coastal areas. About 92 percent is Christian (83 percent Roman Catholic, 9 percent Protestant); 5 percent is Muslim; and 3 percent is Buddhist or follows other religions. The population growth rate was 2.23 percent in 1995 (CIA 1995). The Philippines has an extremely long coastline, stretching for 36,289 kilometers (22,549 miles). The country claimed a 100-nautical-mile territorial sea in its irregularly polygonal-shaped coastal areas as defined in 1898 and a 285-nautical-mile territorial sea off its coastal areas in the South China Sea. The country also claimed the continental shelf to the depth of exploitation in 1968 and a 200-nautical-mile Exclusive Economic Zone in 1978 (CIA 1995; Churchill and Lowe 1988).

The Philippines' economy has been hampered by several factors: a burden of external debt and a public budget deficit, an average annual inflation rate of 9.2 percent during the past decade, natural disasters, and a severe shortage of electricity (*Europa* 1995). In 1994, the Philippines' GDP was $161.4 billion, with a GDP growth rate of 4.3 percent (CIA 1995). Agriculture contributed about 22 percent of the GDP in 1993, employing 45.8 percent of the labor force. The principal crops are rice, coconuts, corn, sugarcane, bananas, and pineapples (*Europa* 1995). Industry contributed 33 percent of the GDP in 1992, employing 15.5 percent of the labor force. Remittances from Filipino workers abroad ($2.1 billion in 1993) and revenue from tourism ($1.7 billion in 1992) are the country's most important sources of foreign exchange. In 1993, as a result of the Ramos administration's liberalization of laws regarding foreign investment and a dramatic improvement in the supply of electricity, the country's economic performance improved, with a GDP growth rate of 44.3 percent (*Europa* 1995). In 1994 per capita GNP in the Philippines was $950 (World Bank 1996)

Spain colonized the Philippines in the sixteenth century. On June 12, 1898, during the Spanish-American War, General Emilio Aguinaldo, leader of the Philippines revolutionary movement, declared the country's independence. Under the Treaty of Paris, Spain ceded the Philippines to the United States. The country's first constitution was ratified in May 1935, giving the Philippines internal self-government. On July 4, 1946, after Japan's occupation of the islands during World War II, the Philippines became an independent republic. The country was governed by a single dictator, Ferdinand Marcos, from the time he became president in 1965 until he was forced to leave the country in 1986. In accordance with the new constitution approved in February 1987, legislative power is vested in a bicameral Congress composed of the Senate, with 24 members generally elected for five-year terms, and the House of Representatives, with 204 members directly elected and another 50 appointed by the president from minority groups. The president is the chief of state, head of government, and commander in chief of the armed forces and appoints a cabinet and other officials. The president is elected by the people for a six-year term and is not eligible for reelection. The Philippines, a republic with a powerful centralized government, has seventy-two provinces and sixty-one municipalities. Local government is carried out by Barangays (citizens' assemblies), and autonomy is granted to any region the government of which is supported in a referendum (*Europa* 1995).

Description of Coastal and Ocean Problems

Fisheries are a traditionally important economic sector in the Philippines. The country produced about 2.6 million metric tons (3 million short tons) of fish catch in 1994. About 44,000 square kilometers (17,000 square miles) of coral reefs off the Philippines' long coastal contour and 450,000 hectares (1.1 million acres) of the original cover of mangrove forest contributed to abundant fish reproduction and productive nursery grounds (Isidro 1996). However, the Philippines' rich coastal natural resources have been heavily damaged or have gradually declined as a result of numerous economic activities since the 1970s.

Over the past decade, the Philippines' total fishery production has grown minimally, but municipal and coastal fishery production has declined from 1.045 billion metric tons (1.150 billion short tons) in 1985 to 1.009 billion metric tons (1.110 billion short tons) in 1994 (Isidro 1996). This phenomenon is basically the result of overfishing by a growing population of fishermen, which increased by 50 percent over the past decade (Isidro 1996), and destruction of traditional fish habitats. About 70 percent of the country's total coral reefs have been heavily damaged by the use of dynamite in fishing and by mining of coral for use in construction. Only 5 percent of the reefs remain undamaged, and they are in urgent need of protection from destructive human activities (Isidro 1996; Munoz 1994).

Tidal wetlands and mangrove forests are being altered or degraded through continuing expansion of human settlements. Seventy-five percent of the original 450,000 hectares of mangrove forests has been cut for timber, fuelwood, and charcoal making or for land, which is converted to other purposes (Alcala 1996).

Rapid population growth in the Philippines has exacerbated the mangrove deforestation problem. Industrial waste discharges and municipal sewage outfalls into
coastal waters cause red tides, deteriorating coastal fisheries (Philippines 1992).
In spite of all these adverse effects on fisheries, small-scale fishermen who face
economic hardship have undermined efforts to enforce fishery laws continuing to
fish with finemesh nets, dynamite, and sodium cyanide (Isidro 1996). As fishery
and environmental law enforcement methods are strengthened, it is important that
a new ICM strategy in the Philippines also take fishermen's socioeconomic conditions into consideration.

Evolution of Government Responses

National concern for CZM in the Philippines began with creation of the National
Environmental Protection Council (NEPC) in 1976. A Coastal Zone Management
Committee was established under the NEPC, later renamed Inter-Agency Committee on Coastal Zone Management. The committee formulated a long-term program for coastal zone management and provided policy recommendations for protection and optimum utilization of the coasts. A basic component of the program
was an inventory of existing coastal zone resources, such as forestry resources,
fisheries and other aquatic resources, and mineral resources, to serve as a rationale for a coastal resources management program (Magdaraos 1996).

Presidential Decree (PD) 704, commonly known as the Fisheries Act of 1975,
gave responsibilities for fisheries management to both national and municipal
governments. However, the government fisheries management measures undertaken under PD 704 were ineffective in promoting rational, sustainable fisheries
management. The government of the Philippines came to recognize the importance of community-based fisheries management given that community authority
and rights over the country's fisheries were traditional. In the early 1990s, the
government began to devolve control over fisheries to local communities under
the Local Government Code (LGC) of 1991. Through the LGC, the government
of the Philippines legally promotes community-based coastal resources management (Abregana et al. 1996; Pomeroy 1995).

This government awareness and a $162 million loan from the Asian Develop-
ment Bank and the Overseas Economic Cooperation Fund (OECF) of Japan
resulted in a comprehensive, integrated five-year fisheries management strategy,
known as the Fisheries Sector Program (FSP) of 1990–1994, under the Department of Agriculture (DA) (Isidro 1996). The FSP is composed of six complementary components that have been implemented by various agencies under the
Department of Agriculture, NGOs, and research and academic institutions. A Program Management Office (PMO) was established under the DA to oversee integration of the agencies in program implementation.

Six components of the FSP are (1) coastal resources management, (2) resource and ecological assessment, (3) a research and extension component, (4) a credit component, (5) an infrastructure component, and (6) fishery law enforcement. The centerpiece of the FSP is its coastal resources management component. It is to be implemented by the DA's regional offices and provincial fisheries management units of local governments for twelve selected bays and gulfs, with national-level assistance from the Bureau of Fisheries and Aquatic Resources and the PMO. The major objectives of this component are (1) regeneration, conservation, and sustainable management of coastal resources; (2) rehabilitation and protection of the coastal zone; and (3) alleviation of poverty among coastal fishermen through diversification of their sources of livelihood (Isidro 1996).

In the 1970s, the Natural Resources Management Center (NRMC) of the Department of Environment and Natural Resources attempted to create coral reef parks and reserves, employing a top-down approach. Neglecting to involve local stakeholders in the decision-making process, the NRMC had difficulty establishing even a single park or reserve (Alcala 1996). Learning from this government experience, Silliman University, with a written agreement with a local government in the central Visayas, designed Sumilon Island primarily as a research reserve. Research was conducted there over a twenty-year period (1974–1994). The Marine Conservation and Development Program (MCDP) of Silliman University was set up in 1984 to help correct coral reef destruction using community-based management programs. The program developed four other research sites: on Apo, Negros, Pamilacan, and Balicasag Islands in the Visayas (White 1989). As of 1995, after the importance of community and education involvement in marine resources management programs was recognized, twenty-six protected areas (reserves, parks, and seascapes) had been established through national laws and local government ordinances, led by various entities such as government agencies (national and local), universities, NGOs, and owners of small islands (Alcala 1996).

In 1986, the government of the Philippines began a four-year pilot study in the Lingayen Gulf area, on the northwestern coast of Luzon to develop an area-specific coastal management plan. Guidance and funding for the project were provided by the ASEAN-USAID Coastal Resources Management Project (CRMP) (McManus and Chua 1996). The Lingayen Gulf is an important fishing ground, with approximately 12,000 municipal fishermen and 80 commercial trawlers operating in the gulf waters (Quitos and Domingo 1993). Major management issues include depletion of fisheries, conflicts over resource use, extreme economic poverty on the part of coastal inhabitants, degradation of the coral reefs, pollution, and weak institutional arrangements (Philippines 1992; Quitos and Domingo 1993).

Regional coastal area management problems such as these led to the preparation of an integrated gulf area management plan called the Lingayen Gulf Coastal Area Management Plan (LGCAMP). The plan, which was finalized in 1992, contained twenty proposed projects, grouped in eight major categories (Philippines 1992):

1. A fisheries management plan
2. Rehabilitation and enhancement of critical habitats
3. Rehabilitation of linked habitats
4. Environmental quality management
5. Coastal zonation
6. Establishment of alternative livelihood for fishing families
7. Aquaculture development
8. Institutional development

The LGCAMP was designed to be implemented in two phases. During Phase I, an interim period for the first two implementing years, the National Economic and Development Authority's (NEDA's) Regional Office (NRO) was to serve as technical secretariat to oversee implementation of the program. In Phase II, during the program's final three years, a newly created Project Management Office was to take over the NRO's functions (Philippines 1992). However, until the Lingayen Gulf Coastal Area Management Commission was created in 1994, there was no formal implementation of the plan (McManus and Chua 1996).

Two legislative measures in 1993 and 1994 acted as turning points that prompted this hiatus. On March 25, 1993, President Fidel Ramos declared the Lingayen Gulf an environmentally critical zone through Presidential Proclamation 156. A year later, on April 20, 1994, the president issued Executive Order 171, creating the Lingayen Gulf Coastal Area Management Commission (Gulf Commission), composed of eight cabinet-level members, the chairman of the Regional Development Council (Region I), the governors of the provinces of Pangasinan and La Union, seventeen town mayors, and one city mayor (McManus and Chua 1996).

The Gulf Commission is directly under the Office of the President and serves as an advisory body to the president. After creating a Technical Secretariat, the commision incorporated the LGCAMP prepared by the NEDA into a new Ten-Year Master Implementation Plan for the Gulf (Gulf Plan), to operate from 1995 to 2004. The Gulf Plan aims to ensure environmental protection, alleviate poverty, encourage local autonomy, promote ecotourism, and maintain sustainable development of the Lingayen Gulf (McManus and Chua 1996).

Although implementation of the Gulf Plan had been delayed until creation of the Gulf Commission in 1994, the plan was successfully launched in 1995 by the Gulf Commission. Media-based support for environmental protection, reduction of illegal fishing, resource rehabilitation, and local community environmental education were all successful outputs of the Gulf Plan's first year of implementation. In further phases, the Gulf Commission will focus on institution building and resource-specific management (McManus and Chua 1996).

Assessment of Government Responses

The Philippines' initial strategy for coastal resources management was a fragmented, sectoral issue–oriented approach, implemented largely through the national government. The country experienced difficulty, inefficiency, and some degree of failure in the management of coastal resources, particularly regarding protection of fisheries and coral reefs. Lessons learned from Silliman University's successful MCDP, with its community-based coastal resources management, helped the government of the Philippines realize the need for ICM that takes into account the socioeconomic needs of local users of coastal resources.

As a result, institutional reforms were incorporated into the Fisheries Sector Program: decentralization of authority to achieve local fisheries management; municipally based enforcement of interagency fishery laws; community-based initiation of coastal resources management; NGO participation; and limited access to fisheries areas (Pomeroy 1995). The devolution of management of nearshore fisheries to municipalities in local fishing communities was the key element of reform under the Local Government Code (LGC) of 1991. Under the LGC, local government units (LGUs) may consolidate or coordinate themselves to manage coastal resources for commonly beneficial purposes (Pomeroy 1995). However, in order for LGUs to fully assume their staffing mandate under the code and actually operationalize devolution in law, a document titled *Legal Challenges for Local Management of Marine Resources: A Philippine Case Study* recommends clarification of the exact extent of jurisdiction of LGUs and the national government over coastal resources (Abregana et al. 1996).

On one hand, experience with ICM efforts in the Philippines during the past two decades has shown that community-based coastal resources management is an effective strategy in addressing local issues of depletion of open-access resources such as fisheries and coral reef habitats. On the other hand, when fisheries management becomes successful to the point that fisheries are enhanced, fishing efforts rapidly expand, creating an overfishing problem. For example, in Panguil Bay, on the northern coast of Mindanao, and in Carigara Bay, in the Eastern Visayas, fish catch rates increased by 50 to 100 percent between 1994 and 1995 because of successful coastal resources management within these areas. The result was a tendency to attract more fishing effort, contrary to the tenets of good fisheries management (Isidro 1996). One of the new management issues is how to reduce fishing effort when fisheries are enhanced.

When coastal resources management programs are not well implemented, or when their progress reaches a plateau, a strong interest in these issues on the part of high-level officials is crucial to the successful continuation of these programs.

For example, with the LGCAMP, President Ramos declared through executive orders his interest in the Lingayen Gulf as an environmentally critical area. This alleviated problems in implementing the LGCAMP and led to the prompt creation of the Gulf Commissions and the Ten-Year Master Implementation Plan for the Gulf.

The Philippines' ICM capacity building has been supported by international organizations such as the Asian Development Bank, the Organization for Economic Cooperation and Development and the ASEAN-USAID Coastal Resources and Management Project. The latter's technical and administrative support initiated the site-specific ICM study in the Lingayen Gulf. Implementation of ICM in the Lingayen Gulf could be further strengthened by the Gulf Commission's "learning by doing" philosophy. The first and foremost goal of ICM efforts in the Lingayen Gulf is to learn from the practice of ICM. Accordingly, the Gulf Commission's staff members and their LGU counterparts obtain ICM know-how through technical training in ICM methods and tools used in the LGCAMP. The LGUs then facilitate ICM environmental education in local communities (Isidro 1996).

The second phase of ICM efforts in the Lingayen Gulf involves efforts to increase institution-building capacity. At the national level, the Gulf Commission leads sectoral agencies toward institutional integration. At the local level, the Municipal Development and Planning Offices under the local government units alongside the Lingayen Gulf can encourage NGOs to help, just as the Gulf Commission works with organizations of local fishermen (Isidro 1996).

Moreover, the Fisheries Sector Program's three-year local community educational programs, conducted to obtain the support for the program, were well taught, explaining the concept of ICM in rational fisheries management to members of the local communities. The FSP also provided a series of planning seminars in the fisheries sector, facilitating community participation in the seminars in order to promote the mutual understanding between the fisheries sector and local communities.

Sources Consulted

Abregana, B., et al. 1996. *Legal Challenges for Local Management of Marine Resources: A Philippines Case Study*. Philippines: Environment and Resource Management Project.

Alcala, A. C. 1996. Coastal resource management as practiced in the Philippines. Paper presented at international workshop, Integrated Coastal Management in Tropical Developing Countries: Lessons Learned from Successes and Failures, May 24–28, Xiamen, People's Republic of China.

Churchill, R. R., and A. V. Lowe. 1988. *The Law of the Sea*. Manchester, England: Manchester University Press.

CIA (Central Intelligence Agency). 1995. *The World Fact Book*. Washington, D.C.: Government Printing Office.

The Europa World Year Book. 1995. 36th ed. London: Europa publications.

Gomez, E. D. 1996. President, Coastal Management Center, Metro-Manila, Philippines. Response to ICM cross-national survey. Newark: University of Delaware, Center for the Study of Marine Policy.

Isidro, A. 1996. Lessons in coastal resources management: The experience of the Philippines fisheries sector program. Paper presented at international workshop, Integrated Coastal Management in Tropical Developing Countries: Lessons Leaned from Successes and Failures, May 24–28, Xiamen, People's Republic of China.

Magdaraos, C. C. 1996. Head Executive Assistant, Department of Environment and Natural Resources, Quezon City, Metro-Manila, Philippines. Response to ICM cross-national survey. Newark: University of Delaware, Center for the Study of Marine Policy.

McManus L. T., and T. E. Chua. 1996. The Lingayen Gulf case: If we were to do it over again. Paper presented at international workshop, Integrated Coastal Management in Tropical Developing Countries: Lessons Learned from Successes and Failures, May 24–28, Xiamen, People's Republic of China.

Munoz, J. C. 1994. Planning for the ecologically sustainable harvest of marine resources: A Philippines experience. In *Coastal Zone Canada '94*, ed. R. G. Wells and P. J. Ricketts, 320–329. Halifax, Nova Scotia: Coastal Zone Canada Association.

Philippines. National Economic Development Authority. 1992. *The Lingayen Gulf Coastal Area Management Plan*. ICLARM Technical Report No. 30. Manila: International Centre for Living Aquatic Resources Management.

Pomeroy, R. S. 1995. Community-based and co-management institutions for sustainable coastal fisheries management in Southeast Asia. *Ocean & Coastal Management* 27 (3): 143–162.

Quitos, L. N., Jr., and F. D. Domingo. 1993. Coastal resource management planning: The Lingayen Gulf experience. In *World Coast Conference 1993: Proceedings* vols. 1 and 2. CZM-Centre Publication No. 4. The Hague: Ministry of Transport, Public Works, and Water Management, National Institute for Coastal Managment, Coastal Zone Managment Centre.

White, A. T. 1989. The Marine Conservation and Development Program of Silliman University as an example for Lingayen Gulf. In *Towards Sustainable Development of the Coastal Resources of Lingayen Gulf, Philippines,* ed. G. Silvestre et al., 119–123. ICLARM Conference Proceedings No. 1. Manila: International Centre for Living Aquatic Resources Management.

···

Cross-National Survey and Respondents

Respondents

Developed Countries

Canada

Larry Hildebrand, Head, Coastal Liaison, Environment Conservation Branch, Environment Canada, Atlantic Region, Dartmouth, Nova Scotia, Canada. (Survey completed together with Chris Morry, Department of Fisheries and Oceans, Ottawa.)

United States

Jack H. Archer, Professor, Environmental Sciences Program, University of Massachusetts, Boston, Massachusetts, United States.

United Kingdom

Hance D. Smith, Senior Lecturer, Department of Maritime Studies and International Transport, University of Wales, Cardiff, Wales.

Susan Gubbay, Coastal Management Specialist and NGO Advisor, Ross-on-Wye, England.

France

Alain Miossec, Professor, Institut de Géographie et d'Aménagement Régional de l'Université de Nantes (IGARUN), Nantes, France.

Netherlands

Leo de Vrees, Senior Project Manager, Coastal Zone Management Centre, The Hague, Netherlands.

Spain

Juan Luis Suarez de Vivero, Professor and Head of Department, Facultad de Geografía e Historia, Universidad de Sevilla, Sevilla, Spain.

Francisco Montoya, Head of Coastal Service, Ministry of Public Works, Transport, and Environment, Tarragona, Spain.

Japan

Hiroyuki Nakahara, Managing Director, Research Institute for Ocean Economics, Tokyo, Japan.

Republic of Korea

Ji Hyun Lee, Senior Researcher, Marine Policy Center, Korea Maritime Institute (KMI), Seoul, Korea.

Australia

Karen Anutha, Program Manager, Coastal and Marine Planning Division, Department of Environment and Land Management, Hobart, Tasmania, Australia.

Marcus Haward, Senior Lecturer, Department of Political Science, University of Tasmania, Hobart, Tasmania, Australia.

Robert Kay, Coastal Planning Coordinator, State Goverment of Western Australia, Perth, Western Australia, Australia.

New Zealand

Wayne Hastie, Resource Policy Manager, Wellington Regional Council, Wellington, New Zealand.

Middle Developing Countries

Mexico

Alejandro Yanez-Arancibia, Director Científíco del Programa EPOMEX Universidad Autónoma de Campeche, Campeche, Mexico. (Survey completed together with Ana Laura Lara Dominguez, Universidad Autónoma de Campeche).

Carlos Valdes-Casillas, Director, Research and Graduate Program, Instituto Tecnologico y de Educación Superior de Monterrey, Campus Guaymas, Guaymas, Sonora, Mexico.

Brazil

Renato Herz, Associate Professor, Instituto Oceanografíco, Universidade de São Paulo, São Paulo, Brazil.

Chile

Armando G. Sanchez Rodriguez, Secretario Técnico, Comisión Nacional Uso del Borde Costero, Subsecretaria de Marina, Ministerio de Defensa Nacional, Santiago, Chile.

Gonzalo A. Cid, Researcher, Centro EULA–Chile, Universidad de Concepción, Concepción, Chile. (Survey completed together with Professor Victor A. Gallardo, Researcher, EULA.)

Turkey

Erdal Ozhan, Professor, Department of Civil Engineering, Middle East Technical University, Chairman, MEDCOAST, Ankara, Turkey.

Thailand

Sirichai Roungrit, Environment Officer, Office of Environmental Policy and Planning, Bangkok, Thailand.

Surophol Sudara, Head, Department of Marine Science, Chulalongkorn University, Bangkok, Thailand.

Ampai Harakunarak, Center for the Study of Marine Policy, University of Delaware, Newark, Delaware, United States.

Barbados

No response.

Malaysia

Abdul Aziz Ibrahim, Director General, National Hydraulics Research Institute of Malaysia, Nahrim, Kuala Lumpur, Malaysia.

Datin Paduka Fatimah, Abdullah, Director, State Economics Planning Unit, Bangunan Sultan Ibrahim, Bahru, Malaysia.

Ir. Ooi Choon Ann, Director General, Coastal Engineering Division, Department of Irrigation and Drainage, Jalan Sultan Salahuddin, Kuala Lumpur, Malaysia.

Mohd Nizam Basiron, Senior Analyst, Malaysian Institute of Maritime Affairs (MIMA), Kuala Lumpur, Malaysia.

Fiji

Joeli Veitayaki, Lecturer, Ocean Resources Management Programme, University of the South Pacific, Suva, Fiji.

Sefa Nawadra, Project Coordinator, Coastal Management, Department of the Environment, Suva, Fiji.

Republic of South Africa

No response.

Developing Countries

Ecuador

Luis Arriaga, Country Advisor, Coastal Resources Center, University of Rhode Island, Programa de Manejo de Recursos Costeros, Guayaquil, Ecuador.

Stephen B. Olsen, Director, Coastal Resources Center; Consultant, CRC-USAID, Ecuador Pilot Program, University of Rhode Island, Narragansett, Rhode Island, United States.

Republic of Bulgaria

Konstantin Galabov, Director of CZM Office, Ministry of Regional Development and Construction, Sofia, Bulgaria.

Professor Jack H. Archer, Consultant to the Government of Bulgaria and World Bank, University of Massachusetts, Boston, Massachusetts, United States.

People's Republic of China

Ying Wang, Dean, School of Geo Sciences; Director, State Laboratory of Coast and Island Exploitation, Nanjing University, Nanjing, People's Republic of China.

Huming Yu, Former Deputy Director General, Department of Integrated Marine Management, State Oceanic Administration, People's Republic of China.

Republic of Indonesia

Triono Soendoro, Bureau Chief, Marine, Aerospace, Environment, Science, and Technology, BAPPENAS (National Development Planning Board), Jakarta, Indonesia.

Rokhmin Dahuri, Head of Coastal Zone Management, Bogor Agricultural University, Bogor, Indonesia.

Etty Agoes, Law Lecturer, Center for Archipelago, Law, and Development Studies, Bandung, Indonesia.

Islamic Republic of Pakistan

Shahid Amjad, Director General, National Institute of Oceanography, Karachi, Pakistan.

Syed Mazhar Haq, Former Professor and Director, Centre of Excellence in Marine Biology, University of Karachi, Karachi, Pakistan.

Republic of India

R. Krishnamoorthy, Project Scientist, Centre for Water Resources and Ocean Management, Anna University, Madras, India.

Raj Murthy, Research Scientist, Consultant to the Indian Government, CCIW–Environment Canada, Burlington, Ontario, Canada.

Republic of the Philippines

Clarissa C. Magdaraos, Head Executive Assistant, Department of Environment and Natural Resources, Quezon City, Metro-Manila, Philippines.

Edgardo D. Gomez, President, Coastal Management Center, Metro-Manila, Philippines.

Democratic Socialist Republic of Sri Lanka

R. A. D. B. Samaranayake, Manager, Coastal Resources Development, Coast Conservation Department, Maligawatte, Colombo, Sri Lanka.

Kem Lowry, Professor, Consultant to Sri Lanka coastal program, Department of Urban and Regional Planning, University of Hawaii, Honolulu, Hawaii, United States.

Federated States of Micronesia

Asher Edward, Sea Grant Office, Kolonia, Pohnpei, Federated States of Micronesia.

Christopher Dahl, Program Assistant, Sea Grant Extension Pacific Program, Honolulu, Hawaii, United States.

United Republic of Tanzania

Magnus Ngoile, Director and Program Coordinator, Marine Affairs, World Conservation Union, Gland, Switzerland, and former Director, Institute of Marine Science, University of Dar es Salaam, Dar es Salaam, United Republic of Tanzania.

Cross-National Survey

Center for the Study of Marine Policy • Graduate College of Marine Studies • University of Delaware
Newark, DE 19716, USA • Telephone: 1-302-831-8086 • Fax: 1-302-831-3668

RESPONDENT'S NAME: _____ COUNTRY: _____

POSITION: _____ ADDRESS: _____

MULTICOUNTRY SURVEY OF EFFORTS IN INTEGRATED COASTAL MANAGEMENT

INSTRUCTIONS:
This survey pertains to integrated coastal management (ICM) in your country. We define ICM as a dynamic process by which decisions are taken for use, development, and protection of coastal/marine areas. As no country can claim to have a fully integrated system of ICM, we consider, as a minimum definition of ICM, the management of multiple uses (two or more activities) in coastal/marine areas.

Please note that survey questions are of three types: close-ended questions in which we ask you to CHECK or RANK your response(s); FILL IN THE BLANK questions; and open-ended questions for further elaboration (PLEASE ELABORATE). If you are uncertain about the response to a particular question, please check I AM UNCERTAIN or OTHER as appropriate. Feel free to use additional paper if needed.

(Please note: we use "marine" and "ocean" interchangeably in the survey.)

PART 1. THE INITIATION OF INTEGRATED COASTAL MANAGEMENT (ICM) EFFORTS

1. Describe the origin of ICM in your country. Were there political, environmental, or other events that gave rise to ICM?

_____

_____

_____

_____

2. What gave rise to the need for ICM in your country?
   (PLEASE RANK THE RELEVANT REASONS FROM 1 TO 7 BELOW—WITH "1" BEING THE MOST IMPORTANT)
   _____ Desire to increase economic benefits flowing from the use of coastal/marine resources
   _____ Serious resource depletion problems
   _____ Increasing pollution of the coastal and ocean environment
   _____ Loss of or damage to productive coastal ecosystems (e.g., mangroves, coral reefs)
   _____ Perceived economic opportunities associated with new forms of development in the coastal zone
   _____ Other—please specify:

   _____ I am uncertain

SURVEY OF EFFORTS IN INTEGRATED COASTAL MANAGEMENT IN VARIOUS COUNTRIES

3. Was there a major catalyst(s) that facilitated the initiation of ICM in your country?

_____ An environmental crisis

_____ A proposal for new coastal/marine economic development project

_____ National-level government initiative

_____ State- (provincial-) or local-level initiative

_____ Initiative from nongovernmental organization

_____ Initiative from international organization

_____ Initiative from regional entity

_____ External funding

_____ Prescriptions of the Earth Summit (e.g., Chapter 17 of Agenda 21)

_____ Other—please specify:

_____ Uncertain

PART 2. DESCRIPTION OF THE ICM EFFORT(S) IN YOUR COUNTRY

4. Please indicate which is the PRIMARY level of government responsible for ICM in your country (PLEASE CHECK):

_____ Federal (or national)   _____ Other—please specify:

_____ State (or provincial)

_____ Local   _____ I am uncertain

5. Please provide a brief summary (FILL IN) of the major specific actions taken/processes started/ programs started regarding integrated coastal management in your country. (Please indicate approximate time frame of actions/processes/programs.)

Actions taken for <u>coastal area</u>   Approximate time
(coastal lands and nearshore waters)   frame (year)

By federal (national) level _____   _____

By state (provincial) level _____   _____

By local level _____   _____

Actions taken for <u>ocean waters</u>

By federal (national) level _____   _____

By state (provincial) level _____   _____

By local level _____   _____

(PLEASE ATTACH OR MAIL COPIES OF RELEVANT LAWS OR PROGRAM DESCRIPTIONS, IF EASILY AVAILABLE.)

SURVEY OF EFFORTS IN INTEGRATED COASTAL MANAGEMENT IN VARIOUS COUNTRIES

6. Below is a list of activities that ICM efforts may undertake. Please check the main types of activities that the ICM effort in your country has involved (PLEASE CHECK ALL THAT APPLY):

◆ Conduct of studies on:

_____ Physical character of the shoreline and of coastal processes

_____ Coastal/marine resources (beaches, wetlands, mangroves, estuaries, coral reefs, etc.)

_____ Socioeconomic studies of multiple coastal and ocean uses and their interactions

_____ Character and needs of coastal settlements

_____ Special perspectives of indigenous peoples and their traditional activities

_____ Economic potential of coastal/ ocean resources (aquaculture, offshore oil and gas, etc.)

_____ Policy options for governance of ICM

◆ _____ Preparation of coastal inventories atlases, GIS

◆ _____ Preparation of an ICM plan

◆ _____ Formal government adoption of ICM policies, goals, management measures

◆ _____ Enactment of a new coastal/ocean law

◆ _____ Implementation of an ICM plan (including staffing and organizational changes)

◆ _____ Approval of funds for the ICM program

◆ _____ Establishment of a monitoring and evaluation program

◆ _____ Other—please specify

7. Please specify below the legal boundaries of the area included in ICM in your country. In answering this question, please refer to domestic practice on ICM, not the international legal boundaries (i.e., we want to know what area is thought to be the ICM area by national-level authorities in your country) (PLEASE FILL IN):

Definition of the scope of the ICM area:

Landward boundary

_____ (Number of _____ mi/km inland)

_____ Not yet determined

_____ Varies according to use

_____ Other:

_____ I am uncertain

Seaward boundary

_____ (Number of _____ mi/km offshore)

_____ Not yet determined

_____ Varies according to use

_____ Other

_____ I am uncertain

SURVEY OF EFFORTS IN INTEGRATED COASTAL MANAGEMENT IN VARIOUS COUNTRIES

8. Please check which ICM activities listed below are part of your country's ICM effort
(PLEASE CHECK):

____ AREA PLANNING

    ____ Studies of coastal environments and their uses

    ____ Zoning of uses

    ____ Anticipation and planning of new uses

    ____ Regulation of coastal developments and their proximity to the shoreline

    ____ Public education to raise awareness about the value of coastal/marine areas

    ____ Provision of public access to coastal/marine areas

    ____ Other—please specify:

____ PROMOTION OF ECONOMIC DEVELOPMENT IN COASTAL/MARINE AREA

    ____ Encouragement of coastal-dependent enterprises (PLEASE CHECK):

        ____ Industrial fisheries

        ____ Artisanal fisheries

        ____ Mass tourism

        ____ Ecotourism

        ____ Aquaculture

        ____ Marine transportation

        ____ Port development

        ____ Marine recreation (boating, beaching, surfing, tourist submarines, etc.)

        ____ Other offshore minerals (sands, placer deposits, etc.)

        ____ Ocean research

        ____ Access to marine genetic resources

        ____ Other—please specify:

____ STEWARDSHIP OF RESOURCES

    ____ Conduct of environmental assessments

    ____ Assessments of relative risk

____ Establishment and enforcement of environmental standards

____ Protection and improvement of coastal water quality

    ____ Point sources

    ____ Nonpoint sources

____ Establishment and management of coastal/marine protected areas

____ Conservation and restoration of coastal/marine environments, e.g., mangroves, sea grass beds, other wetlands, coral reefs

____ Other—please specify

____ CONFLICT RESOLUTION

    ____ Studies of multiple uses and their interactions

    ____ Application of methods of conflict resolution

    ____ Mitigation for unavoidable adverse impacts

    ____ Other—please specify

____ PROTECTION OF PUBLIC SAFETY

    ____ Reduction of vulnerability to natural disasters and to globally induced changes (e.g., sea-level rise)

    ____ Regulation of development in high-risk areas such as through the establishment of "set-back lines"

    ____ Coastal defense measures (e.g., seawalls)

    ____ Creation of evacuation plans or other measures in case of coastal emergencies

    ____ Other—please specify

____ PROPRIETORSHIP OF PUBLIC SUBMERGED LANDS AND WATERS

    ____ Establishment of leases and fees for use of publicly held ocean resources and space

    ____ Establishment of joint venture to exploit nonrenewable resources (e.g., offshore oil)

    ____ Other—please specify

SURVEY OF EFFORTS IN INTEGRATED COASTAL MANAGEMENT IN VARIOUS COUNTRIES

9. Among the various functions that your ICM program performs, how important is the management of conflicts among multiple coastal and ocean uses? (PLEASE CIRCLE ONE):

⟵ Very Important — Moderately Important — Moderately Unimportant — Very Unimportant — Uncertain ⟶

10. Which are the most important coastal and ocean conflicts that your country faces?

_____

_____

_____

_____

PART 3. INSTITUTIONAL ASPECTS OF ICM

11. Were any new institutions or processes established in connection with the ICM effort?

_____ At the federal (national) level? (PLEASE DESCRIBE)

_____ At the state (provincial) level? (PLEASE DESCRIBE)

_____ At the local level? (PLEASE DESCRIBE)

_____ Other—please specify:

_____ I am uncertain

SURVEY OF EFFORTS IN INTEGRATED COASTAL MANAGEMENT IN VARIOUS COUNTRIES

12. Were any existing institutions strengthened (or altered) in connection with the ICM effort?

    ____ At the federal (national) level? (PLEASE DESCRIBE)

    ____ At the state (provincial) level? (PLEASE DESCRIBE)

    ____ At the local level? (PLEASE DESCRIBE)

    ____ OTHER—please specify:

    ____ I am uncertain

13. Is there one main institution (government agency) responsible for ICM in your country at the national level, or are several institutions (government agencies) involved?

    ____ One main institution involved at national level (PLEASE NAME THE INSTITUTION)
         (GO TO QUESTION 15)

    ____ Several institutions involved (PLEASE NAME THE MAJOR INSTITUTIONS)
         (GO TO QUESTION 14)

    ____ Other—please specify:

    ____ I am uncertain

SURVEY OF EFFORTS IN INTEGRATED COASTAL MANAGEMENT IN VARIOUS COUNTRIES

14. If several institutions are involved:
How would you characterize the relationship among these various institutions at the national level? (PLEASE CHECK ONE)

_____ Generally cooperative with one another

_____ Have little to do with one another

_____ Generally competitive with one another

_____ It varies, depending on the issue

_____ Other—please specify:

_____ I am uncertain

15. Has some type of coordinating mechanism been set up to coordinate the ICM activities? (PLEASE CHECK ALL THAT APPLY)

_____ An interagency or interministerial committee has been set up

_____ A special coordinating commission or committee has been set up

_____ One of the ministries (or agencies) has been named the "lead" agency to oversee the ICM process

_____ The prime minister's office (or president's office) is charged with ICM coordination

_____ There is no national coordinating mechanism

_____ Other—please specify:

_____ I am uncertain

SURVEY OF EFFORTS IN INTEGRATED COASTAL MANAGEMENT IN VARIOUS COUNTRIES

16. To what extent (if any) was the creation of a coordinating mechanism for ICM in your country related to the prescriptions on this subject of Chapter 17 of Agenda 21? (PLEASE CIRCLE ONE)

◄—— Highly Related —— Somewhat Related —— Little Related —— Not at all Related —— Uncertain ——►

17. Please characterize the relationship between <u>federal</u>- (national-) level institutions concerned with ICM and <u>state/provincial and local institutions</u> concerned with ICM (PLEASE CHECK ONE):

_____ Generally cooperative with one another

_____ Have little to do with one another

_____ Generally competitive with one another

_____ It varies, depending on the issue

_____ Other—please specify:

_____ I am uncertain

Please describe the mechanisms that exist (if any) to coordinate the actions of institutions at the federal (national) and subnational levels (state, provincial, local) regarding ICM:

PART 4. EVALUATION OF ICM EFFORT

18. Concerning your nation's programs in <u>coastal</u> management and <u>ocean</u> management (PLEASE CHECK):

	YES	NO	UNCERTAIN
◆ Are they administered by the same organizational unit?	_____	_____	_____
◆ Do they operate with similar policy mandates, program priorities, and resource levels?	_____	_____	_____
◆ Can it be said that both programs represent an effort at an integrated approach?	_____	_____	_____
◆ In your view, notwithstanding any organizational separations, are the programs adequately coordinated?	_____	_____	_____

SURVEY OF EFFORTS IN INTEGRATED COASTAL MANAGEMENT IN VARIOUS COUNTRIES

19. Please provide your assessment of the extent to which the ICM effort has been successful in integrating activities across jurisdictional boundaries, government agencies, and levels of government and between scientists and policy makers (PLEASE CHECK ONE):

Extent of <u>spatial</u> integration
(land and sea)

_____ Highly successful
_____ Moderately successful
_____ Moderately unsuccessful
_____ Highly unsuccessful
_____ Uncertain

Extent of interagency integration at the
<u>federal</u> (national) level

_____ Highly successful
_____ Moderately successful
_____ Moderately unsuccessful
_____ Highly unsuccessful
_____ Uncertain

Extent of interagency integration at the
<u>state</u> (provincial) level

_____ Highly successful
_____ Moderately successful
_____ Moderately unsuccessful
_____ Highly unsuccessful
_____ Uncertain

Extent of interagency integration at the
<u>local</u> level

_____ Highly successful
_____ Moderately successful
_____ Moderately unsuccessful
_____ Highly unsuccessful
_____ Uncertain

Extent of integration between <u>federal</u>
(national) <u>and subnational</u> levels of
government

_____ Highly successful
_____ Moderately successful
_____ Moderately unsuccessful
_____ Highly unsuccessful
_____ Uncertain

Extent of integration between <u>science and
policy</u> (i.e., to what extent is policy based
on scientific understanding from the
natural and the social sciences?)

_____ Highly successful
_____ Moderately successful
_____ Moderately unsuccessful
_____ Highly unsuccessful
_____ Uncertain

SURVEY OF EFFORTS IN INTEGRATED COASTAL MANAGEMENT IN VARIOUS COUNTRIES

20. Have there been particular problems regarding coastal and ocean management that your
    country has faced vis-à-vis other neighboring countries?

    (PLEASE EXPLAIN)

    (IF YES, HOW HAVE THESE PROBLEMS BEEN ADDRESSED?)

21. Thinking about your country's experience with ICM, are there any particular "successes" the ICM
    effort has achieved that might serve as useful examples for ICM managers in other countries?

    (PLEASE EXPLAIN)

22. Thinking about your country's experience with ICM, are there any particular "difficulties" (or
    obstacles) the ICM effort has faced that might be of interest to ICM managers in other countries?

    (PLEASE EXPLAIN)

PART 5. CAPACITY BUILDING IN ICM

23. Capacity building (CHECK ALL APPROACHES THAT HAVE BEEN USED):

    _____ Specialized in-country training of existing staff

    _____ Specialized training at overseas institutions

    _____ Participation in UNDP, UNEP, FAO, IOC, or other training programs

    _____ Hiring of new staff with appropriate qualifications/training

    _____ Establishment of new graduate programs in ICM at the university level

    _____ Other—please specify

SURVEY OF EFFORTS IN INTEGRATED COASTAL MANAGEMENT IN VARIOUS COUNTRIES

PART 6. OTHER COMMENTS

24. Please indicate the extent of participation of the following types of nongovernmental organizations in the development and implementation of your ICM program:

   ◆ Environment/conservation groups
   _____ Major involvement   _____ Minor involvement
   _____ No involvement   _____ Uncertain

   ◆ Industry/development-oriented groups
   _____ Major involvement   _____ Minor involvement
   _____ No involvement   _____ Uncertain

   ◆ Commercial/recreational fishing interests
   _____ Major involvement   _____ Minor involvement
   _____ No involvement   _____ Uncertain

   ◆ Local community organizations
   _____ Major involvement   _____ Minor involvement
   _____ No involvement   _____ Uncertain

   ◆ Academic institutions
   _____ Major involvement   _____ Minor involvement
   _____ No involvement   _____ Uncertain

   Other—please specify:

   ◆ I am uncertain

25. Please suggest the name(s) of a government official or officials (at the appropriate level, given where the primary responsibility for ICM lies) as a possible contact for us to get further information on ICM laws and programs. Please provide the address, phone, and fax, if available.

THANK YOU VERY MUCH FOR YOUR TIME AND COOPERATION.
PLEASE CHECK HERE ☐ IF YOU WOULD LIKE TO OBTAIN A SUMMARY OF THE SURVEY RESULTS.
PLEASE FAX COMPLETED SURVEY TO:
GREGORY W. FISK
CENTER FOR THE STUDY OF MARINE POLICY
UNIVERSITY OF DELAWARE
ROBINSON HALL 301
NEWARK, DELAWARE 19716, USA

FAX 1-302-831-3668
TELEPHONE 1-302-831-8086

SURVEY OF EFFORTS IN INTEGRATED COASTAL MANAGEMENT IN VARIOUS COUNTRIES

Glossary

Adaptive management. A type of management approach that is flexible and is modified from time to time on the basis of new data related to changing circumstances.

Agenda 21. The forty-chapter action plan emanating from the 1992 United Nations Conference on Environment and Development (UNCED) that provides guidance to nations on a wide range of matters related to environment and development. Chapter 17 of Agenda 21 addresses oceans and coasts.

Biological diversity. The genetic variety of faunal and floral species living in the biosphere. Biological diversity is critical to maintaining the biophere's life-sustaining systems.

Bottom-up approach to ICM. The formulation of an ICM program by starting first at the local community level, in contrast to the top-down approach.

Capacity building. Enhancement of the skills of people and the capacity of institutions in resources management through education and training.

Coastal hazards. Natural phenomena such as storms, hurricanes, flooding, and erosion that cause damage to property, resources, and populations in coastal regions.

Coastal zone. The area at the interface between land and sea, where the sea influences the land and vice versa. Coastal zone boundaries vary depending on biogeographical conditions, the mix of uses and problems present, and the legal system. In some cases, the coastal zone may extend from the top of the watershed (as the landward boundary) out to the edge of the nation's 200-nautical-mile exclusive economic zone (as the seaward boundary). In other cases, it may be more narrowly defined, encompassing smaller areas on both the land and the sea sides. Entire islands are often defined as coastal zones.

Coastal zone management. Term used initially in the United States in its

459

1972 Coastal Zone Management Act. The term was meant to encompass management of all uses in the coastal zone, but it does not necessarily emphasize integration among uses and policies. As implemented in the United States, CZM programs initially emphasized management of land, not ocean, uses in the coastal zone.

Co-management. A type of management whereby responsibility for resource management is shared between the government and resource users' groups.

Common heritage of mankind. A concept referring to natural resources considered by the community of nations to be held in common for the benefit of all humankind. For example, the 1982 United Nations Convention on the Law of the Sea placed the mineral resources of the deep seabed in this category.

Consensus building. The building of agreement regarding ICM decisions among government agencies, users' groups, and local communities through informed discussion, negotiation, and public participation.

Contiguous zone. A zone adjacent to a nation's territorial sea, not extending beyond 24 nautical miles of the baseline from which the breadth of the territorial sea is measured (usually between 12 and 24 nautical miles offshore). In this zone, a coastal nation may exercise the control necessary to prevent infringement of its custom, fiscal, immigration, or sanitary laws.

Continental shelf. The seabed and subsoil of the submerged areas that extend beyond a coastal nation's territorial sea throughout the natural prolongation of its land territory and to the outer edge of the continental margin, or to a distance of 200 nautical miles from the baseline from which the breadth of the territorial sea is measured.

Cross-national. Comparing or dealing with two or more nations.

Easement. A right, vested by an implied or express agreement, of an owner of land (or the public as a whole) to make lawful and beneficial use of land owned by another party.

Ecosystem. The linked system of interactive relationships among organisms and between organisms and their physical environment in a given geographical unit.

Ecotourism. Tourism focusing on environmental and cultural resources and usually based on a conservation theme.

Environmental impact assessment. A process whereby a detailed prediction is made of the effects of a proposed development project on the environment and natural resources. Such assessments generally include a consideration of options for reducing or mitigating adverse environmental effects and of alternative courses of action.

Exclusionary zones. Coastal area where building construction or other activities is prohibited because of the hazardous or sensitive nature of the area.

Exclusive economic zone. The maritime zone beyond and adjacent to the territorial sea but not exceeding 200 nautical miles from the baseline from which the territorial sea is measured. In the exclusive economic zone, the coastal nation has sovereign rights to explore and exploit, conserve, and manage the natural resources, whether living or nonliving, of the waters superjacent to the seabed and of the seabed and its subsoil.

Geographic information system (GIS). Computer-based tool to aid in the display and analysis of geographically based information.

Habitat. The place of residence of an animal species or community of species.

High seas. All parts of the sea that are not included in the exclusive economic zone, the territorial sea, or the internal waters of individual nations. On the high seas, nations have such freedoms as freedom of navigation, overflight, laying of submarine cables and pipelines, construction of artificial islands and other installations, fishing, and scientific research.

Integrated coastal management (ICM). A continuous and dynamic process by which decisions are made for the sustainable use, development, and protection of coastal and marine areas and resources. ICM acknowledges the interrelationships that exist among coastal and ocean uses and the environments they potentially affect and is designed to overcome the fragmentation inherent in the sectoral management approach. ICM is multipurpose oriented; it analyzes implications of development, conflicting uses, and interrelationships between physical processes and human activities; and it promotes linkages and harmonization among sectoral coastal and ocean activities.

Interdependence. A term describing the growing economic, environmental, and sociopolitical interrelationships among nations.

Intergovernmental integration. Harmonization of goals and policies among different levels of government (national, provincial or state, and local).

Intersectoral integration. Harmonization of goals and policies among different coastal and marine sectors (e.g., oil and gas development, fishing operations, coastal tourism, marine mammal protection, and port development).

Land-based sources of marine pollution. Marine pollution originating from the land through rivers, estuaries, pipelines, runoff, outfall structures, and so forth.

Land reclamation. A type of coastal construction activity aimed at gaining land from the sea for agriculture, urban development, and industrial sites by filling in wetlands or semienclosed bays, constructing dikes, or building dams and other barriers to exclude coastal waters.

Mariculture. The farming of marine finfish, mollusks, crustaceans, and sea-weeds, as well as ocean ranching of anadromous fish such as salmon.

Marine protected areas. Areas of coastal land or water that are specially designated to protect coastal and marine resources, preserve biological diversity, increase public awareness, and provide sites for recreation, research, and monitoring.

Mitigation. Prevention, elimination, reduction, or control of a project's negative environmental effects by avoiding or minimizing the effects, or by compensating for them by providing substitute resources.

Monitoring. Regular collection of data and information with the goal of analyzing trends over time.

Nonpoint-source marine pollution. Marine pollution orginating from diffuse sources associated with agricultural and urban runoff, runoff from construction activities, atmospheric deposition, and the like.

Permit. A coastal management tool that requires anyone wishing to undertake an activity in the coastal zone to obtain specific permission from the relevant coastal management agency.

Pilot project. A demonstration project generally applying to only a small segment of a nation's coastal area.

Point-source marine pollution. Pollution discharged from a specific, fixed location such as a pipe or an outfall structure.

Police power. The inherent power of state governments in the United States, often delegated in part to local governments, to impose restrictions on private rights that are reasonably related to promotion and maintenance of the health, safety, morals, and general welfare of the public.

"Polluter-pays" principle. The principle that the cost of controlling environmental pollution should be internalized (i.e., borne by the polluter, developer, or consumer) rather than imposed on society as a whole.

Precautionary principle. The principle that preventive or remedial action should be taken, on the basis of the best available scientific evidence, to avoid making policy decisions that have irreversible adverse effects on the environment. In the precautionary approach, when full scientific information is not available, the burden of proof (for demonstrating lack of irreversible adverse consequences) lies with the developer of a proposed coastal project, not with the government.

Public trust doctrine. A doctrine prevalent in countries with a common law tradition wherein the country is said to own lands lying under navigable waters and to hold such lands in trust for the benefit of the people of the country. According to this doctrine, these submerged lands may not be sold or otherwise alienated by the country except in a manner that promotes the public interest.

Remote sensing. A technology involving the use of satellites or aircraft to observe the characteristics of broad areas of the earth's surface in relatively short periods of time. This technology can play a role both in the initial mapping of resources and in the monitoring of trends over time.

Restoration. Repair or rehabilitation of an environmental resource to restore its structure and ecological processes.

Risk assessment. Technique to quantify risks and thus provide some guidance as to which environmental problems need the most prompt attention and how they might be addressed.

Scaling up. Expanding the experiences and know-how gained from a pilot area project to other coastal regions or to a country's entire coastal zone.

Sea-level rise. An increase in elevation of the sea. It may be associated with local land subsidence as a result of tectonic factors or the withdrawal of hydrocarbons or water, or it may be associated with global warming phenomena, which can cause heat expansion of ocean waters and melting of glaciers and polar ice caps.

Sectoral management. A type of coastal management approach whereby each different use of the ocean (e.g., fisheries operations, coral mining, oil and gas development, tourism) is managed separately.

Set-back line. A legally prescribed boundary seaward of which new construction in the coastal zone is generally prohibited.

Spatial integration. Harmonization of goals and policies between the land and ocean sides of the coastal zone.

Special area management. A type of coastal management in which an area in the coastal zone is designated and managed as a unit, regardless of government jurisdictions.

Stakeholders. Users who have particular interests in a coastal area, such as oil and gas developers, fishermen, hotel owners, port developers, aquaculture farmers, environmental groups, and local government authorities.

Subnational. Level of government below the national level, such as regional, state or provincial, or local, etc.

Sustainable development. A process in which the exploitation of resources, the direction of investments, the orientation of technological development, and institutional change are made consistent with the needs of future generations as well as those of present generations.

"Takings" issue. According to the Fifth Amendment of the United States Constitution, no person shall be deprived of his or her property without just compensation. Recently, some U.S. courts have decided that some forms of environmental regulation constitute a "taking" of private property.

Territorial sea. The area of sea where the sovereignty of a coastal nation extends beyond its land territory and internal waters and, in the case of an archipelagic nation, its archipelagic waters, to an adjacent belt of sea described as the territorial sea. Every nation has the right to establish the breadth of its territorial sea to a limit not exceeding 12 nautical miles, measured from baselines. The sovereignty of a coastal nation extends to the airspace over the territorial sea and the territorial sea's water column, seabed, and subsoil.

Top-down approach to ICM. In contrast to the bottom-up approach, the formulation of an ICM program from a higher level of government, such as the national level.

Traditional management. Management practices based on the traditional knowledge of indigenous peoples.

Transboundary. A term describing environmental effects or other processes that cross jurisdictional boundaries.

Transparency. A quality of decision making that is conducted in an open manner with participation from all major groups potentially affected by the outcome of the decision, including women, children, youth, indigenous peoples, non-governmental organizations, local authorities, and others, as appropriate.

Upland. A term describing land areas sufficiently inland from the shoreline to have limited interaction with the sea.

User fee. A fee charged by a government agency for use of a particular coastal resource.

Zoning or zonation. A regulatory process that divides a given geographical area into subareas, each of which is designated for a particular use or uses.

Acronyms

ACAP	Atlantic Coastal Action Program
ADB	Asian Development Bank
AOSIS	Alliance of Small Island States
ASEAN	Association of Southeast Asian Nations
BCDC	San Francisco Bay Conservation and Development Commission
CAMP	Coastal area management program
CAP	Coastal Action Plan (Australia)
CBD	Convention on Biological Diversity
CCA	Coast Conservation Act (Sri Lanka)
CCD	Coast Conservation Department (Sri Lanka)
CCP	Commonwealth Coastal Policy (Australia)
CFCs	Chlorofluorocarbons
CIDA	Canadian International Development Agency
CITES	Convention on International Trade in Endangered Species of Wild Fauna and Flora
COA	Canada Oceans Act
COE	U.S. Army Corps of Engineers
COREMAP	Coral Reef Rehabilitation and Management Program (Indonesia)
CORIN	Coastal Resources Institute (Prince of Songkla University, Thailand)
CRMP	ASEAN-USAID Coastal Resources Management Project
CRMPSJ	Coastal Resources Management Plan for South Johore (Malaysia)
CSD	United Nations Commission on Sustainable Development
CSO	Combined sewer overflow
CZM	Coastal zone management
CZMA	Coastal Zone Management Act (United States)
DDT	Dichloro-diphenyl-trichloroethane
DGIP	Division for Global and Interregional Programmes (UNDP)
DOD	Department of Ocean Development (India)
EEZ	Exclusive economic zone

EIA	Environmental impact assessment
FAO	Food and Agriculture Organization of the United Nations
FCCC	Framework Convention on Climate Change
FCZ	Fishery conservation zone (United States)
GBR	Great Barrier Reef (Australia)
GBRMPA	Great Barrier Reef Marine Park Authority (Australia)
GDP	Gross domestic product
GEF	Global Environment Facility
GESAMP	Group of Experts on the Scientific Aspects of Marine Environmental Protection
GIS	Geographic information system
GNP	Gross national product
IAEA	International Atomic Energy Agency
IBAMA	Institute for the Environment and Renewable Natural Resources (Brazil)
ICAM	Integrated coastal area management
ICCOPS	International Centre for Coastal and Ocean Policy Studies (Italy)
ICLARM	International Centre for Living Aquatic Resources Management (Philippines)
ICM	Integrated coastal management
ICRI	International Coral Reef Initiative
ICSU	International Council of Scientific Unions
ICZM	Integrated coastal zone management
IFREMER	Institut Français de Recherches sur l'Exploitation de la Mer (France)
IMCAM	Integrated marine and coastal area management
IMO	International Maritime Organization
IMPSA	Integrated Management Plan for Segara Anakan–Cilacap (Indonesia)
INC	International negotiating committee
INC/BIODIV	Intergovernmental Negotiating Committee for Biological Diversity
INC/FCCC	Intergovernmental Negotiating Committee of the Framework Convention on Climate Change
IOC	Intergovernmental Oceanographic Commission (UNESCO)
IOI	International Ocean Institute
IPCC	Intergovernmental Panel on Climate Change
ITQ	Individual transferable quota
IUCN	World Conservation Union (formerly International Union for the Conservation of Nature and Natural Resources)
JICA	Japan International Cooperation Agency

JNCC	Joint Nature Conservation Committee (United Kingdom)
KMI	Korea Maritime Institute
KORDI	Korea Ocean Research and Development Institute
LARI	Louisiana Artificial Reef Initiative (United States)
LGCAMP	Lingayen Gulf Coastal Area Management Plan (Philippines)
LOS	Law of the Sea
MAP	Mediterranean Action Plan
MAPA	Marine Affairs and Policy Association (United States)
MARPOL	International Convention for the Prevention of Pollution from Ships
MEDCOAST	International Conference on the Mediterranean Coastal Environment
MOMAF	Ministry of Maritime Affairs and Fisheries (Republic of Korea)
MOSTE	Ministry of Science, Technology, and Environment (Malaysia)
MREP	Marine Resource Evaluation and Planning Project (Indonesia)
NEAP	National environmental action plan
NGO	Nongovernmental organization
NIO	National Institute of Oceanography (India)
NOAA	National Oceanic and Atmospheric Administration (United States)
NRC	National Research Council (United States)
OCRM	Office of Ocean and Coastal Resource Management (NOAA, United States)
OECD	Organization for Economic Cooperation and Development
OSP	Office of State Planning (Hawaii, United States)
PMRC	Coastal Resources Management Program (Ecuador)
PrepCom	Preparatory Committee (UNCED)
RACE	Rapid appraisal of coastal environment
SAM	Special area management
SAREC	Swedish Agency for Research Cooperation with Developing Countries (now part of SIDA)
SIDA	Swedish International Development Authority
SIDS	Small island developing states
SMA	Special management area
SOA	State Oceanic Administration (China)
SPA	Specially protected area (Turkey)
SPREP	South Pacific Regional Environment Programme

STAPS	Science, Technology, and Private Sector Division (UNDP)
TEMA	Committee for Training, Education, and Mutual Assistance in the Marine Sciences (IOC)
TS	Territorial sea
UNCED	United Nations Conference on Environment and Development (also Earth Summit, Rio Conference)
UNCLOS III	Third United Nations Convention on the Law of the Sea
UNDOALOS	United Nations Division for Ocean Affairs and Law of the Sea
UNDP	United Nations Development Programme
UNDPCSD	United Nations Department for Policy Coordination and Sustainable Development
UNEP	United Nations Environment Programme
UNESCO	United Nations Educational, Scientific, and Cultural Organization
UNGA	United Nations General Assembly
USAID	United States Agency for International Development
WCC	World Coast Conference (1993)
WCED	World Commission on Environment and Development
WHO	World Health Organization
WMO	World Meteorological Organization
WMU	World Maritime University
WRI	World Resources Institute
ZEM	Zonas especiales de manejo (special area management zones) (Ecuador)

References

Abdullah, D. 1996. Director, State Economic Planning Unit, Bangunan Sultan Ibrahim, Johor Bahru, Malaysia. Response to ICM cross-national survey. Newark: University of Delaware, Center for the Study of Marine Policy.

Abregana, B., et al. 1996. *Legal Challenges for Local Management of Marine Resources: A Philippines Case Study.* Philippines: Environment and Resource Management Project.

Agate, A. D., S. D. Bonde, and K. P. N. Kumaran, eds. 1990. *Proceedings of the Symposium on the Significance of Mangroves.* March 16, Pune, Maharashtra, India. Pune, Maharashtra, India: Maharashtra Association for the Cultivation of Science Research Institute.

Agoes, E. R. 1996. Law Lecturer, Center for Law and Archipelago Studies. Response to ICM cross-national survey. Newark: University of Delaware, Center for the Study of Marine Policy.

Alcala, A. C. 1996. Coastal resource management as practiced in the Philippines. Paper presented at international workshop, Integrated Coastal Management in Tropical Developing Countries: Lessons Learned from Successes and Failures, May 24–28. Xiamen, People's Republic of China.

Alder, J., N. Sloan, and H. Uktolseya. 1994. Advances in marine protected area management in Indonesia: 1988–1993. *Ocean & Coastal Management* 25: 63–75.

Amjad, S. 1996. Director General, National Institute of Oceanography, Karachi, Pakistan. Response to ICM cross-national survey. Newark: University of Delaware, Center for the Study of Marine Policy.

Andresen, S., and B. Fløistad. 1988. Sea use planning in Norwegian waters: National and international dimensions. *Coastal Management* 16 (3): 183–200.

Andriguetto-Filho, A. M. 1993. Institutional prospects in managing coastal environmental conservation units in Paraná State, Brazil. In *Coastal Zone '93*, ed. O. T. Magoon, W. S. Wilson, and H. Converse, 2354–2368. New York: American Society of Civil Engineers.

Ann, O. C., Jr. 1996. Director General, Department of Irrigation and Drainage, Kuala Lumpur, Malaysia. Response to ICM cross-national survey. Newark: University of Delaware, Center for the Study of Marine Policy.

471

Annex. 1994. An outline on the implementation of strategy and programme of action adopted by regional leadership. Seminar on marine and ocean affairs in Africa, March 28–April 2, Addis Ababa, Ethiopia.

Antenucci, J. C., K. Brown, P. L. Croswell, M. J. Kevany, and H. Archer. 1991. *Geographic Information Systems: A Guide to the Technology.* New York: Van Nostrand Reinhold.

Anutha, K. 1996. Program Manager, Coastal and Marine Planning Division, Department of Environment and Land Management, Hobart, Tasmania, Australia. Response to ICM cross-national survey. Newark: University of Delaware, Center for the Study of Marine Policy.

Aquarone, M. C. 1988. French marine policy in the 1970's and 1980's. *Ocean Development and International Law* 19: 267–285.

Archer, J. H. 1988. *Coastal Management in the United States: A Selective Review and Summary.* Kingston: University of Rhode Island, Center for Ocean Management Studies, Coastal Resources Center, International Coastal Resources Management Project.

———. 1989. Resolving intergovernmental conflicts in marine resources management: The U.S. experience. *Ocean and Shoreline Management* 12: 253–269.

Archer, J. H., and M. C. Jarman. 1992. Sovereign rights and responsibilities: Applying public trust principles to the management of EEZ space and resources. *Ocean & Coastal Management* 17 (1): 251–270.

Archer, J. H., D. Connors, K. Lawrence, S. C. Columbia, and R. Bowen. 1994. *The Public Trust Doctrine and the Management of America's Coasts.* Amherst, Mass.: University of Massachusetts Press.

Arriaga, L. M. 1996. Country Advisor, University of Rhode Island Coastal Resources Center, Programa de Manejo de Recursos Costeros, Guayaquil, Ecuador. Response to ICM cross-national survey. Newark: University of Delaware, Center for the Study of Marine Policy.

ASEAN-USAID CRMP and DGF (Association of Southeast Asian Nations–United States Agency for International Development Coastal Resources Management Project and Directorate-General of Fisheries). 1992. *The Integrated Management Plan for Segara Anakan–Cilacap, Central Java, Indonesia.* ICLARM Technical Report No. 34. Manila: International Centre for Living Aquatic Resources Management.

Australia. DEST (Department of the Environment, Sport, and Territories). 1995. *Living on the Coast: The Commonwealth Coastal Policy.* Canberra, Australia: Department of the Environment, Sport, and Territories.

Australia. RAC (Resource Assessment Commission). 1993. *Resource Assessment Commission Coastal Zone Inquiry—Final Report.* Canberra: Australian Government Publishing Service.

Bailey, R. 1997. Ocean management by coastal states: The Oregon case. *Ocean & Coastal Management* 34 (3): 205–224.

Bandora, C. M. M. 1995. The network for environmental training at tertiary-level in

Asia and the Pacific (NETTLAP): An experience in environmental networking. Paper presented at the International Workshop on Educating Coastal Managers, March 4–10. University of Rhode Island.

Barua, D. K. 1993. Practices, possibilities, and impacts of land reclamation activities in the coastal areas of Bangladesh. In *International Perspectives on Coastal Ocean Space Utilization*, ed. P. M. Grifman and J. A. Fawcett, 343–356. Los Angeles: University of Southern California, Sea Grant Program.

Basiron, M. N. 1996. Senior Analyst, Malaysian Institute of Maritime Affairs, Kuala Lumpur, Malaysia. Response to ICM cross-national survey. Newark: University of Delaware, Center for the Study of Marine Policy.

BCDC (Bay Conservation and Development Commission). 1985. *Applying for Permits: BCDC Permit Application Information Booklet.* San Francisco: Bay Conservation and Development Commission.

Beck, R. E. 1994. The movement in the United States to restoration and creation of wetlands. *Natural Resources Journal* 34: 781–822.

Bennett, E. C., and L. F. Curtis. 1992. *Introduction to Environmental Remote Sensing.* 3rd ed. New York: Chapman and Hall.

Bergin, A., and D. Lawrence. 1993. Aboriginal and Torres Strait Islander Interests in the Great Barrier Reef Marine Park. In *Turning the Tide: Indigenous Sea Rights.* Northern Territory University Law School International Conference, July. Townsville, Queensland, Australia: Great Barrier Reef Marine Park Authority.

Berg-Schlosser, D. 1990. Typologies of Third World political systems. In *Contemporary Political Systems: Classifications and Typologies*, ed. A. Bebler and J. Seroka, 173–203. Rienner.

Bingham, G. 1986. *Resolving Environmental Disputes: A Decade of Experience.* Washington, D.C.: Conservation Foundation.

Boelaert-Suominen, S., and C. Cullinan. 1994. *Legal and Institutional Aspects of Integrated Coastal Area Management in National Legislation.* Rome: Food and Agriculture Organization of the United Nations, Development Law Service, Legal Office.

Boesch, D., and S. A. Macke. 1995. Bridging the gap: What natural scientists and policy-makers need to know about each other. In *Improving Interactions between Coastal Science and Policy: Proceedings of the California Symposium*, 33–48. Washington, D.C.: National Academy Press.

Borgese, E. M. 1994. Training and education in ocean management for the 21st century: Regional and global aspects. Paper presented at the Second International Conference on Oceanography: Toward Sustainable Use of Oceans and Coastal Zones, November 14–19. Lisbon, Portugal.

———. 1995. *Ocean Governance and the United Nations.* Halifax, Nova Scotia: Dalhousie University, Center for Foreign Policy Studies.

Bower, B. T. 1992. *Producing Information for Integrated Coastal Management*

Decisions: An Annotated Seminar Outline. Silver Spring, Md.: U.S. Department of Commerce, National Oceanic and Atmospheric Administration, National Ocean Service, Office of Ocean Resources Conservation and Assessment.

Brazil. 1997. *Towards Brazil's Agenda 21: Principles and Actions, 1992–1997.* Prepared by the Ministry for the Environment, Water Resources, and the Amazon, with contributions from the other ministries and state agencies. Brasília: Gráfica Valci Editorial Brazil.

Brown, K. R. 1996. Coastal resource management: Interdisciplinary, multidisciplinary, or transdisciplinary—The role of universities in capacity building. Paper presented at international workshop, Integrated Coastal Management in Tropical Developing Countries: Lessons Learned from Successes and Failures, May 24–28. Xiamen, People's Republic of China.

Buhat, D. Y. 1994. Community-based coral reef and fisheries management, San Salvador Island, Philippines. In *Collaborative and Community-Based Management of Coral Reefs: Lessons from Experience*, ed. A. T. White, L. Z. Hale, Y. Renard, and L. Cortesi, 33–50. West Hartford, Conn.: Kumarian Press.

Bunpapong, S., and S. Ausavajitanon. 1991. Saving what's left of tourism development at Patong Beach, Phuket, Thailand. In *Coastal Zone '91*, ed. O. T. Magoon, H. Converse, V. Tippie, L. Tobin, and D. Clark, 1688–1697. New York: American Society of Civil Engineers.

Caldwell, L. K. 1990. *International Environmental Policy: Emergence and Dimensions.* 2nd ed. Durham, N.C.: Duke University Press.

California Coastal Zone Conservation Commission. 1975. *California Coastal Plan.* Sacramento: California Coastal Zone Conservation Commission.

Canada. Department of Fisheries and Oceans. 1996a. Integrated coastal area management: A Canadian retrospective and update. Presentation to the United Nations Commission on Sustainable Development, April, United Nations, New York.

———. 1996b. Minister Mifflin welcomes passage of the Oceans Act. *News Release.* NR-HQ-96-100E. Ottawa, Ontario: Fisheries and Oceans Canada.

Carey, J. J., and R. B. Mieremet. 1992. Reducing vulnerability to sea level rise: International initiatives. *Ocean & Coastal Management* 18: 161–177.

Chalapan, K. 1996. Coastal area management in the Pacific island nations. Paper presented to the Commission on Sustainable Development. United Nations, New York.

Chape, S. P. 1990. Coastal tourism and environmental management in Fiji. In *Proceedings of the 1990 Congress on Coastal and Marine Tourism*, vol. 1, ed. M. Miller and J. Auyong, 67–77. Newport, Oreg.: National Coastal Resources Research and Development Institute.

Chape, S. P., and D. Watling, ed. 1992. *Fiji National Report to UNCED.* Apia, Western Samoa: South Pacific Regional Environment Programme.

Ch'ng, L. K. 1987. Coastal zone management plan development in Malaysia: Issues and possible solutions. In *Coastal Zone '87*, ed. O. T. Magoon, H. Converse, D.

Miner, and L. T. Tobin, 4601–4615. New York: American Society of Civil Engineers.

Chua, T. E. 1993. Essential elements of integrated coastal management. *Ocean & Coastal Management* 21 (1–3): 81–108.

———, ed. 1996. *Integrated Coastal Management in Tropical Developing Countries: Lessons Learned from Successes and Failures.* MPP-EAS Technical Report No. 4. Quezon City, Philippines: GEF/UNDP/IMO Regional Programme for the Prevention and Management of Marine Pollution in the East Asian Seas and Coastal Management Center.

Chua, T. E., L. M. Chou, and M. S. M. Sadorra, eds. 1987. *The Coastal Environmental Profile of Brunei Darussalem: Resource Assessment and Management Issues.* ICLARM Technical Reports on Coastal Area Management Series No. 18. Manila: International Centre for Living Aquatic Resources Management.

Chua, T. E., and J. N. Paw. 1996. A dichotomy of the planning process used in the ASEAN-US project on coastal resources management. Paper presented at international workshop, Integrated Coastal Management in Tropical Developing Countries: Lessons Learned from Successes and Failures, May 24–28, Xiamen, People's Republic of China.

Chua, T. E., and L. F. Scura, eds. 1992. *Integrative Framework and Methods for Coastal Area Management.* ICLARM Conference Proceedings No. 37. Manila: International Centre for Living Aquatic Resources Management.

Chua, T. E., and H. Yu. 1996. From sectoral to integrated coastal management: A case study in Xiamen, China. Paper presented at international workshop, Integrated Coastal Management in Tropical Developing Countries: Lessons Learned from Successes and Failures, May 24–28, Xiamen, People's Republic of China.

Churchill, R. R., and A. V. Lowe. 1988. *The Law of the Sea.* Manchester, England: Manchester University Press.

CIA (Central Intelligence Agency). 1995. *The World Factbook.* Washington, D.C.: Government Printing Office.

Cicin-Sain, B. 1982. Managing the ocean commons: U.S. marine programs in the 70s and 80s. *Marine Technology Society Journal* 16 (4): 6–18.

———. 1992. Multiple use conflicts and their resolution: Toward a comprehensive research agenda. In *Ocean Management in Global Change*, ed. P. Fabbri, 280–307. New York: Elsevier Applied Science.

———, ed. 1992. *Ocean Governance: A New Vision.* Newark: University of Delaware, Center for the Study of Marine Policy.

———. 1993. Sustainable development and integrated coastal management. *Ocean & Coastal Management* 21 (1–3), 11–43.

———, ed. 1993. Special issue: Integrated coastal management. *Ocean & Coastal Management* 21 (1–3), 1–377.

———. 1996. Earth Summit implementation: Progress since Rio. *Marine Policy* 20 (2): 123–143.

Cicin-Sain, B., C. N. Ehler, R. Knecht, S. South, and R. Weiher. 1997. *Guidelines for Integrating Coastal Management Programs and National Climate Change Action Plans.* International Workshop on Planning for Climate Change through Integrated Coastal Management, February 24–28, Taipei, Taiwan.

Cicin-Sain, B., M. Hershman, and J. Isaacs. 1990. *Improving Ocean Management Capacity in the Pacific Coast Region: State and Regional Perspectives.* Newport, Oreg.: National Coastal Resources Research and Development Institute.

Cicin-Sain, B., and R. W. Knecht. 1985. The problem of governance of U.S. ocean resources and the new exclusive economic zone. *Ocean Development and International Law* 15 (3–4): 289.

———. 1987. Federalism under stress: The case of offshore oil development and California. In *Perspectives on Federalism*, ed. H. Scheiber, 149–176. Berkeley: University of California, Institute for Governmental Studies.

———. 1992. Research agenda on ocean governance. In *Ocean Governance: A New Vision*, ed. B. Cicin-Sain, 9–16. Newark: University of Delaware, Center for the Study of Marine Policy.

———. 1993. Implications of the earth summit for ocean and coastal governance. *Ocean Development and International Law* 24: 121–153.

———. 1995. Measuring Progress on UNCED Implementation. In Special issue: Earth Summit implementation—Progress achieved on oceans and coasts. *Ocean & Coastal Management* 29 (1–3): 1–11.

Cicin-Sain, B., R. W. Knecht, and G. Fisk. 1995. Growth in capacity for integrated coastal management since UNCED: An international perspective. *Ocean & Coastal Management* 29 (1–3): 93–123.

Cicin-Sain, B., and A. Tiddens. 1989. Private and public approaches to solving oil/fishing conflicts offshore California. *Ocean and Shoreline Management* 12: 233–251.

Clark, J. R. 1991. *The Status of Integrated Coastal Management: A Global Assessment.* Miami: University of Miami, Coastal Area Management and Planning Network (CAMPNET), Rosentiel School of Marine and Atmospheric Science.

———. 1992. *Integrated Management of Coastal Zones.* FAO Fisheries Technical Paper No. 327. Rome: Food and Agriculture Organization of the United Nations.

———. 1996. *Coastal Zone Management Handbook.* New York: Lewis.

Cole-King, A. 1995. Marine protected areas in Britain: A conceptual problem? *Ocean & Coastal Management* 27 (1–2): 109–120.

Commission Océanographique Intergouvernementale. 1997. *Guide méthodologique d'aide à la gestion intègree de la zone côtiere.* UNESCO.

Cort, R. P. 1995. A practical guide to environmental impact assessment. *Quarterly Review of Biology* 70 (4): 533.

Couper, A. D., ed. 1983. *Atlas of the Oceans.* London: Times Books.

Crawford, B. R., J. S. Cobb, and A. Friedman. 1993. Building capacity for integrated coastal management in developing countries. *Ocean & Coastal Management* 21 (1–3): 311–338.

Cullen, P. 1982. Coastal zone management in Australia. *Coastal Zone Management Journal* 10 (3): 183–212.

Dahl, C. 1995. Micronesia. In *Coastal Management in the Asia Pacific Region: Issues and Approaches*, ed. K. Hotta and I. Dutton, 337–358. Tokyo: Japan International Marine Science and Technology Federation.

———. 1996. Program Assistant, Sea Grant Extension Pacific Program, Honolulu, Hawaii, United States. Response to ICM cross-national survey. Newark: University of Delaware, Center for the Study of Marine Policy.

———. In press. Integrated coastal resources management and community participation in a small island setting. *Ocean & Coastal Management.*

Dahuri, R. 1996a. Coastal zone management and transmigration in Indonesia. Paper presented at international workshop, Integrated Coastal Management in Tropical Developing Countries: Lessons Learned from Successes and Failures, May 24–28. Xiamen, People's Republic of China.

———. 1996b. Head of Coastal Zone Management, Bogor Agricultural University, Bogor, Indonesia. Response to ICM cross-national survey. Newark: University of Delaware, Center for the Study of Marine Policy.

Darus, M., and H. Haron. 1988. The management of Matang mangrove forest reserves in peninsular Malaysia. In *Coastal Area Management in Southeast Asia: Policies, Management Strategies and Case Studies*, ed. T. E. Chua and D. Pauly, 77–84. ASEAN-USAID Coastal Resources Management Project. ICLARM Conference Proceedings No. 2. Manila: International Centre for Living Aquatic Resources Management.

de Fontaubert, C. A., D. R. Downes, and T. S. Agardy. 1996. *Biodiversity in the seas: implementing the convention on biological diversity in marine and coastal habitats.* IUCN Environmental Policy and Law Paper No. 32. Gland, Switzerland: World Conservation Union.

Degong, C. 1989. Coastal zone development, utilization, legislation, and management in China. *Coastal Management* 17 (1): 55–62.

de Vrees, L. 1996. Senior Project Manager, Coastal Zone Management Centre, The Hague, Netherlands. Response to ICM cross-national survey. Newark: University of Delaware, Center for the Study of Marine Policy.

Dixon, J., and M. Hufschmidt, eds. 1986. *Economic Valuation Techniques for the Environment: A Case Study Workbook.* Baltimore: Johns Hopkins University Press.

Dobias, R. J. 1989. Beaches and tourism in Thailand. In *Coastal Area Management in Southeast Asia: Policies, Management Strategies and Case Studies*, ed. T. E. Chua and D. Pauly, 43–55. ASEAN-USAID Coastal Resources Management Project. ICLARM Proceedings No. 2. Manila: International Centre for Living Aquatic Resources Management.

Donalson, C. 1994. An unholy alliance: working with coastal communities—A practitioner's perspective. In *Coastal Zone Canada '94*, ed. R. G. Wells and P. J. Ricketts, 696–705. Halifax, Nova Scotia: Coastal Zone Canada Association.

DuBois, R., and E. L. Towle. 1985. Coral harvesting and sand mining management practices. In *Coasts: Coastal resources management—Development case studies*, ed. J. R. Clark, 443–508. Washington, D.C.: U.S. Department of the Interior, National Park Service, and U.S. Agency for International Development.

Dulvy, N., and W. R. T. Darwell. 1996. Tanzania, Mafia Island: Participatory control of coral mining. In *Coastal Zone Management Handbook*, ed. J. R. Clark, 587–590. New York: Lewis.

Edward, A. 1996. Sea Grant Office, Kolonia, Pohnpei, Federated States of Micronesia. Response to ICM cross-national survey. Newark: University of Delaware, Center for the Study of Marine Policy.

Edwards, S. F. 1987. *An Introduction to Coastal Zone Economics: Concepts, Methods, and Case Studies*. New York: Taylor and Francis.

Eichenberg, T., and J. Archer. 1987. The federal consistency doctrine: Coastal zone management and "new federalism." *Ecology Law Quarterly* 14 (1): 9–68.

Elkington, N. 1996. The effects of human impact on some mangrove ecosystems in Karnataka, India. Paper presented at international workshop, Integrated Coastal Management in Tropical Developing Countries: Lessons Learned from Successes and Failures, May 24–28. Xiamen, People's Republic of China.

Ellsworth, J. P. 1994. Closing the gap between community expectations and service delivery: Canada's Atlantic Coastal Action Program. In *Coastal Zone Canada '94*, ed. R. G. Wells and P. J. Ricketts, 687–695. Halifax, Nova Scotia: Coastal Zone Canada Association.

Ellsworth, J. P., L. P. Hildebrand, and E. A. Glover. 1997. Canada's coastal action program: A community-based approach to collective governance. *Ocean & Coastal Management* 36 (1–3): 121–142.

Englehardt, R., ed. 1994. Remote sensing for marine and coastal environments. *Marine Technology Journal* 28 (2): 83.

EPA (Environmental Protection Agency). 1992. *Framework for Ecological Risk Assessment*. EPA/630/R92/001. Washington, D.C.: Environmental Protection Agency.

———. Science Advisory Board. 1990. *Reducing Risk: Setting Priorities and Strategies for Environmental Protection*. Washington, D.C.: Environmental Protection Agency.

Etchart, G. 1995. Mitigation banks: A strategy for sustainable development. *Coastal Management* 23: 223–237.

The Europa World Year Book. 1995. 36th ed. London: Europa.

Fabbri, P., ed. 1992. *Ocean Management in Global Change*. New York: Elsevier Applied Science.

FAO (Food and Agriculture Organization of the United Nations) 1991. *Development of coastal areas and enclosed seas*. UN Conference on Environment and Development Research Paper No. 4. Rome: Food and Agriculture Organization of the United Nations.

Field, B. C. 1994. *Environmental Economics*. New York: McGraw-Hill.

Fiji. National Report to the United Nations Conference on the Environment and Development (UNCED). 1992. Apia, Western Samoa: South Pacific Regional Environment Programme.

Fisk, G. W. 1996. Integrated coastal management in developed countries: The case of Australia. Master's thesis, Graduate College of Marine Studies, University of Delaware.

Foster, N. M., and J. H. Archer. 1988. Introduction: The National Marine Sanctuary Program—Policy, education, and research. *Oceanus* 31 (1): 5.

Freeman, M. A. 1993. *The Measurement of Environmental and Resource Values.* Washington, D.C.: Resources for the Future.

Friedheim, R. L. 1993. *Negotiating the New Ocean Regime.* Columbia: University of South Carolina Press.

———, ed. 1979. The political, economic, and legal ocean. In *Managing Ocean Resources: A Primer*, ed. R. L. Friedheim, 26–42. Boulder, Colo.: Westview Press.

GBRMPA (Great Barrier Reef Marine Park Authority). 1992. *Cairns Section Zoning Plan.* Townsville, Queensland, Australia: Great Barrier Reef Marine Park Authority.

George, S., and K. E. Nichols. 1994. *Conflict resolution: A case study of the coastal town of Soufriere, St. Lucia.* In *Coastal Zone Canada '94*, ed. P. G. Wells and P. J. Ricketts, 466–471. Halifax, Nova Scotia: Coastal Zone Canada Association.

GESAMP (Group of Experts on the Scientific Aspects of Marine Environmental Protection). 1996. *Report of the task force on integrated coastal management.* Rome: Food and Agriculture Organization of the United Nations.

Gilpin, A. 1995. *Environmental Impact Assessment: Cutting Edge for the Twenty-First Century.* Cambridge: Cambridge University Press.

Goldberg, E. D. 1994. *Coastal Zone Space: Prelude to Conflict?* Paris: United Nations Educational, Scientific, and Cultural Organization.

Gomez, E. D. 1996. President, Coastal Management Center, Metro-Manila, Philippines. Response to ICM cross-national survey. Newark: University of Delaware, Center for the Study of Marine Policy.

Gregory, R., T. C. Brown, and J. L. Knetsch. 1996. Valuing risks to the environment. *Annals of the American Academy of Political and Social Science* 545: 54–63.

Griffith, M. D. 1992. The South and the United Nations conference on environment and development: The dawn of a probable turning point in international relations between states. *Ocean & Coastal Management* 18 (1): 55–77.

Gubbay, S. 1990. *A Future for the Coast: Proposals for a U.K. Coastal Zone Management Plan.* Report prepared for the World Wide Fund for Nature and the Marine Conservation Society. London: World Wide Fund for Nature.

———. 1995. Integrated coastal zone management—Opportunities for Scotland. Briefing for the World Wide Fund for Nature, Scotland.

———. 1996. Coastal Management Specialist and NGO Advisor, Ross-on-Wye, England. Response to ICM cross-national survey. Newark: University of Delaware, Center for the Study of Marine Policy.

Guochen, Z. 1993. Coastal zone management in China. In *World Coast Conference 1993: Proceedings*, vols. 1 and 2. CZM-Centre Publication No. 4. The Hague: Ministry of Transport, Public Works, and Water Management, National Institute for Coastal and Marine Management, Coastal Zone Management Centre.

Haas, P. M. 1990. *Saving the Mediterranean: The Politics of International Environmental Cooperation.* New York: Columbia University Press.

Hale, L. Z., and M. H. Lemay. 1994. Coral reef protection in Phuket, Thailand: A step toward integrated coastal management. In *Collaborative and Community-Based Management of Coral Reefs: Lessons from Experience*, ed. A. T. White, L. Z. Hale, Y. Renard, and L. Cortesi, 68–79. West Hartford, Conn.: Kumarian Press.

Hanley, N. 1993. *Cost-Benefit Analysis and the Environment.* Aldershot, England: Elgar.

Haq, S. M. 1996. Former Director, Center of Excellence in Marine Biology, University of Karachi, Pakistan. Response to ICM cross-national survey. Newark: University of Delaware, Center for the Study of Marine Policy.

Hatziolos, M. E. In press. A World Bank framework for ICM with special emphasis on Africa. *Ocean & Coastal Management.*

Hatziolos, M. E., and I. Trumbic. 1997. Preliminary results of an assessment of coastal zone management initiatives in the Mediterranean. *Memorandum*, Washington, D.C.: World Bank.

Hawaii. 1990. *Hawaii Coastal Zone Management Program.* Honolulu: Hawaii Office of State Planning.

Haward, M. 1989. The Australian offshore constitutional settlement. *Marine Policy* 13 (4): 334–348.

———. 1996. Senior Lecturer, Department of Political Science, University of Tasmania, Hobart, Australia. Response to ICM cross-national survey. Newark: University of Delaware, Center for the Study of Marine Policy.

Hennessey, T. M. 1994. Governance and adaptive management for estuarine ecosystems: The case of Chesapeake Bay. *Coastal Management* 22 (2): 119–146.

Hershman, M., T. Bernd-Cohen, J. W. Good, B. Goodwin, M. Gordon, V. Lee, and P. Pogue. 1997. National CZM effectiveness study—Findings and recommendations. Paper presented at conference, Coastal Zone '97, July 22–24, Boston, Massachusetts.

Herz, R. 1989. Coastal ocean space management in Brazil. In *Coastal Ocean Space Utilization*, ed. S. D. Halsey and R. B. Abel, 29–52. New York: Elsevier Applied Science.

Herz, R., and A. S. Mascarenhas Jr. 1993. Political and planning actions on the Brazilian Coastal Management Program. In *Coastal Zone '93*, ed. O. T. Magoon, W. S. Wilson, and H. Converse, 1084–1091. New York: American Society of Civil Engineers.

Hildebrand, L. P. 1989. *Canada's Experience with Coastal Zone Management.* Halifax, Nova Scotia: Oceans Institute of Canada.

———. 1996a. Canada: National experience with coastal zone management. In *Coastal Zone Management Handbook*, ed. J. R. Clark, 506–508. New York: Lewis.

———. 1996b. Head Coastal Liaison, Environment Conservation Branch, Environment Canada, Atlantic Region, Dartmouth, Nova Scotia, Canada. Response to ICM cross-national survey. Newark: University of Delaware, Center for the Study of Marine Policy.

———, ed. 1997. Special issue on community-based coastal management. *Ocean & Coastal Management* 36 (1–3).

Hiscock, K., ed. 1996. *Marine Nature Conservation Review: Rationale and Methods.* Summary report. Petersborough, England: Joint Nature Conservation Committee.

Hodgson, G., and J. A. Dixon. 1988. Measuring economic losses due to sediment pollution: Logging versus tourism and fisheries. *Tropical Coastal Area Management* (April): 5–7.

Hollick, A. 1981. *U.S. Foreign Policy and the Law of the Sea.* Princeton, N.J.: Princeton University Press.

HOMRC (Hawaii Ocean and Marine Resources Council). 1991. *Hawaii Ocean Resources Management Plan,* vols. 1 and 2. (Technical Supplement.) Honolulu: Hawaii Ocean and Marine Resources Council.

Hong, S. Y. 1991. Assessment of coastal zone issues in the Republic of Korea. *Coastal Management* 19: 391–415.

———. 1995. A framework for emerging new marine policy: The Korean experience. *Ocean & Coastal Management* 25 (2): 77–101.

———. 1996. Ministry of Maritime Affairs and Fisheries Letter to Professors Biliana Cicin-Sain and Robert W. Knecht, June 25.

Hong, S. Y., and J. H. Lee. 1995. National level implementation of Chapter 17: The Korean example. *Ocean & Coastal Management* 29 (1–3): 231–249.

Hooten, A. J., and M. E. Hatziolos, eds. *Sustainable Financing Mechanisms for Coral Reef Conservation.* Environmentally Sustainable Development Proceedings Series No. 9. Washington, D.C.: The World Bank.

ICLARM (International Centre for Living Aquatic Resources Management). 1991. *The Coastal Environmental Profile of South Johore, Malaysia.* ASEAN-USAID Coastal Resources Management Project. ICLARM Technical Series No. 6. Manila: International Centre for Living Aquatic Resources Management.

ICRI (International Coral Reef Initiative). 1995. *Final Report—The International Coral Reef Initiative Workshop.* May 29–June 2. ICR Secretariat, Washington, D.C.: International Coral Reef Initiative.

IOC (International Oceanographic Commission). 1994a. *International Workshop on Integrated Coastal Zone Management (ICZM).* October 10–14, Karachi, Pakistan. Intergovernmental Oceanographic Commission Workshop Report No. 114. Paris: United Nations Educational, Scientific, and Cultural Organization.

————. 1994b. Second International Conference on Oceanography. *Towards Sustainable Use of Oceans and Coastal Zones.* November 14–19, Lisbon, Portugal.

IPCC (Intergovernmental Panel on Climate Change). 1990. *Climate Change: The IPCC Scientific Assessment,* ed. J. T. Houghton, G. J. Jenkins, and J. J. Ephraums. Cambridge: Cambridge University Press.

————. 1992. *Global Climate Change and the Rising Challenge of the Sea.* Report of the Coastal Zone Management Subgroup, Response Strategies Working Group, March. The Hague, The Netherlands: Ministry of Transport, Public Works, and Water Management.

————. 1994. *Preparing to Meet the Coastal Challenges of the 21st Century: Report of the World Coast Conference 1993.* November 1–5, Noordwijk, Netherlands. The Hague: Ministry of Transport, Public Works, and Water Management, National Institute for Coastal and Marine Management, Coastal Zone Management Centre.

————. 1995. *Economic and Social Dimensions of Climate Change: Contribution of Working Group III to the Second Assessment Report of the IPCC,* ed. J. P. Bruce, H. Lee, and E. F. Haites. Cambridge: Cambridge University Press.

Isidro, A. 1996. Lessons in coastal resources management: The experience of the Philippines fisheries sector program. Paper presented at international workshop, Integrated Coastal Management in Tropical Developing Countries: Lessons Learned from Successes and Failures. May 24–28, Xiamen, People's Republic of China.

IUCN (World Conservation Union). 1993. *The National Environment Strategy Fiji.* Gland, Switzerland: World Conservation Union.

Jang, D. 1997. *Dalian Report.* Report prepared for the Intergovernmental Oceanographic Commission and State Oceanic Administration's International Training Workshop on the Integration of Marine Science into the Process of Integrated Coastal Management (ICM), May 19–24, Dalian, People's Republic of China.

Jentoft, S., and B. McCay. 1995. User participation in fisheries management: Lessons drawn from international experiences. *Marine Policy* 19 (3): 227–246.

Jernelov, A. 1988. *EIA—A Practical Approach: Proceedings of the ROPME Workshop on Coastal Area Development.* UNEP Regional Seas Reports and Studies No. 90, 143–169. Nairobi, Kenya: United Nations Environment Programme.

Johnson, A. I., C. B. Patterson, and J. L. Fulton, eds. 1992. *Geographic Information Systems and Mapping: Practices and Standards.* Philadelphia: ASTM.

Jorge, M. A. 1997. Developing capacity for coastal management in the absence of the public sector: A case study in the Dominican Republic. *Ocean & Coastal Management* 36 (1–3): 47–72.

Juda, L., and R. H. Burroughs. 1990. The prospects for comprehensive ocean management. *Marine Policy* 14 (1): 23–35.

Kahawita, B. S. 1993. Coastal zone management in Sri Lanka. In *World Coast Conference 1993: Proceedings,* 669–675. CZM-Centre Publication No. 4. The Hague: Ministry of Transport, Public Works, and Water Management, National Institute for Coastal and Marine Management, Coastal Zone Management Centre.

Kay, R. 1996. Coastal Planning Coordinator, State Government of Western Australia, Perth, Australia. Response to ICM cross-national survey. Newark: University of Delaware, Center for the Study of Marine Policy.

Kenchington, R. 1990. *Managing Marine Environments*. New York: Taylor and Francis.

Kenchington, R., and D. Crawford. 1993. On the meaning of integration in coastal zone management. *Ocean & Coastal Management* 21 (1–3): 109–128.

Khalimonov, O. 1995. Report on GESAMP and its role for the protection and sustainable development of the marine environment. *Ocean & Coastal Management* 29 (1–3): 297–302.

Kimball, L. A. 1995. An international regime for managing land-based activities that degrade marine and coastal environments. *Ocean & Coastal Management* 29 (1–3): 187–206.

King, G., and L. Bridge. 1994. *Directory of Coastal Planning and Management Initiatives in England*. Compiled on behalf of the National Coasts and Estuaries Advisory Group.

Kingdon, J. W. 1984. *Agendas, Alternatives, and Public Policies*. Boston: Little, Brown.

Kitsos, T. R. 1981. Ocean policy and the uncertainty of implementation in the 80s: A legislative perspective. *Marine Technology Society Journal* 15 (3): 3–11.

———. 1994. Troubled waters: A half dozen reasons why the federal offshore oil and gas program is failing—A political analysis. In *Moving Ahead on Ocean Governance*, ed. B. Cicin-Sain and K. A. Leccese, 36–45. Newark: University of Delaware, Center for the Study of Marine Policy, Ocean Governance StudyGroup.

Klomp, R. 1993. Lessons learned from the experience with policy decisions and supporting tools in coastal water management. Paper presented at MEDCOAST 93, the First International Conference on the Mediterranean Coastal Environment, November 2–5, Ankara, Turkey.

Knecht, R. W. 1979. Coastal zone management: The first five years and beyond. *Coastal Zone Management Journal* 6: 259–272.

———. 1986. The exclusive economic zone: A new opportunity in federal-state ocean relations. In *Ocean Resources and U.S. Intergovernmental Relations in the 1980s*, ed. M. Silva, 263–275. Boulder: Westview.

———. 1992. National ocean policy in the United States: Less than the sum of its parts. In *Ocean Management in Global Change*, ed. P. Fabbri, 184–208. New York: Elsevier Applied Science.

———. 1993. A perspective on the relationship between the local and national levels of government in coastal zone management. Paper presented at MEDCOAST 1993, the First International Conference on the Mediterranean Coastal Environment, November 2–5, Ankara, Turkey.

Knecht, R. W., and J. Archer. 1993. Integration in the U.S. Coastal Zone Management Program. *Ocean & Coastal Management* 21: 183–200.

Knecht, R. W., B. Cicin-Sain, and J. Archer. 1988. National ocean policy: A window of opportunity. *Ocean Development and International Law* 19: 113–142.

Knecht, R. W., B. Cicin-Sain, and G. W. Fisk. 1996. Perceptions of the performance of state coastal zone management programs in the United States. *Coastal Management* 24: 141–163.

Knecht, R. W., J. M. Broadus, M. E. Silva, R. E. Bowen, H. S. Marcus, and S. B. Peterson. 1984. *The Management of Ocean and Coastal Resources in Colombia: An Assessment.* Technical Report. Woods Hole, Mass.: Woods Hole Oceanographic Institution, Marine Policy and Ocean Management Center.

Koekebakker, P., and G. Peet. 1987. Coastal zone planning and management in the Netherlands. *Coastal Management* 15: 121–134.

Koh, T. B. 1983. Statement by Tommy B. Koh, president of the Third United Nations Conference on the Law of the Sea. In *The Law of the Sea,* xxxiii–xxxvii. New York: United Nations.

Kongsangchai, J. 1986. The conflicting interests of mangrove resources use in Thailand. Paper presented at UNDP/UNESCO Regional Project RAS/79/002, Workshop for Mangrove Zone Managers, September 9–10, Phuket, Thailand.

Kullenberg, G. 1995. Reflections on marine science contributions to sustainable development. *Ocean & Coastal Management* 29 (1–3): 1–11.

Kusuma-Atmadja, M., and T. H. Purwaka. 1996. Legal and institutional aspects of coastal zone management in Indonesia. *Marine Policy* 20 (1): 63–86.

Lay, G. F. T., M. C. Rockwell, G. B. J. Fader, and R. O. Miller. 1993. The potential for fisheries habitat creation through marine aggregate mining in the Bay of Fundy, Nova Scotia, Canada. In *International Perspectives on Coastal Ocean Space Utilization,* ed. P. M. Grifman and J. A. Faucet, 253–264. Los Angeles: University of Southern California, Sea Grant Program.

Lee, J. H., Y. Lee, S.-Y Hong, and M. Kwon. 1993. Coastal zone utilization of the Republic of Korea: Status and prospects. In *International Perspectives on Coastal Ocean Space Utilization,* ed. P. M. Grifman and J. A. Fawcett, 253–264, Los Angeles: University of Southern California, Sea Grant Program.

Lee, J. H. 1996a. Policy issues and management framework of Chinhae Bay in the Republic of Korea. Paper presented at international workshop, Integrated Coastal Management in Tropical Developing Countries: Lessons Learned from Successes and Failures, May 24–28, Xiamen, People's Republic of China.

———. 1996b. Research Scientist, Korea Ocean Research and Development Institute, Marine Policy Center, Ansan, Republic of Korea. Response to ICM cross-national survey. Newark: University of Delaware, Center for the Study of Marine Policy.

Lee, K. N. 1993. *Compass and Gyroscope: Integrating Science and Politics for the Environment.* Washington, D.C: Island Press.

Lemay, M. H., S. Ausavajitanon, and L. Zeitlin-Hale. 1991. A national coral reef management strategy for Thailand. In *Coastal Zone '91,* ed. O. T. Magoon, H. Converse, V. Tippie, L. T. Tobin, and D. Clark, 1698–1712. New York: American Society of Civil Engineers.

Lennard, D. E. 1988. Marine science and technology in the United Kingdom. *MTS Journal* 24 (1): 72–75.

Lester, C. 1996. Cumulative impact management in the coastal zone: Regulation and planning in the Monterey Bay Region of California. Paper presented at the Fifth Annual Symposium of the Ocean Governance Study Group, July 21–23, Boston, Massachusetts.

Levitus, S., and R. D. Gelfeld. 1992. *National Oceanographic Data Center Inventory of Physical Oceanographic Profiles: Global Distributions by Year for All Countries*. Silver Spring, Md.: U.S. Department of Commerce, National Oceanic and Atmospheric Administration, National Environmental Satellite, Data, and Information Service and National Ocean Data Center.

Levy, J. P. 1988. Towards integrated marine policy in developing countries, *Marine Policy* (July): 326–342.

Li, H. 1996. Intergovernmental Oceanographic Commission Secretariat, United Nations Educational, Scientific, and Cultural Organization, Paris. 1996. Personal review of and comments on manuscript for *Integrated Coastal and Ocean Management: Concepts and Practices,* August 27.

Lijphart, A. 1990. Democratic political systems. In *Contemporary Political Systems: Classifications and Typologies*, ed. A. Bebler and J. Seroka, 173–203. Boulder: Colo.: Rienner.

Loi, H. K. 1993. Coastal zone management in Malaysia. In *World Coast Conference 1993: Proceedings*, vols. 1 and 2. CZM-Centre Publication No. 4. The Hague: Ministry of Transport, Public Works, and Water Management, National Institute for Coastal and Marine Management, Coastal Zone Management Centre.

Lowry, K., and H. J. M. Wickremeratne. 1988. Coastal area management in Sri Lanka. In *Ocean Yearbook 7*, ed. E. M. Borgese, N. Ginsberg, and J. R. Morgan, 263–293. Chicago: University of Chicago Press.

Lowry, K. 1996. Professor, Department of Urban and Regional Planning, University of Hawaii, Honolulu, Hawaii, United States. 1996. Response to ICM cross-national survey. Newark: University of Delaware, Center for the Study of Marine Policy.

Lyon, J., and J. McCarthy, eds. 1995. *Wetland and Environmental Applications of GIS*. Boca Raton, Fla.: CRC Press.

McAlister, I. J. and R. A. Nathan. 1987. Malaysian national coastal erosion study. In *Coastal Zone '87*, ed. O. T. Magoon, H. Converse, D. Miner, and L. T. Tobin, 45–55. New York: American Society of Civil Engineers.

Macilwain, C. 1996. Risk: A suitable case for analysis? *Nature* 380: 10–11.

Mackay, G. A. 1981. Offshore oil and gas policy: United Kingdom. In *Comparative Marine Policy: Perspectives from Europe, Scandinavia, Canada, and the United States,* Center for Ocean Management Studies, University of Rhode Island, 103–116. New York: Praeger.

McManus, L. T., and T. E. Chua. 1996. The Lingayen Gulf case: If we were to do it over again. Paper presented at international workshop, Integrated Coastal Management in Tropical Developing Countries: Lessons Learned from Successes and Failures, May 24–28, Xiamen, People's Republic of China.

Magdaraos, C. C. 1996. Head Executive Assistant, Department of Environment and Natural Resources, Quezon City, Metro-Manila, Philippines: Response to ICM cross-national survey. Newark: University of Delaware, Center for the Study of Marine Policy.

Malaysia. Ministry of Science, Technology, and the Environment. Coastal Resources Study Team. 1992. *The Coastal Resources Management Plan for South Johore, Malaysia.* ASEAN-USAID Coastal Resources Management Project. ICLARM Technical Series No. 11. Manila: International Centre for Living Aquatic Resources Management.

Matson, P., and L. Ustin. 1991. The future of remote sensing in ecological studies. *Ecology* 72: 1917–1945.

Matthews, E., J. Veitayaki, and V. Ram-Bidesi. In press. Evolution of traditional and community-based management of marine resources in a Fijian village. *Ocean & Coastal Management.*

Meltzer, E., et al. 1996. *International Review of Integrated Coastal Zone Management*, vols. I–III. Consultancy report prepared by Meltzer Research and Consulting for the Canadian Department of Fisheries and Oceans, Halifax, to be published by government of Canada in 1998.

Meltzoff, S. K., and E. Lipuma. 1986. The troubled seas of Spanish fishermen: Marine policy and the economy of change. *American Ethnologist* 13 (4): 681–699.

METAP (Mediterranean Environment Technical Assistance Program). 1991. Coastal zone management (CZM) in Turkey. Consultant Report to the World Bank.

Mieremet, R. B. 1995. The International Coral Reef Initiative: A seed from the Earth Summit tree which now bears fruit. *Ocean & Coastal Management* 29 (1–3): 303–328.

Miles, E. L. 1991. Lecture, Marine Policy Program, Graduate College of Marine Studies, University of Delaware, November 14.

———. Future challenges in ocean management: Towards integrated national ocean policy. In *Ocean Management in Global Change*, ed. P. Fabbri, 594–620. New York: Elsevier Applied Science.

Miossec, A. 1996. Professor, Institut de Géographie et d'Aménagement Regional de l'Université de Nantes (IGARUN), Nantes, France. Response to ICM cross-national survey. Newark: University of Delaware, Center for the Study of Marine Policy.

Mlot, C. 1989. Global risk assessment. *BioScience* 39: 428–430.

Montoya, F. 1996. Head of Coastal Service, Ministry of Public Works, Transport and the Environment, Tarragona, Spain. Response to ICM cross-national survey. Newark: University of Delaware, Center for the Study of Marine Policy.

MOPT (Ministerio de Obras Públicas y Transportes). 1992. *Informe Nacional UNCED, Brazil '92.* (National Report for UNCED 1992). Madrid: Ministerio de Obras Públicas y Transportes.

MOPU (Ministerio de Obras Públicas y Urbanismo). 1988a. *Actuaciónes en la Costa* (Coastal Action). Madrid: Ministerio de Obras Públicas y Urbanismo.

———. 1988b. *Ley de Costas* (The Shores Act). Madrid: Ministerio de Obras Públicas y Urbanismo.

Morris, P., and R. Therivel, eds. 1995. *Methods of Environmental Impact Assessment.* Vancouver, B.C.: UCL Press.

Msangi, J. P. 1991. The integrated utilization of the sea off the coast of Tanzania. In *The Development of Integrated Sea-Use Management*, ed. H. D. Smith and A. Vallega. 230–237. London: Routledge.

MTE (Multidisciplinary Team of Experts). 1996. *The Coastal Environmental Profiles of the Batangas Bay Region.* MPP-EAS Technical Report No. 5. GEF/UNDP/IMO Regional Programme for the Prevention and Management of Marine Pollution in the East Asian Seas, Metro-Manila, Phillipines.

Muñoz, J. C. 1994. Planning for the ecologically sustainable harvest of marine resources: A Philippines experience. In *Coastal Zone Canada '94*, ed. R. G. Wells and P. J. Ricketts, 320–329. Halifax, Nova Scotia: Coastal Zone Canada Association.

Murthy, R. 1996. Research Scientist, Consultant to the Indian Government, CCIW–Environment Canada. Response to ICM cross-national survey. Newark: University of Delaware, Center for the Study of Marine Policy.

Nandan, S. N. 1995. Closing statement by Ambassador Satya N. Nandan, Chairman of the Conference on Straddling Fish Stocks and Highly Migratory Fish Stocks, August 4, United Nations, New York.

NAS (National Academy of Sciences) National Research Council. Committee on Science and Policy for the Coastal Ocean. 1995. *Science, Policy, and the Coast: Improving Decisionmaking.* Washington, D.C.: National Academy Press.

Nawadra, S. 1996. Project Coordinator, Coastal Management, Department of the Environment, Suva, Fiji. Response to ICM cross-national survey. Newark: University of Delaware, Center for the Study of Marine Policy.

Nayak, B. U., P. Chandra-Mohan, and B. N. Desai. 1992. Planning and manage-ment of the coastal zone in India: A perspective. *Coastal Management* 20: 365–375.

NetCoast. 1996. *A Guide to Integrated Coastal Management.* World Wide Web site operated by the Ministry of Transport, Public Works, and Water Management, National Institute for Coastal and Marine Management, Coastal Zone Management Centre, The Hague, Netherlands. Internet: http://www.minvenw.nl/projects/netcoast/.

Ngoile, M. A. K., and O. Linden. In press. Lessons learned from eastern Africa: The development policy on ICZM at national and regional levels. *Ocean & Coastal Management.*

Ngoile, M. A. K. 1996, Director, Institute of Marine Science, University of Dar es Salaam, Dar es Salaam, Tanzania; Director and Program Coordinator, Marine Affairs, World Conservation Union, Gland, Switzerland. Response to ICM cross-national survey. Newark: University of Delaware, Center for the Study of Marine Policy.

North Carolina. 1985. *A Handbook for Development in North Carlina's Coastal Area* Raleigh: North Carolina Department of Natural Resources and Community Development, Division of Coastal Management.

OECD (Organization for Economic Co-operation and Development). 1991. *Report on CZM: Integrated Policies and Draft Recommendation of the Council on Integrated Coastal Zone Management*. Washington, D.C.: Organization for Economic Co-operation and Development.

———. 1992. *Environmental Policies in Turkey*. Proceedings from MEDCOAST. Washington, D.C.: Organization for Economic Co-operation and Development.

Olsen, S. B. 1993. Will integrated coastal management programs be sustainable? The constituency problem. *Ocean & Coastal Management* 21 (1–3): 201–225.

———. 1996. Director, Coastal Resources Center, University of Rhode Island, Narragansett, Rhode Island, United States. 1996. Response to ICM cross-national survey. Newark: University of Delaware, Center for the Study of Marine Policy.

Olsen, S., J. Tobey, and M. Kerr. In press. A common framework for learning from ICM experience. *Ocean & Coastal Management*.

Orbach, M. 1995. Social scientific contributions to coastal policy making. In *Improving Interactions between Coastal Science and Policy: Proceedings of the California Symposium* 49–59. Washington, D.C.: National Academy Press.

OSIR (Oil Spill International Report). 1995. Korea officials report widespread aquaculture damage from tanker spill. *OSIR* 18 (28) (August 3).

Ozhan, E. 1996a. Coastal zone management in Turkey. *Ocean & Coastal Management* 30 (2–3): 153–176.

———. 1996b. Professor, Civil Engineering Department, Middle East Technical University, Ankara, Turkey. Response to ICM cross-national survey. Newark: University of Delaware, Center for the Study of Marine Policy.

Ozhan, E., and E. B Culhaogie. 1995. MEDCOAST Institute: A Training Program on Coastal Zone Management in the Mediterranean and the Black Sea. In *Proceedings of the Workshop on Educating Coastal Managers*, ed. B. R. Crawford, J. S. Cobb, and L. M. Chou, 84–109. July 5–10. Kingston: University of Rhode Island, Coastal Resources Center.

Ozhan, E., A. Engin, and U. Aktas. 1993. Turkish legislation pertinent to coastal zone management. Paper presented at MEDCOAST 1993, the First International Conference on the Mediterranean Coastal Environment, November 2–5. Ankara, Turkey.

Peet, G. 1987. Sea use management for the North Sea. In *The UN Convention on the Law of the Sea: Impact and Implementation*, ed. E. D. Brown and R. R. Churchill; 430–440. Honolulu: University of Hawaii, Law of the Sea Institute.

Pernetta, J. C., and D. L. Elder. 1993. *Cross-Sectoral, Integrated Coastal Area Planning (CICAP): Guidelines and Principles for Coastal Area Development*. Marine Conservation and Development Report. Gland, Switzerland: World Conservation Union.

Philippines. National Economic Development Authority. 1992. *The Lingayen Gulf Coastal Area Management Plan*. ICLARM Technical Report 30. Manila: International Centre for Living Aquatic Resources Management.

Pido, M. D., and T. E. Chua. 1992. A framework for rapid appraisal of coastal environments. In *Integrative Framework and Methods for Coastal Area Management,* ed. T. E. Chua and L. F. Scura, 144–147. ICLARM: Conference Proceedings No. 37. Manila: International Centre for Living Aquatic Resources Management.

Pintukanok, A., and S. Borothanarat. 1993. National coastal resources management in Thailand. In *World Coast Conference 1993: Proceedings,* vols. 1 and 2. CZM-Centre Publication No. 4. The Hague: Ministry of Transport, Public Works, and Water Management, National Institute for Coastal and Marine Management, Coastal Zone Management Centre.

Pires-Filho, I. A., and D. E. Cycon. 1987. Planning and managing Brazil's coastal resources. *Coastal Management* 15 (1): 61–74.

Piyakarnchana, T., N. Paphavasit, J. Suchareekul, T. Rochana Buranon, S. Suwannodom, and S. Panich. 1991. Environmental education curricula at the tertiary levels in Thailand: Case study of marine science and marine affairs programs. In *Coastal Area Management Education in the ASEAN Region,* ed. T. E. Chua, N. Paphavasit, J. Suchareekul, T. Rochana Buranon, S. Suwannodom, and S. Panich. ASEAN-USAID, Coastal Resources Management Project. ICLARM Conference Proceedings No. 8. Manila: International Centre for Living Aquatic Resources Management.

Pomeroy, R. S. 1995. Community-based and co-management institutions for sustainable coastal fisheries management in Southeast Asia. *Ocean & Coastal Management* 27 (3): 143–162.

Premaratne, A. 1991. Difficulties of coastal resources management in developing countries: Sri Lanka experiences. In *Coastal Zone '91,* ed. O. T. Magoon, H. Converse, V. Tippie, L. T. Tobin, and D. Clark, 3026–3041. New York: American Society of Civil Engineers.

Quitos, L. N., Jr., and F. D. Domingo. 1993. Coastal resource management planning: The Lingayen Gulf experience. In *World Coast Conference 1993: Proceedings,* vols. 1 and 2. CZM-Centre Publication No. 4. The Hague: Ministry of Transport, Public Works, and Water Management, National Institute for Coastal and Marine Management, Coastal Zone Management Centre.

Rais, J. 1993. Marine Resource Evaluation and Planning Project for an integrated coastal zone planning and management in Indonesia. In *World Coast Conference '3: Proceedings,* vols. 1 and 2. CZM-Centre Publication No. 4. The Hague: inistry of Transport, Public Works, and Water Management, National Institute for Co..al and ..rine Management, Coastal Zone Management Centre.

Ramos-.spia, A. A., and S. E. McNeill. 1994. The status of marine conservation in Spain. *Ocean & Coastal Management* 24 (2): 125–138.

Rivas, V., E. Frances, D. de Teran Jr., A. Cendrero, J. Hildago, and A. Serrando. 1994. Conservation and restoration of endangered coastal areas: The case of small estuaries in northern Spain. *Ocean & Coastal Management* 23 (2): 129–148.

Rivera, R., and G. F. Newkirk. 1997. Power from the people: A documentation of NGO experience in community-based coastal resource management in the Philippines. *Ocean & Coastal Management* 36 (1–3): 73–95.

Rizvi, S. H. N. 1993. Coastal zone management issues in Pakistan—An overview. Paper presented at the IPCC Eastern Hemisphere Workshop on the Vulnerability of Sea-Level Rise and Coastal Zone Management, August 3–6. Tsukuba, Japan.

Robadue, D. D., Jr., ed. 1995. *Eight Years in Ecuador: The Road to Integrated Coastal Management.* Kingston: University of Rhode Island, Coastal Resources Center.

Robadue, D. D., Jr., and L. Arriaga. 1993. Policies and programs toward sustainable coastal development in Ecuador's special area management zones: Creating vision, consensus, and capacity. In *Coastal Zone '93*, ed. O. T. Magoon, W. S. Wilson, and H. Converse, 2668–2682. New York: American Society of Civil Engineers.

Rose, R. 1992. What is lesson-drawing? *Journal of Public Policy* 2: 3–30.

Saigal, K., and T. Fuse. 1994. National case-studies: India and Japan. In *Ocean Governance: Sustainable Development of the Seas*, ed. P. B. Payoyo, 119–135 United Nations University Press.

Samaranayake, R. A. D. B. 1996. Manager, Coastal Resources Development, Coast Conservation Department, Maligawatte, Colombo, Sri Lanka. Responses to ICM cross-national survey. Newark: University of Delaware, Center for the Study of Marine Policy.

Sample, V. A., ed. 1994. *Remote Sensing and GIS in Ecosystem Management.* Washington, D.C.: Island Press.

SAREC (Swedish Agency for Research Cooperation with Developing Countries). Marine Science Program. 1994. *Technical Recommendations of the Workshop on Integrated Coastal Zone Management in Eastern Africa Including the Island States.* April 21–22, Arusha, Tanzania. Stockholm: SAREC Marine Science Program.

Scura, L. F., T.-E. Chua, M. D. Pido, and J. N. Paw. 1992. Lessons for integrated coastal zone management: The ASEAN experience. In *Integrative Framework and Methods for Area Management*, ed. T. E. Chua and L. F. Scura, 1–70. ICLARM Conference Proceedings 37. Manila: International Centre for Living Aquatic Resources Management.

Simpson, J. 1994. Remote sensing in fisheries: A tool for better management in the utilization of a renewable resource. *Canadian Journal of Fisheries and Aquatic Sciences* 51: 743–771.

Sloan, N. A., and A. Sugandhy. 1994. An overview of Indonesian coastal environmental management. *Coastal Management* 22: 215–233.

Smith, H., 1996. Senior Lecturer, Department of Maritime Studies and International Transport, University of Wales, Cardiff. Response to ICM cross-national survey. Newark: University of Delaware, Center for the Study of Marine Policy.

Smith, L. G. 1996. Introduction to environmental impact assessment: Principles and procedures. *Environment and Planning* 28 (2): 373.

Smith, R. A. 1992. Conflicting trends of beach resort development: A Malaysian case. *Coastal Management* 20: 167–187.

Soegiarto, A. 1996. Integrated coastal zone management in Indonesia: Problems and plans of action. Paper presented at international workshop, Integrated Coastal

Management in Tropical Developing Countries: Lessons Learned from Successes and Failures, May 24–28, Xiamen, People's Republic of China.

Soendoro, T. 1994. The Indonesian experience in planning and management of coastal zone through MREP project. Paper presented at IOC-WESTPAC Third International Scientific Symposium, November 22–26. Bali, Indonesia.

———. 1996. Bureau Chief, Marine Aerospace, Environment, Science, and Technology, Bappenas, Jakarta, Indonesia. Responses to ICM cross-national survey. Newark: University of Delaware, Center for the Study of Marine Policy.

Sorensen, J. C. 1993. The international proliferation of integrated coastal zone management efforts. *Ocean & Coastal Management* 21 (1–3): 45–80.

———. 1997. National and international efforts at integrated coastal management: Definitions, achievements, and lessons. *Coastal Management* 25: 3–41.

Sorensen J. C. and S. T. McCreary. 1990. *Institutional Arrangements for Managing Coastal Resources and Environments.* Renewable Resources Information Series No. 2. Washington, D.C.: U.S. Department of the Interior, National Park Service.

South, G. R., D. Goulet, S. Tuqiri, and M. Church, eds. 1994. *Traditional Marine Tenure and Sustainable Management of Marine Resources in Asia and the Pacific: Proceedings of the International Workshop.* July 1–8. Suva, Fiji: University of South Pacific.

SPREP (South Pacific Regional Environment Programme). 1994a. *Assessment of Coastal Vulnerability and Resilience to Sea Level Rise and Climate Change: Case Study Yasawa Islands, Fiji. Phase 2: Development of Methodology.* Apia, Western Samoa: South Pacific Regional Environment Programme.

———.1994b. *Coastal Protection in the Pacific Islands: Current trends and future prospects.* Apia, Western Samoa: South Pacific Regional Environment Programme.

———. 1995. *Coastal Area Management in the Pacific Island Nations.* Apia, Western Samoa: South Pacific Regional Environment Programme.

SPREP, IUCN, and ADB (South Pacific Regional Environment Programme, World Conservation Union, and Asian Development Bank). 1993. *National Environmental Management Strategy: Federated States of Micronesia.* Apia, Western Samoa: South Pacific Regional Environment Programme.

Sri Lanka. Coast Conservation Department. 1987. *Coastal Zone Management Plan.* Colombo, Sri Lanka: Coast Conservation Department.

Stanley, D. R., and C. A. Wilson. 1989. Utilization of offshore platforms by recreational fishermen and scuba divers off the Louisiana coast. In *Petroleum Structures as Artificial Reefs: A Compendium,* comp. by V. C. Reggio, Jr., U.S. Department of the Interior, Minerals Management Service, Gulf of Mexico OCS Regional Office.

Suarez de Vivero, J. 1992. The Spanish Shores Act and its implications for regional coastal management. *Ocean & Coastal Management* 18: 307–317.

———. 1996. Professor, Facultad de Geografía e Historia, Universidad de Sevilla, Sevilla, Spain. Response to ICM cross-national survey. Newark: University of Delaware, Center for the Study of Marine Policy.

Susskind, L., and J. Cruikshank. 1987. *Breaking the Impasse: Consensual Approaches to Resolving Public Disputes*. New York: Basic Books.

Swaney, J. A. 1996. Comparative risk analysis: Limitations and opportunities. *Journal of Economic Issues* 30 (2): 463–473.

Tabucanon, M. S., 1991. State of coastal resource management strategy in Thailand. *Marine Pollution Bulletin* 23: 579–586.

Tanzania. Region of Tanga. 1996. *Tanga Coastal Zone Conservation and Development Programme*. Pamphlet. Tanzania, Region of Tanga.

Thailand. Office of the National Environment Board and Ministry of Science, Technology, and Environment. 1992. *The Integrated Management Plan for Ban Don Bay and Phangnga Bay, Thailand*. ICLARM Technical Report No. 30. Manila: International Centre for Living Aquatic Resources Management.

Thomas, J. C. 1995. *Public Participation in Public Decisions: New Skills and Strategies for Public Managers*. San Francisco: Jossey-Bass Publishers.

Trends and challenges: The new environmental landscape. 1996. *Environmental Science and Technology* (special issue), 30: 24–44.

Underdahl, A. 1980. Integrated marine policy—What? Why? How? *Marine Policy* (July): 159–169.

UNDIESA (United Nations Department of International Economic and Social Affairs). Ocean Economics and Technology Branch. 1982. *Coastal Area Management and Development*. Elmsford, N.Y.: Pergamon Press.

UNDP, ADB, and SPREP (United Nations Development Programme, Asian Development Bank, and South Pacific Regional Environment Programme). 1992. *Country Report for UNCED*. Apia, Western Samoa: South Pacific Regional Environment Programme.

UNDP and GEF (United Nations Development Programme and Global Environment Facility). 1993. *Support for regional oceans training programs*. Project Document. July.

UNDPCSD (United Nations Department for Policy Coordination and Sustainable Development). 1996. *Report of the Expert Group Meeting on Identification of Principles of International Law for Sustainable Development, Geneva, Switzerland*. September 26–28, 1995. Background Paper #3.

UNEP (United Nations Environment Programme). 1994. *Integrated Management Study for the Area of Izmir*. MAP Technical Reports Series No. 84. Regional Activity Center for Priority Actions Programme.

———. 1995a. *Guidelines for Integrated Management of Coastal and Marine Areas with Special Reference to the Mediterranean Basin*. UNEP Regional Seas Reports and Studies No. 161. Nairobi, Kenya: United Nations Environment Programme.

———. 1995b. *Report of the Second Meeting of the Conference of Parties to the Convention on Biological Diversity*. November 6–17. Jakarta, Indonesia. U.N. Doc. UNEP/CBD/COP/2/19, November 30. Nairobi, Kenya: United Nations Environment Programme.

———. 1995c. *Review of the Draft Global Programme of Action to Protect the Marine Environment from Land-Based Activities*. Intergovernmental Conference to

Adopt a Global Programme of Action to Protect the Marine Environment from Land-Based Activities, Washington, D.C., October 23–November 3, 1995. U.N. Doc. UNEP (OCA)/LBA/IG.2/L.3/Add.1., November.

United Kingdom. DOE (Department of the Environment). 1992. *Coastal Zone Protection and Planning: The Government's Response to the Second Report from the House of Commons Select Committee on the Environment.* London: Department of the Environment.

———. 1995. *Policy Guidelines for the Coast.* London: Department of the Environment.

United Kingdom. DOE (Department of the Environment) and Scottish Office. 1996. *Scotland's coasts. A discussion paper.* HMSO. March. London: Stationery Office.

United Kingdom. DOE (Department of the Environment) and Welsh Office. 1992. *Planning Policy Guidance Note on Guidance Planning.* September 20. London: Stationery Office.

———. 1993a. *Development Below Low Water Mark: A Review of Regulations in England and Wales.* October. London: Stationery Office.

———. 1993b. *Managing the Coast: A Review of Coastal Management Plans in England and Wales and the Powers Supporting Them.* October. London: Stationery Office.

United Nations. 1983. *The Law of the Sea: The United Nations Convention on the Law of the Sea with Index and Final Act of the Third United Nations Conference on the Law of the Sea.* New York: United Nations.

———. 1994a. *Report of the Global Conference on the Sustainable Development of Small Island Developing States, Bridgetown, Barbados, 26 April–6 May 1994.* U.N. Sales No. E. 94.I.8 and Corrigendum. New York: United Nations.

———. 1994b. *Report of the Secretary-General on the Law of the Sea.* U.N. Doc. A/49/631, November 16.

Untawale, A. G. 1993. Significance of mangroves for the coastal communities of India. In *Proceedings of the Seventh Pacific Science Inter-Congress Mangrove Session.* July 1, Okinawa, Japan. Okinawa, Japan: International Society for Mangrove Ecosystems.

Untawale, A. G., and S. Wafar. 1990. Socio-economic significance of mangroves in India. In *Proceedings of the Symposium on the Significance of Mangroves,* ed. A. D. Agate, S. D. Bonde, K. P. N. Kumaran, March 16, Pune, Maharashtra, India. Pune, Maharashtra, India: Maharashtra Association for the Cultivation of Science Research Institute.

URI (University of Rhode Island, Coastal Resources Center). 1996. Database of integrated coastal management efforts worldwide. Internet: http://brooktrout.gso.uri.edu.

U.S. Department of Commerce. National Oceanic and Atmospheric Administration. 1979. *Final Environmental Impact Statement: Proposed CZMP for the Virgin Islands.* Silver Spring, Md.: U.S. Department of Commerce, National Oceanic and Atmospheric Administration, Office of Ocean and Coastal Resource Management.

———. 1993a. *Biennial Report to Congress on the Administration of the U. S.*

Coastal Zone Management Act. Silver Spring, Md.: U.S. Department of Commerce, National Oceanic and Atmospheric Administration.

———. 1993b. *Draft Description of Environmental Impacts of Management Alternatives.* Silver Spring, Md.: U.S. Department of Commerce, National Oceanic and Atmospheric Administration, Office of Ocean and Coastal Resource Management, Sanctuaries and Reserves Division.

———. 1995. *Economic Valuation of Natural Resources: A Handbook for Coastal Resource Policymakers.* NOAA Coastal Ocean Program Decision Analysis Series No. 5. Silver Spring, Md.: U.S. Department of Commerce, National Oceanic and Atmospheric Administration, Coastal Ocean Program Office.

Vallega, A. 1992. The Management of the Mediterranean Sea: The Role of Regional Complexity. *Ocean & Coastal Management* 18 (2–4): 279–290.

———. 1992. *Sea Management: A Theoretical Approach.* New York: Elsevier Applied Science.

———. 1993. A conceptual approach to integrated coastal management. *Ocean & Coastal Management* 21: 149–162.

———. 1995. Regional level implementation of Chapter 17: The UNEP approach to the Mediterranean. In Special issue: Earth summit implementation—Progress achieved on oceans and coasts, *Ocean & Coastal Management* 29 (1–3): 1–11.

———. 1996. *The Agenda 21 of Ocean Geography: The Epistemological Challenge.* Twenty-Eighth International Geographical Union International Geographical Congress, Land, Sea and Human Effort, August 4–10, The Hague, Netherlands. Geneva: International Geographical Union.

Vallejo, S. M. 1993. The integration of coastal zone management into national development planning. *Ocean & Coastal Management* 21: 163–181.

———. 1995. The Train-Sea-Coast programme: A decentralized, cooperative training network for the systematic development of human resources. Paper presented at the International Workshop on Educating Coastal Managers, March 4–10. University of Rhode Island.

Vanclay, F., and D. Bronstein. 1985. *Environmental and Social Impact Assessment.* New York: Wiley.

Van der Weide, J. 1993. A systems view of integrated coastal management. *Ocean & Coastal Management* 21 (1–3): 129–148.

Van Dyke, J. M. 1992. Substantive principles for a constitution for the U.S. oceans. In *Ocean Governance: A New Vision.* Newark: University of Delaware, Center for the Study of Marine Policy, Ocean Governance Study Group.

———. 1996. The Rio principles and our responsibilities of ocean stewardship. *Ocean & Coastal Management* 31 (1): 1–23.

Van Horn, H., G. Peet, and K. Wieriks. 1985. Harmonizing North Sea policy in the Netherlands. *Marine Policy* 9 (1): 53–61.

Veitayaki, J. 1996. Ocean Resources Management Programme, University of the South Pacific, Suva, Fiji. Response to ICM cross-national survey. Newark: University of Delaware, Center for the Study of Marine Policy.

Vellinga, P., and R. J. T. Klein. 1993. Climate change, sea level rise and integrated coastal zone management: An IPCC approach. *Ocean & Coastal Management* 21 (1–3): 245–268.

Waite, C. 1981. Coastal management in England and Wales. In *Comparative Marine Policy: Perspectives from Europe, Scandinavia, Canada, and the United States, Center for Ocean Management Studies, University of Rhode Island*, 57–64. New York: Praeger.

Walker, K. J. 1992. *Australian Environmental Policy: Ten Case Studies.* Sydney: University of New South Wales Press.

Wang, Y. 1992. Coastal management in China. In *Ocean Management in Global Change*, ed. P. Fabbri, 460–469. New York: Elsevier Applied Science.

―――. Director, State Pilot Laboratory of Coast and Inland Exploitation, Nanjing University, Nanjing, China. Response to ICM cross-national survey. Newark: University of Delaware, Center for the Study of Marine Policy.

WCC (World Coast Conference). 1993. *How to Account for Impacts of Climate Change in Coastal Zone Management: Concepts and Tools for Approach and Analysis.* Versions 1 and 2. World Coast Conference 1993, November 1–5, Noordwijk, Netherlands. The Hague: Ministry of Transport, Public Works, and Water Management, National Institute for Coastal and Marine Management, Coastal Zone Management Centre.

―――. 1994. *Preparing to Meet the Coastal Challenges of the 21st Century: Report of the World Conference 1993.* November 1–5, Noordwijk, Netherlands. The Hague: Ministry of Transport, Public Works, and Water Management, National Institute for Coastal and Marine Management, Coastal Zone Management Centre.

WCED (World Commission on Environment and Development). 1987. *Our Common Future.* Oxford: Oxford University Press.

Wehr, P. 1979. *Conflict Regulation.* Boulder, Colo.: Westview Press.

Weiss, J. A. 1987. Pathways to cooperation among public agencies. *Journal of Policy Analysis and Management* 7 (1): 94–117.

Wessel, A. E., and M. J. Hershman. 1988. Mitigation: Compensating the environment for unavoidable harm. In *Urban Ports and Harbor Management*, ed. M. J. Hershman. New York: Taylor and Francis.

White, A. T. 1989. The Marine Conservation and Development Program of Silliman University as an example for Lingayen Gulf. In *Towards Sustainable Development of the Coastal Resources of Lingayen Gulf, Philippines*, ed. G. Silvestre et al., 119–123. ICLARM Conference Proceedings No. 1. Manila: International Centre for Living Aquatic Resources Management.

White, A. T., L. Z. Hale, Y. Renard, and L. Cortesi. 1994. *Collaborative and Community-Based Management of Coral Reefs: Lessons from Experience.* West Hartford, Conn.: Kumarian Press.

White, A. T., P. Martosubroto, and M. S. M. Sadorra, eds. 1989. *The Coastal Environmental Profile of Segara Anakan–Cilacap, South Java, Indonesia.* ICLARM Technical Reports on Coastal Area Management Series No. 25. Manila: International Centre for Living Aquatic Resources Management.

Wiggerts, H. 1981. Coastal management in the Netherlands. In *Comparative Marine*

Policy: Perspectives from Europe, Scandinavia, Canada, and the United States, Center for Ocean Management Studies, University of Rhode Island, 75–81. New York: Praeger.

Williams, A. T. 1992. The quiet conservators: Heritage coasts of England and Wales. *Ocean & Coastal Management* 17 (2): 151–169.

Wood, M. J. 1995. The International Ocean Institute Network. Paper presented at workshop, Educating Coastal Managers, March 4–10. University of Rhode Island.

World Bank. 1993. Noordwijk guidelines for integrated coastal zone management. Document presented at World Coast Conference 1993, November 1–5, Noordwijk, Netherlands. Republished as Post, J. C., and C. G. Lundin, eds. 1996. *Guidelines for Integrated Coastal Zone Management.* Environmentally Sustainable Development Studies and Monographs Series No. 9. Washington, D.C.: World Bank.

———. 1994. Coastal zone management and environmental assessment. *Environmental Assessment Source Book Update,* Environment Department World Bank No. 7. March. Washington, D.C.: World Bank.

World Bank. 1995a. *Africa: A Framework for Integrated Coastal Zone Management.* Washington, D.C.: World Bank, Land, Water, and Natural Habitats Division, African Environmentally Sustainable Development Division.

———. 1995b. *Mainstreaming the Environment: The World Bank Group and the Environment Since the Rio Earth Summit.* Washington, D.C.: World Bank.

———. 1996. *World Development Report 1996: From Plan to Market.* New York: Oxford University Press.

WRI (World Resources Institute). 1994. *World Resources 1994–95.* New York: Oxford University Press.

Wright, D. 1994. The approach of the Department of Fisheries and Oceans to land use planning in Canada's Arctic coastal zone. In *Coastal Zone Canada '94*, ed. R. G. Wells and P. J. Ricketts, 19: 31–45. Halifax, Nova Scotia: Coastal Zone Canada Association.

WRI, IUCN, and UNEP (World Resources Institute, World Conservation Union, and United Nations Environment Programme). 1992. *Global Biodiversity Strategy: Guidelines for Action to Save, Study, and Use Earth's Basic Wealth.* Washington, D.C.: World Resources Institute.

WWF (World Wide Fund for Nature). 1995. Integrated coastal zone management: U.K. and European initiatives. *Marine Update* (Surrey, England: World Wide Fund for Nature).

Yu, H. 1991. Marine fishery management in the People's Republic of China. *Marine Policy* 15 (1): 23–32.

———. 1993. A new stage in China's marine management. *Ocean & Coastal Management* 19: 185–198.

———. 1994. China's coastal ocean uses: Conflicts and impacts. *Ocean & Coastal Management* 25 (3): 161–178.

———. Former Deputy Director General, Department of Integrated Marine Management, State Oceanic Administration, Beijing, People's Republic of China.

Response to ICM cross-national survey. Newark: University of Delaware, Center for the Study of Marine Policy.

Zann, L. P. 1995. *Our Sea, Our Future: Major Findings of the State of the Marine Environment Report (SOMER) for Australia.* Canberra, Australia: Department of the Environment, Sport, and Territories and Great Barrier Reef Marine Park Authority.

Zhijie, F., and R. P. Cote. 1990. Coastal zone of the People's Republic of China: Management approaches and institutions. *Marine Policy* 14 (4): 305–314.

———. 1991. Population, development, and marine pollution in China. *Marine Policy* 15 (3): 210–219.

Index

Abregana, B., 434, 437
Accidents, marine, 73
Accountability, 247–48
Acquisition of coastal land, 227, 229
Acronyms, list of, 467–70
Activities and uses of the coastal zone and ocean, 19,
   21–22, 40
   see also specific activities/uses
Adaptive management, 172–73, 195
Adler, J., 402, 404, 405
Adoption of polices, formal, 59–60, 211–13
Advice giving public participation, 238, 239
Affected interests in the process, involving the,
   220–21
Africa, 34–35
Agardy, T. S., 99
Agate, A. D., 431
Agencies/institutional bodies/projects:
   (Australia) *Australian Coastal Zone Management
     Report* of 1980, 349
   (Australia) Department of the Environment, Sport,
     and Territories (DEST), 351
   (Australia) *Injured Coastline Protection of the
     Coastal Environment* of 1991, 349
   (Australia) *Resource Assessment Commission
     Coastal Zone Inquiry Final Report* of 1993,
     349
   (Bangladesh) Coastal Embankment Project (CEP),
     26
   (Brazil) Coastal Pollution Management Project,
     112
   (Brazil) Environmental Institute of Parana, 357
   (Brazil) Fishing Sector Executive Group (GESPE),
     356
   (Brazil) Institute for the Environment and Natural
     Resources (IBAMA), 357
   (Brazil) Interministerial Commission for Marine
     Resources (CIRM), 355–57
   (Brazil) State Council for Environment
     (CONSEMA), 357
   (Canada) Atlantic Coastal Action Program
     (ACAP), 313
   (Canada) Canadian International Development
     Agency (CIDA), 314

(Canada) Fraser River Estuary Management
   Program (FREMP), 312–13
(China) Chinese Academy of Sciences, 397–98
(China) Chinese Society of Oceanography, 397
(China) Comprehensive Coastal Environment and
   Resources Survey, 394
(China) Comprehensive Environment and
   Resources Survey on Islands, 394
(China) "Integrated Coastal Management in
   Tropical Developing Countries: Lessons
   Learned from Successes and Failures," 398
(China) Integrated Management Task Force
   (IMTF), 396
(China) Interagency Executive Committee, 396
(China) Marine Management and Coordination
   Working Committee/Task Force, 396
(China) National Marine Development Plan, 395
(China) State Oceanic Administration (SOA), 394
(Ecuador) Coastal Resources Management
   Program (PMRC), 388–89
(Ecuador) National Coastal Resources Commission
   (NMRC), 389
(Fiji) Department of Town and Country Planning,
   383
(Fiji) Environmental Management Committee
   (EMC), 383
(Fiji) Environmental Management Unit (EMU),
   383
(Fiji) *Fiji National Report to UNCED,* 384
(Fiji) National Environmental Strategy (NES), 383,
   384
(Fiji) National Sustainable Development Council,
   383
(Fiji) Native Lands Conservation and Preservation
   Committee, 383
(Fiji) Native Land Trust Board (NLTB), 382
(France) Institut Francais de Recherches sur l'Ex-
   ploitation de la Mer of 1984 (IFREMER), 331
(France) Ministry of the Sea (MOS), 330
(India) Central Beach Erosion Board, 429
(India) Central Water Power Research Station
   (CWPRS), 428
(India) Department of Ocean Development (DOD),
   276, 429

499